Reference Manual for Magnetic Resonance Safety, Implants, and Devices: 2004 Edition

Frank G. Shellock, Ph.D.
Adjunct Clinical Professor of Radiology
Keck School of Medicine
University of Southern California

Founder
Institute for Magnetic Resonance
Safety, Education, and Research

President
Magnetic Resonance Safety Testing Services
Los Angeles, California

Biomedical Research Publishing Group
Los Angeles, CA

Made in the United States of America

Library of Congress Cataloging-in-Publication Data

Shellock, Frank G.
    Reference Manual for Magnetic Resonance Safety, Implants, and Devices: 2004 Edition
    Frank G. Shellock
        p. cm.
        Includes bibliographical references.
        ISBN 0-9746410-0-6
    1. Magnetic resonance imaging-Complications-Handbooks, manuals, etc. 2. Metals in medicine-Magnetic properties-Handbooks, manuals, etc. 3. Implants, Artificial-Magnetic properties-Handbooks, manuals, etc. I. Title.
        [DNLM: 1. Magnetic Resonance Imaging-contraindications-handbooks. 2. Nuclear Magnetic Resonance-handbooks. 3. Metals-handbooks. 4. Implants

---

Disclaimer

This textbook was designed to provide a desk reference for radiologists, MRI technologists, MRI facility managers, and others. The materials are current through the publication date of this textbook. The content of this book is designed for general informational purposes only and is not intended to be nor should it be construed to be technical or medical advice or opinion on any specific facts or circumstances.

The author and publisher of this work disclaim any liability for the acts of any physician, individual, group, or entity acting independently or on behalf of any organization that utilizes any information for any medical procedure, activity, service, or other situation through the use of this textbook.

The content of this book makes no representations or warranties of any kind, expressed or implied, as to the information content, materials or products, included in this book. The author and publisher assume no responsibilities for errors or omissions that may include technical or other inaccuracies, or typographical errors.

The author and publisher of this work specifically disclaim all representation and warranties of any kind, expressed or implied, as to the information, content, materials, or products included or referenced in this book. The publisher and author assume no responsibilities for errors or omissions, which may include technical or other inaccuracies or typographical errors.

The author and publisher disclaim responsibility for any injury and/or damage to persons or property from any of the methods, products, instructions, or ideas contained in this publication.

The author and publisher disclaim liability for any damages of any kind arising from the use of the book, including but not limited to direct, indirect incidental, punitive and consequential damages.

DEDICATION

To Jaana Shellock, a truly extraordinary person,
a gentle soul, and kindred spirit.
Mina rakastan sinua!

To Jaan Cohan, Baby Dumpling, Pikachu, Nala,
Ren, Domino, Ugly Puss, Scarface, Runt, and
all of my friends at the Boo Boo Zoo.

Contents

SECTION II
MR PROCEDURES AND IMPLANTS, DEVICES, AND MATERIALS

Acknowledgments

Special thanks to the members of the
Medical, Technology, and Scientific Advisory Board
and Corporate Advisory Board of the
Institute for Magnetic Resonance
Safety, Education, and Research (www.IMRSER.org).

Preface

Magnetic resonance (MR) procedures continue to expand with regard to usage and complexity. Safety and patient management issues are important aspects of this diagnostic modality. Because MR technology continuously evolves, it is necessary to update, revise, and add to MR safety topics in consideration of the latest information.

Accordingly, similar to prior editions, the **Reference Manual for Magnetic Resonance Safety, Implants, and Devices: 2004 Edition** includes updated guidelines and recommendations from the latest information in the pertinent peer-reviewed literature as well as documents developed by the International Society for Magnetic Resonance in Medicine (ISMRM), the American College of Radiology (ACR), the Food and Drug Administration (FDA), the National Electrical Manufacturers Association (NEMA), the International Electrotechnical Commission (IEC), and the Medical Devices Agency (MDA).

This textbook is intended to be a comprehensive yet concise information resource on MR safety for healthcare professionals. *Section I* presents safety guidelines and recommendations. *Section II* has the latest information for implants, devices, and materials tested for safety in the MR environment. **"The List"** continues to grow and now has a tabular listing of data for more than 1,200 objects.

The clinical use of 3.0-Tesla MR systems for brain, musculoskeletal, body, cardiovascular, and other applications is increasing worldwide. Because previous investigations performed to determine MR safety for implants and devices used mostly MR systems operating at 1.5-Tesla or less, it is crucial to perform ex vivo testing at 3.0-Tesla to characterize safety for these objects, especially with regard to magnetic field interactions. Importantly, metallic objects that displayed "weakly" ferromagnetic qualities in association with 1.5-Tesla MR systems may exhibit substantial magnetic field interactions during exposure to 3.0-Tesla MR systems, posing possible hazards to patients. Therefore, recent investigations have evaluated MR safety for implants and devices at 3.0-Tesla. This book contains information for over 200 objects tested at 3.0-Tesla.

Frank G. Shellock, Ph.D.

SECTION I

GUIDELINES AND RECOMMENDATIONS FOR MR SAFETY

ACOUSTIC NOISE AND MR PROCEDURES*

Various types of acoustic noise are produced during the operation of MR systems. The problems associated with acoustic noise for patients and healthcare workers include simple annoyance, difficulties in verbal communication, heightened anxiety, temporary hearing loss and, potential permanent hearing impairment.

Acoustic noise may pose a particular hazard to specific patient groups who may be at increased risk. Patients with psychiatric disorders, elderly and pediatric patients may be confused or suffer from heightened anxiety. Sedated patients may experience discomfort due to high noise levels. Certain drugs are known to increase hearing sensitivity. Neonates with immature anatomical development may have an increased response to acoustic noise. For example, significant alterations in vital signs of newborns have been reported during MR examinations, which may be attributed to acoustic noise.

Aside from issues of safety, acoustic noise levels also pose problems for the increasing numbers of researchers involved in functional MRI (fMRI) studies of brain activation. For example, MR system-related acoustic noise will interfere with communication of activation task instructions that are typically given during the MR procedure.

One area of particular interest is the study of auditory and language function. In this work, the response to pure tone stimuli is analyzed, and any background levels of unwanted or uncontrolled acoustic noise can interfere with the delivery of these sound stimuli and the integrity of the experimental findings that may be attributed to the presence of acoustic noise.

Acoustic noise levels during echo planar imaging (EPI) have been reported to significantly increase pure tone hearing thresholds in the optimal frequency hearing range (i.e., 0.1-8 kHz). These effects vary across the fre-

2

quency range. The threshold changes according to the characteristics of the sequence-generated acoustic noise. It may be possible to take into account the MR system-induced auditory activation by using a control series of scans in task paradigms. Experimental results have been reported for mapping auditory activation induced by MR system-related acoustic noise.

HEARING AND THE IMPACT OF ACOUSTIC NOISE

The ear is a highly sensitive wide-band receiver, with the typical frequency range for normal hearing being between 20 Hz to 20,000 Hz. The human ear does not tend to judge sound powers in absolute terms, but assesses how much greater one power is than another. Combined with the very wide range of powers involved, the logarithmic decibel scale, dB, is used when dealing with sound power.

Noise is defined in terms of frequency spectrum (in Hz), intensity (in dB), and duration. Noise may be steady-state, intermittent, impulsive, or explosive. Transient hearing loss may occur following loud noise, resulting in a temporary threshold shift (shift in audible threshold). Brummett et al. have reported temporary shifts in hearing thresholds in 43% patients scanned without ear protection and also in those with improperly fitted protection. Recovery from the effects of noise should be exponential and occur quickly. However, if the noise insult is severe, full recovery can take up to several weeks. If the noise is sufficiently injurious, this may result in a permanent threshold shift (i.e., permanent hearing loss) at specific frequencies.

MRI-RELATED ACOUSTIC NOISE

The gradient magnetic field is the primary source of acoustic noise associated with MR procedures. This noise occurs during the rapid alterations of currents within the gradient coils. These currents, in the presence of a strong static magnetic field of the MR system, produce significant (Lorentz) forces that act upon the gradient coils.

Acoustic noise, manifested as loud tapping, knocking, or chirping sounds, is produced when the forces cause motion or vibration of the gradient coils as they impact against their mountings which, in turn, also flex and vibrate.

Alteration of the gradient output (rise time or amplitude) caused by modifying the MR imaging parameters will cause the level of gradient-induced acoustic noise to vary. This noise is enhanced by decreases in section thickness, field of view, repetition time, and echo time. The physical features of the MR system, especially whether or not it has special sound insulation, and the material and construction of coils and support structures, also affect the transmission of the acoustic noise and its subsequent perception by the patient and MR system operator.

CHARACTERISTICS OF MR SYSTEM-RELATED ACOUSTIC NOISE

Gradient magnetic field-induced noise levels have been measured during a variety of pulse sequences for MR systems with static magnetic field strengths ranging from 0.35 to 1.5-Tesla and one 2.0-Tesla research system. Hurwitz et al. reported that the sound levels varied from 82 to 93 dB on the A-weighted scale and from 84 to 103 dB on the linear scale. The report concluded that gradient magnetic field-induced noise was an annoyance but well within recognized safety guidelines.

Later studies performed using other MR parameters including "worst-case" pulse sequences showed that, unsurprisingly, fast gradient echo pulse sequences produced the greatest noise during MR imaging and 3-D sequences (e.g., MP-RAGE), where multiple gradients are applied simultaneously, are among the loudest sequences. Acoustic noise levels in these investigations did not exceed a range of 103 to 113 dB (peak) on the A-weighted scale.

Additional studies have been performed to include measurement of acoustic noise generated by echo planar and fast spin echo sequences. Echo planar sequences, in collecting a complete image in one RF excitation of the spin system, tend to have extremely fast gradient switching times and high gradient amplitudes. They can generate potentially high levels of acoustic noise, although the duration of the sequence and, thus, patient exposure, is shorter than conventional sequences.

Shellock et al. reported comparable high levels of noise of 114-115 dBA on two different high field strength MR systems tested when running EPI sequences with parameters chosen to represent a 'worst-case' protocol. Although high, the recorded acoustic noise levels were within current permissible limits. Increased interest in the use of diffusion-weighted and fMRI techniques has meant an increased utilization of ultra-high-field strength MR systems (e.g., MR systems with ≥ 3.0-Tesla static magnetic fields) with fast gradient capabilities (25 to 30 mT/m switching rates and high amplitudes) to acquire high quality, multi-slice EPI images.

In addition to dependence on imaging parameters, acoustic noise is dependent on the MR system hardware, construction, and the surrounding environment. Furthermore, noise characteristics have a spatial dependence. For example, noise levels have been found to vary by as much as 10 dB as a function of patient position along the magnet bore. The presence and size of the patient may also affect the level of acoustic noise. An increase of from 1 to 3 dB has been measured with a patient present in the bore of the MR system, which may be due to pressure doubling (i.e., an increase in sound pressure) close to a solid object, as sound waves reflect and undergo an in-phase enhancement.

MRI-RELATED ACOUSTIC NOISE AND PERMISSIBLE LIMITS

In general, the acoustic noise levels recorded by various researchers in the MR environment have been below the maximum limit permissible by the Occupational Safety and Health Administration of the United States. This is particularly the case when one considers that the duration of exposure is one of the most important physical factors that determines the effect of noise on hearing.

Other physical factors involved in hearing loss include the sound frequency, temporal pattern, and intensity of the noise. High root-mean-square (RMS) and peak noise levels at low frequency can induce strong vibration of the entire cochlea, which may lead to a temporary shift in hearing threshold or potentially permanent shift in noise sensitive patients.

On July 14, 2003, the U.S. Food and Drug Administration released updated guidelines for acoustic noise levels associated with the operation of MR systems, as follows: Sound Pressure Level - Peak unweighted sound pressure level greater than 140 dB. A-weighted root mean square (rms) sound pressure level greater than 99 dBA with hearing protection in place.

The exposure of staff and other health-workers in the MR system environment is also a concern (e.g., those involved in interventional MR procedures or who remain in the room for patient management reasons). Shellock et al. reported levels of acoustic noise ranging from 108 to 111 dB at the entrances and exits of the magnet bores of MR systems while running EPI sequences. The acceptable duration for exposure to these noise levels is 15 to 30 minutes. This suggests that staff members should wear ear protection if they remain in the MR system room for longer periods of time. In the United Kingdom, guidelines issued by the Department of Health recommend hearing protection be worn by staff exposed to an average of 85 dB over an eight hour day.

While acoustic noise levels suggested for patients exposed during MR procedures on an infrequent and short-term temporal basis are considered to be highly conservative, they may not be appropriate for individuals with underlying health problems that have problems with noise at certain levels or at particular frequencies. The acoustic noise produced during MR procedures represents a potential risk to such patients. As previously-mentioned, the possibility exists that significant gradient magnetic field-induced noise may produce substantial hearing problems in patients who are susceptible to the damaging effects of loud noises.

ACOUSTIC NOISE CONTROL TECHNIQUES

Passive noise control. The simplest and least expensive means of preventing problems associated with acoustic noise during MR procedures is to encourage the routine use of earplugs or headphones. Ear plugs, when properly used, can abate noise by 10 to 30 dB, which is usually an adequate amount of sound attenuation for the MR environment. The use of disposable earplugs has been shown to provide a sufficient decrease in acoustic noise that, in turn, would be capable of preventing the potential temporary hearing loss associated with MR procedures. Therefore, it behooves all MR facilities to recommend that patients undergoing MR examinations wear these protective devices. MR-safe headphones that substantially muffle acoustic noise are also commercially available.

Unfortunately, the passive noise control methods suffer from a number of limitations. For example, these devices hamper verbal communication with patients during the operation of the MR system. In certain circumstances, they can cause discomfort or inhibit the immobilization of the patient's head when optimal immobilization is required for certain studies that are sensitive to patient movement (i.e., diffusion-weighted and fMRI studies). Additionally, standard earplugs are often too large for the ear canal of young infants.

Importantly, passive noise control devices offer non-uniform noise attenuation over the hearing range. While high frequency sound may be well attenuated, attenuation is often poor at low frequency. This is unfortunate because the low frequency range is also where the peak MR-related acoustic noise is generated.

Passive noise control techniques provide poor attenuation of noise transmitted to the patient through bone conduction. The presence of an insulating foam mattress on the patient couch has been found to reduce vibration-related coupling to the patient and noise levels by around 10 dB.

Active Noise Control (ANC). A significant reduction in the level of acoustic noise caused by MR procedures has been accomplished by implementing the use of an active noise cancellation, or "anti-noise", technique with the existing audio system. Controlling the noise from a particular source, by introducing anti-phase noise to interfere destructively with the noise source is not a new idea. For example, Goldman et al. used combined passive and active noise control system (i.e. an active system built into a headphone), achieving an average noise reduction of around 14 dB.

Advances in digital signal processing technology allow efficient ANC systems to be realized at a moderate cost. The anti-noise system involves a continuous feedback loop with continuous sampling of the sounds in the noise environment so that the gradient magnetic field-induced noise is attenuated. It is possible to attenuate the pseudo-periodic scanner noise while allowing the transmission of vocal communication or music to be maintained. Some commercial manufacturers are currently offering ANC systems, based on delivering anti-noise to headphones in a manner similar to airline music systems.

McJury et al. reported the sound attenuation of MR system-generated acoustic noise with a real-time adaptive active noise control (ANC) system. In the adaptive ANC controller, the system attempts to minimize the error signal power using a feedback control algorithm. If successful, a "zone of quiet" appears around the error microphone.

OTHER SOURCES OF MR SYSTEM-RELATED ACOUSTIC NOISE

RF hearing. When the human head is subjected to pulsed radiofrequency (RF) radiation at certain frequencies, an audible sound perceived as a click, buzz, chirp, or knocking noise may be heard. This acoustic phenomenon is referred to as "RF hearing", "RF sound" or "microwave hearing".

Thermoelastic expansion is believed responsible for the production of RF hearing, whereby there is absorption of RF energy that produces a minute temperature elevation (i.e., approximately 1×10^{-6} °C) over a brief time period in the tissue of the head. Subsequently, a pressure wave is induced that is sensed by the hair cells of the cochlea via bone conduction. In this manner, the pulse of RF energy is transferred into an acoustic wave within the human head and sensed by the hearing organs.

With specific reference to the operation of MR scanners, RF hearing has been found to be associated with frequencies ranging from 2.4 to 170 MHz. The gradient magnetic field-induced acoustic noise that occurs during MR procedures is significantly louder than the sounds associated with RF hearing. Therefore, noises produced by the RF auditory phenomenon are effectively masked and not perceived by patients or MR operators.

Currently, there is no evidence of any detrimental health effect related to the presence of RF hearing. However, Roschmann recommends an upper level limit of 30 kW applied peak pulse power of RF energy for head coils and 6 kW for surface coils used during MR imaging or spectroscopy to avoid RF-evoked sound pressure levels in the head increasing above the discomfort threshold of 110 dB.

Noise From Subsidiary Systems. Patient comfort fans and cryogen reclamation systems associated with superconducting magnets of MR systems are the main sources of ambient acoustic noise found in the MR environment. Acoustic noise produced by these subsidiary systems is considerably less than that caused by the activation of gradient magnetic fields during MR procedures. Therefore, this acoustic noise, at the very least, may only be an annoyance to patients or MR system operators.

SUMMARY AND CONCLUSIONS

MR procedures can generate significant levels of acoustic noise. This noise can be an annoyance, hinder communication with staff and, at high levels, presents a potential hazard to patients and MR healthcare workers.

Many investigators have measured and analyzed the acoustic noise associated with MR procedures. Acoustic noise levels increase with changes in imaging parameters including a decrease in section thickness, field-of-view, gradient ramp time, and an increase in the gradient amplitude. Environment, and hardware design, presence of patient in the MR system and patient position will also affect noise levels recorded.

Although many options are available for noise control, the use of simple passive protection in the form of earplugs is typically sufficient to bring noise levels well within permissible limits for the great majority of patients and other individuals. However, care must be taken to ensure passive protection is properly fitted and in good condition.

Passive methods of course have limitations and several more elegant solutions are possible and under investigation by researchers. These methods range from the simply optimizing the MR procedure to be used in terms of gradient parameters, to the utilization of anti-noise systems or redesigning of gradient coils themselves.

As with many aspects of MR procedures, the specifications for achieving high noise attenuation often run counter to fast acquisition of high quality diagnostic images and compromises must be made. Optimizing procedures to lengthen gradient ramp times, lower amplitudes or minimize pulsing can result in sequences with reduced performance. Designing gradient coils that are force-balanced to minimize acoustic resonances will compromise performance in terms in increasing inductance and loss of gradient strength.

The solution that presents a minimal impact on the performance of the MR system is active noise control using anti-noise. This is a promising technique, but an optimal system has yet to be fully implemented clinical MR systems.

Current trends in MR techniques include increasing static magnetic field strengths and improving gradient performance for rapid imaging and other clinical applications. Both of these developments will result in increases in MR-generated acoustic noise levels. This will mean an increasing interest in acoustic noise control methods and warrants continued investigation of these techniques.

[*Portions of this content were excerpted with permission from McJury M. Acoustic Noise and Magnetic Resonance Procedures, In: Magnetic Resonance Procedures: Health Effects and Safety, FG Shellock, Editor, CRC Press, Boca Raton, FL, 2001 and McJury M, Shellock FG. Acoustic noise and MR procedures: A review. Journal of Magnetic Resonance Imaging 12: 37-45, 2000.]

REFERENCES

Bandettini PA, Jesmanowicz A, Van Kylen J, Birn RA and Hyde J. Functional MRI of brain activation induced by scanner acoustic noise. Magn Reson Med 1998;39:410-416.

Brummett RE, Talbot JM, Charuhas P. Potential hearing loss resulting from MR imaging. Radiology 1988;169: 539-540.

Bowtell RW, Mansfield PM. Quiet transverse gradient coils: Lorentz force balancing designs using geometric similitude. Magn Reson Med 1995;34:494-497.

Chaplain GBB. Anti-noise: the Essex breakthrough. Chart Mech Engin 1983;30:41-47.

Chen CK, Chiueh TD, Chen JH. Active cancellation system of acoustic noise in MR imaging. IEEE Trans Biomed Engin 1999;46:186-190.

Cho ZH, Park SH, Kim JH, Chung SC, Chung ST, Chung JY, Moon CW, Yi JH, Wong EK. Analysis of acoustic noise in MRI. Magn Reson Imaging 1997;15:815-822.

Counter SA, Olofsson A, Grahn, Borg E. MRI acoustic noise: sound pressure and frequency analysis. J Magn Reson Imaging 1997;7:606-611.

Elliott SJ, Nelson Active noise control. IEEE Sign Proceed Magn 1983; 12-35.

Elder JA. Special senses. In: United States Environmental Protection Agency, Health Effects Research Laboratory. Biological Effects of Radio frequency Radiation. EPA-600/8-83-026F, Research Triangle Park, 1984; pp. 570-571.

Goldman AM, Gossman WE, Friedlander PC. Reduction of sound levels with anti-noise in MR imaging. Radiology 1989;173:549-550.

Hedeen RA, Edelstein WA. Characteristics and prediction of gradient acoustic noise in MR imagers. Magn Reson Med 1997;37:7-10.

http://www.MRIsafety.com

Hurwitz R, Lane SR, Bell RA, Brant-Zawadzki MN. Acoustic analysis of gradient-coil noise in MR imaging. Radiology 1989;173:545-548.

Kanal E, Shellock FG, Talagala L. Safety considerations in MR imaging. Radiology 1990; 176:593-606.

Mansfield PM, Glover PM. Beaumont J Sound generation in gradient coil structures for MRI. Magn Reson Med 1998;39:539-550.

Mansfield PM, Glover PM, Bowtell RW. Active acoustic screening: design principles for quiet gradient coils in MRI. Meas Sci Technol 1994;5:1021-1025.

McJury M, Blug A, Joerger C, Condon B, Wyper D. Acoustic noise levels during magnetic resonance imaging scanning at 1.5 T. Brit J Radiol 1994;413-415.

McJury MJ. Acoustic noise levels generated during high field MR imaging. Clinical Radiology 1995;50:331-334.

McJury M. Acoustic Noise and Magnetic Resonance Procedures. In: Magnetic Resonance Procedures: Health Effects and Safety. FG Shellock, Editor, CRC Press, Boca Raton, FL, 2001.

McJury M, Shellock FG. Acoustic noise and MR procedures: A review. J Magn Reson Imaging 2000;12: 37-45.

McJury M, Stewart RW, Crawford D, Toma E. The use of active noise control (ANC) to reduce acoustic noise generated during MRI scanning: some initial results. Magn Reson Imaging 1997;15:319-322.

Melnick W. Hearing loss from noise exposure. In: Harris CM, ed. Handbook of Noise Control. New York: McGraw-Hill, 1979; pp. 2.

Miller LE, Keller AM. Regulation of occupational noise. In: Harris CM, ed. Handbook of noise control. New York: McGraw-Hill, 1979; pp. 1-16.

Philbin MK, Taber KH, Hayman LA. Preliminary report: changes in vital signs of term newborns during MR. Am J Neuroradiol 1996;17:1033-6.

Quirk ME, Letendre AJ, Ciottone RA, Lingley JF. Anxiety in patients undergoing MR imaging. Radiology 1989;170:464-466.

Robinson DW. Characteristics of occupational noise-induced hearing loss. In: Effects of Noise on Hearing. Henderson D, Hamernik RP, Dosjanjh DS, Mills JH, Editors. Raven Press, New York, 1976; pp. 383-405.

Roschmann P. Human auditory system response to pulsed radio frequency energy in RF coils for magnetic resonance at 2.4 to 170 MHz. Magn Reson Med 1991;21:197-215.

Shellock FG, Kanal E. Policies, guidelines, and recommendations for MR imaging safety and patient management. J Magn Reson Imaging 1991;1:97-101.

Shellock FG, Litwer CA, Kanal E. Magnetic resonance imaging: bioeffects, safety, and patient management. Magnetic Resonance Quarterly 1992;4:21-63.

Shellock FG, Morisoli SM, Ziarati M. Measurement of acoustic noise during MR imaging: Evaluation of six "worst-case" pulse sequences. Radiology 1994;191:91-93.

Shellock FG, Ziarati M, Atkinson D, Chen D-Y. Determination of gradient magnetic field-induced acoustic noise associated with the use of echo planar and three-dimensional fast spin echo techniques. J Magn Reson Imaging 1998;8:1154-1157.

Strainer JC, Ulmer JL, Yetkin FZ, Haughton VM, Daniels DL, Millen SJ. Functional MR of the primary auditory cortex: an analysis of pure tone activation and tone discrimination. Am J Roentgenol 1997;18: 601-610.

Ulmer JL, Bharat BB, Leighton PM, Mathews VP, Prost RW, Millen SJ, Garman JN, Horzewski D. Acoustic echo planar scanner noise and pure tone hearing thresholds: the effects of sequence repetition times and acoustic noise rates. J Comp Assist Tomogr 1998;22:480-486.

U.S. Department of Health and Human Services, Food and Drug Administration, Center for Devices and Radiological Health, Radiological Devices Branch, Division of Reproductive, Abdominal, and Radiological Devices, Office of Device Evaluation. Guidance for Industry and FDA Staff. Criteria for significant Risk Investigations of Magnetic Resonance Diagnostic Devices, July 14, 2003.

BODY PIERCING JEWELRY

Ritual or decorative body piercing is increasing in popularity. Different types of materials are used to make body piercing jewelry including ferromagnetic metals, nonferromagnetic metals, as well as non-metallic materials. The presence of body piercing jewelry that is ferromagnetic may present a problem for a patient referred for a magnetic resonance (MR) procedure or an individual in the MR environment. Other MR-related hazards may also exist for patients with body piercing jewelry made from electrically conductive materials.

Risks include uncomfortable sensations from movement or displacement that may be mild-to-moderate depending on the site of the body piercing and the ferromagnetic qualities of the piercing jewelry (e.g., mass, degree of magnetic susceptibility, etc.). In extreme cases, a serious injury may occur due to adverse interactions between the ferromagnetic jewelry and the MR system. In addition, for body piercing jewelry made from electrically conducting materials, there is a possibility of MRI-related heating that could cause burns.

Because of potential MR safety issues, metallic body piercing jewelry should be removed prior to entering the MR environment. However, patients or individuals with body piercings are often reluctant to remove their jewelry or other similar objects for a variety of reasons.

Therefore, if it is not possible to remove metallic body jewelry or other similar objects used for piercings, the patient or individual should be informed regarding the potential risks. In addition, if the body piercing jewelry is made from ferromagnetic material, some means of stabilization (e.g., application of adhesive tape or bandage) should be used to prevent movement or displacement.

To prevent potential heating of body piercing jewelry made from conductive materials, gauze, tape, or other similar material should be used to wrap the body piercing jewelry in such a manner as to insulate it (i.e., prevent contact) from the underlying skin. This insulation should be a minimum of 1-cm in thickness. The patient should be instructed to immediately inform the MR system operator if any heating or other unusual sensation occurs in association with the body piercing jewelry during the MR procedure. As always, the patient should be monitored continuously during the MR examination to ensure safety.

REFERENCE

http://www.MRIsafety.com

CLAUSTROPHOBIA, ANXIETY, AND EMOTIONAL DISTRESS IN THE MR ENVIRONMENT*

The increasing availability and capabilities of magnetic resonance (MR) studies to improve medical diagnosis and prognosis has dramatically increased the number of MR procedures performed worldwide. Thus, many more first-time and repeat patients are undergoing these MR examinations for an ever-widening spectrum of medical indications. Increasing proportions of these procedures are performed on patients suffering from unstable medical and psychological illnesses. For many of the patients who undergo MR procedures every year, the experience may cause great emotional distress. Referring physicians, radiologists, and technologists are best prepared to manage affected patients if they understand the etiology of the problem and know the appropriate maneuver or intervention to implement for treatment of the condition.

The experience of "psychological distress" in the MR environment includes all subjectively unpleasant experiences that are directly attributable to the MR procedure. Distress for the patient undergoing an MR procedure can range from mild anxiety that can be managed simply with minimal reassurance to a full-blown panic attack that requires psychiatric intervention. Severe psychological distress reactions to MR examinations, namely anxiety and panic attacks, are typically characterized by the rapid onset of at least four of the following clinical signs: fear of losing control or dying, nausea, paresthesias, palpitations, chest pain, faintness, dyspnea, feeling of choking, sweating, trembling, vertigo, or depersonalization.

Many symptoms of panic attack mimic over activity of the sympathetic nervous system, prompting concern that catecholamine responses may

15

precipitate cardiac arrhythmias and/or ischemia in susceptible patients during the MR procedure. However, this has not been reported in a clinical MR setting or any other similar situation. Nevertheless, it is advisable that, in a medically unstable patient, physiologic monitoring be a routine component of the MR procedure. Pre-emptive efforts to minimize patient distress are the most important factors in preventing or containing a panic attack in susceptible patients.

In the mildest form, distress is the normal amount of anxiety any reasonable person will experience when undergoing a diagnostic procedure. Moderate distress severe enough to be described as a dysphoric psychological reaction has been reported by as many as 65% of the patients examined by MR imaging. The most severe forms of psychological distress described by patients are claustrophobia, and anxiety or panic attacks.

Claustrophobia is a disorder characterized by the marked, persistent and excessive fear of enclosed spaces. In such affected individuals, exposure to enclosed spaces such as the MR environment, but no other situations or stimuli, almost invariably provokes an immediate anxiety response that in it's most extreme form is indistinguishable from a panic attack described above.

The actual incidence of distress in the MR environment is highly variable across studies in part reflecting differences in outcome measures used to measure distress. Some studies indicated that as many as 20% of individuals attempting to undergo an MR imaging procedure can't complete it secondary to serious distress such as claustrophobia or other similar sensation. In contrast, other investigators have reported that as few as 0.7% of individuals have incomplete or failed MR procedures due to distress. An estimate of the number of patients that experience distress that compromises either their own well-being or the diagnostic utility of the MR procedure is 3 to 5% of all studies.

There are no perfect predictors of distress in the MR environment. In fact, different studies cite opposing results such as which gender has greater difficulty tolerating the MR studies. Obviously, these differences may reflect cultural, socioeconomic or other influences.

THE IMPACT OF EMOTIONAL DISTRESS

Patient distress can contribute to adverse outcomes for the MR procedure. These adverse outcomes include unintentional exacerbation of patient distress, a compromise in the quality and, thus, the diagnostic aspects of the imaging study and decreased efficiency of the imaging facility due to delayed, cancelled, or prematurely terminated studies. Patient compliance during an MR procedure, such as the ability to remain in the MR system and hold still long enough to complete the study, is of paramount importance to achieving a high quality, diagnostic examination.

If a good quality study can't be obtained, the patient may require an invasive diagnostic examination in place of the safer, less painful and risky MR procedure. Thus, for the distressed patient unable to undergo an MR procedure, there are typically clinical, medico-legal, and economic related considerations implications.

Increasing pressure to use MR system time efficiently to cover the costs of expensive diagnostic imaging equipment puts greater stress on both staff and patients. The ability of referring physicians, radiologists, and MRI technologists to detect patient distress at the earliest possible time, to discover the source of the distress, and then provide appropriate intervention can greatly improve patient comfort, quality of imaging studies and efficiency of the MRI facility.

FACTORS THAT CONTRIBUTE TO DISTRESS

Many factors contribute to distress experienced by certain patients undergoing MR procedures. Most commonly cited are concerns about the physical environment of the MR system. Also well documented are the anxieties associated with the underlying medical problem necessitating the MR procedure. Certain individuals, such as those with psychiatric illnesses, may be predisposed to suffer greater distress due to MR procedures.

The physical environment of the MR system is clearly one important source of distress to patients. Sensations of apprehension, tension, worry, claustrophobia, anxiety, fear, and even panic attacks have been directly attributed to the confining dimensions of the interior of the MR system. For example, for certain types of MR systems, the patient's face may be three to ten inches from the inner portion of the MR system, prompting feelings of uncontrolled confinement and detachment. Importantly, newer short-bore and wide-bore high field strength MR systems tend to alleviate these sensations.

Similar distressing sensations have been attributed to the other aspects of the MR environment including the prolonged duration of the MR examination, the gradient magnetic field-induced acoustic noise, the temperature and humidity within the MR system, and the distress related to the restriction of movement. Noise in of itself can be a source of stress and, thus, particularly troublesome to certain patients undergoing MR procedures. Other studies have reported stress related to the administration of an intravenous MRI contrast agent. Additionally, the MR system may produce a feeling of sensory deprivation, which is also known to be a precursor of severe anxiety states.

MR systems that have an architecture that utilizes a vertical magnetic field offer a more open design that is presumed to reduce the frequency of distress associated with MR procedures. The latest versions of these "open" MR systems, despite having static magnetic field strengths of 0.3-Tesla or lower have improved technology (i.e., faster gradient fields, optimized surface coils, etc.) that permit acceptable image quality for virtually all types of standard, diagnostic imaging procedures. "High-field open" MR systems operating at 0.7-Tesla and 1.0-Tesla are now commercially available and these systems may be more acceptable to patients with feelings of distress. Also, the latest generation, high-field-strength (1.5-Tesla and 3.0-Tesla) MR systems have shorter-and wider bore configurations that likely mitigate feelings of being enclosed that may be experienced by patients undergoing MR procedures.

In 1993, a specially designed, low-field-strength (0.2-Tesla) MR system (Artoscan, Lunar Corporation/General Electric Medical Systems, Madison, WI and Esaote, Genoa, Italy) became commercially available for MR imaging of extremities. The use of this dedicated extremity MR system provides an accurate, reliable, and relatively inexpensive means (i.e., in comparison to the use of a whole-body MR system) of evaluating various types of musculoskeletal abnormalities. Therefore, utilization of the extremity MR system to assess musculoskeletal pathology is a viable and acceptable alternative to the use of whole-body MR systems. This is particularly the case since the image quality and diagnostic capabilities for the evaluation of the knee and other extremities has been reported to be comparable to mid- or high-field-strength MR systems for certain musculoskeletal applications.

The architecture of the extremity MR system has no confining features or other aspects that would typically create patient-related problems. This is because only the body part that requires imaging is placed inside the magnet bore during the MR examination.

A preliminary study reported that 100% of the MR examinations that were initiated were completed without being interrupted or cancelled for patient-related problems. The unique design of the extremity MR system likely contributed to the totally successful completion of MR procedures in the patients of this study. Furthermore, these findings represent a dramatic improvement compared with the published incidence of patient distress that tends to interrupt or prevent the completion of MR procedures using whole-body MR systems. Another dedicated-extremity MR system with an open design permits MR imaging of the shoulder. Finally, there is also a high-field-strength extremity MR system (OrthOne, ONI) that has a configuration that is conducive to preventing psychological distress or claustrophobia.

Adverse psychological reactions are sometimes associated with the MR procedures simply because the examination may be perceived by the patient as a "dramatic" medical test that has an associated uncertainty of outcome, such that there may be a fear of the presence of disease or other abnormality. In fact, any type of diagnostic imaging procedure may produce a certain amount of anxiety for the patient.

Patients with pre-existing psychiatric disorders may be at greater risk for experiencing distress in the MR environment. One problem that arises more often in this population is the refusal of the prescribed MR procedure by the patient. Frequently, the cause for refusal is an inadequate understanding of why the procedure was ordered and what the actual procedure involves.

No specific reports of the differential frequency of distress or adverse outcomes for MR procedures in these patients compared to non-psychiatrically impaired patients has been published to date. However, specific inquiry should be made to identify patients with pre-existing anxiety disorders including claustrophobia, generalized anxiety disorder, post-traumatic stress disorder, and obsessive-compulsive disorder in order to increase anxiety minimizing efforts in these patients (see below).

Patients with other psychiatric illnesses such as depression and any illness complicated by thought disorder such as schizophrenia and manic-depressive disorder may also be at increased risk for distress in the MR environment. Patients with psychiatric illnesses may, under normal circumstances, be able to tolerate the MR environment without a problem, as is clear form the thousands who participate in clinical neuroimaging research studies each year. However, the increased stress due to their medical illness or fear of medical illness may exacerbate their psychiatric symptoms to such an extent that they may have difficulty complying with MR procedures. At the very least, patients with psychiatric illnesses may require more time and patience to provide the appropriate level of preparatory information for the MR examination.

TECHNIQUES TO MINIMIZE PATIENT DISTRESS

A stepwise set of procedures exists for minimizing subjective distress for patients undergoing MR procedures. Certain measures to alleviate patient distress should be employed for all studies. A number of other measures will be required if the patient is experiencing significant distress due to factors as described above. Finally, other distress-alleviation techniques will only be necessary for patients with co-existing psychiatric illness or other special problems. Coordination of these efforts among the referring physician, the radiologist, the MR technologist and the MRI facility support staff is crucial. These methods have been described in the literature and are summarized in Table 1.

Table 1. Recommended techniques for managing patients with distress related to MR procedures.

1. Prepare and educate the patient concerning specific aspects of the MR examination (e.g., MR system dimensions, gradient noise, intercom system, etc.).

2. Allow an appropriately-screened relative or friend to remain with the patient during the MR procedure.

3. Maintain physical or verbal contact with the patient during the MR procedure.

4. Use an MR-compatible stereo system to provide music to the patient and to minimize gradient magnetic field-induced noise.

5. Use an MR-compatible monitor to provide a visual distraction to the patient.

6. Use a virtual reality environment system to provide audio and visual distraction.

7. Place the patient in a prone position inside the MR system.

8. Position the patient feet-first instead of head-first into the MR system.

9. Use special mirrors or prism glasses to redirect the patient's line of sight.

10. Use a blindfold so that the patient is not aware of the close surroundings.

11. Use bright lights inside of the MR system.

12. Use a fan inside of the MR system to provide adequate air movement.

13. Use vanilla scented oil or other similar aroma therapy.

14. Use relaxation techniques such as controlled breathing or mental imagery.

15. Use systematic desensitization.

16. Use medical hypnosis.

17. Use a sedative or other similar medication.

For All Patients Undergoing MR Procedures

Referring clinicians should take the time to explain the rationale for the MR procedure and what he/she expects to learn from the results with respect to the implications for treatment and prognosis. Importantly, the clinician should schedule time with the patient to communicate the results of the MR procedure.

The single most important step is to educate the patient about the specific aspects of the MR examination that are known to be particularly difficult. This includes conveying, in terms that are understandable to the patient, the internal dimensions of the MR system, the level of gradient magnetic field-induced acoustic noise to expect, and the estimated time duration of the examination.

Studies have documented a decrease in the incidence of premature termination of MR procedures when patients are provided with more detailed information regarding the examination. This may be effectively accomplished by means of providing the patient time to view an educational video tape or written brochure supplemented by a question and answer session with an MR-trained healthcare worker prior to the MR procedure.

Some authors have proposed adding a pre-scan "fear assessment" to help predict patients who will experience psychological problems related to the MR procedure. Such a brief questionnaire could be used to help elicit questions and concerns from patients and to provide guidance to staff about which distress minimization strategies are most likely to be effective for the patient.

Upon entering MRI facility, patients treated with respect and are welcomed into a calm environment experience less distress. Many details of patient positioning in the MR system can increase comfort and minimize distress. Taking time to ensure comfortable positioning with adequate padding and blankets to alleviate undue discomfort or pain is also important. Adequate ear protection should be provided routinely to decrease acoustic noise from the MR system, as needed (i.e., this is typically not required for low-field-strength MR systems). Demonstration of the two-way intercom system or other monitoring technique to reassure the patient that the MR staff is readily available during the examination is also important.

For Mildly to Moderately Distressed Patients

If a patient that continues to experience distress after the afore-mentioned measures are implemented, additional interventions are required. Frequently, all that is necessary to successfully complete an MR examination is to allow an appropriately-screened relative or friend to remain with the patient during the procedure. A familiar person in the MR system room often helps anxious patients because they develop an increased sense of security. If a supportive companion is not present, then simply having the MR staff maintain verbal contact via the intercom system or physical contact by having a staff person remain in the MR system room with the patient during the examination will frequently decrease psychological distress.

Placing the patient in a prone position inside the MR system so that the patient can visualize the opening of the bore provides a sensation of being inside a device that is more spacious and alleviates the "closed-in" feeling associated with the supine position. Prone positioning of the patient may not be a practical alternative if MR imaging requires the use of flat local coils or if the patient has underlying medical conditions (e.g., shortness of breath, the presence of chest tubes, etc.) that preclude the use of this position. Another method of positioning the patient that may help is to place the individual feet-first instead of head-first into the MR system.

MR system-mounted mirrors or prism glasses can be used to permit the patient to maintain a vertical view of the outside of the MR system in order to minimize phobic responses. Using a blindfold so that the patient is unaware of the close surroundings has also been suggested to be an effective technique for enabling anxious patients to successfully undergo MR procedures.

The environment of the MR system may be changed to optimize the management of apprehensive patients. For example, the presence of higher lighting levels tends to make most individuals feel less anxious. Therefore, the use of bright lights at either end and inside of the MR system can produce a less imposing environment for the patient. In addition, using a fan inside of the MR system to provide more air movement will help reduce the sensation of confinement and lessen any tissue heating that may result when high levels of RF power absorption are used for MR imaging. Some MR staff members have reported that placing a cotton pad moistened with a few drops of essential lemon or vanilla oil or other similar form of aroma therapy in the MR system for the patient to receive olfactory stimulation can also reduce distress.

Electronic devices that utilize compressed air or other appropriate means to transmit music or audio communication through headphones have been developed specifically for use with MR systems. MR-compatible music/audio systems may be acquired from a commercial vendor or can be made by adapting an airplane pneumatic earphone headset. MR-compatible music systems can be used to provide calming music to the patient and, with the proper design, help to minimize exposure to gradient magnetic field-induced acoustic noise. Reports have indicated that the use of these devices has been successful in reducing symptoms of anxiety in patients during MR procedures. In addition, it is now possible to provide visual stimulation to the patient via monitors or special goggles. Use of visual stimuli to distract patients may also reduce distress. Finally, a system has been developed to provide a virtual reality environment for the patient that may likewise serve as an acceptable means of audio and visual distraction from the MR procedure (this device is also used for fMRI studies).

For Severely Distressed or Claustrophobic Patients

Patients who are at high risk for severe distress in the MR environment and can be identified as such by their referring clinician or by the scheduling MR staff person could be offered the opportunity to have pre-MR procedure behavioral therapy. MR procedures that were conducted in patients that previously refused or were unable to tolerate the MR environment have been reported to be successful as a result of treatment with relaxation techniques, systematic desensitization, and medical hypnosis.

In the majority of MR facilities, patients that are severely affected by claustrophobia, anxiety, or panic attacks in response to MR procedures are usually need sedation when other attempts to counteract their distress fail. Using short-acting sedatives such as lorazepam, diazepam, alprazolam, or intranasal midazolam or one of the newer anxiolytic medications may be the only means of managing the patient with a high degree of anxiety related to the MR procedure. However, the use of sedatives in patients prior to and during MR procedures may not be required in all instances, nor is it always practical.

Of special note, anxious patients with a history of substance abuse who are in recovery programs may not be willing to take mind altering medications because this is typically absolutely contraindicated in their treatment. These patients should be referred for behavioral therapy before MR procedures. In all cases, one or more of the recommended, non-medication-related techniques indicated in Table 1 should be attempted before immediately electing to use a sedative in distressed patients in the MR environment.

Sedation in the MR environment is not a totally benign procedure. Confusion and respiratory compromise as well as other untoward reactions have been reported in response to relatively modest doses of commonly employed sedatives. If a sedative is used in a patient in preparation for an MR procedure, it should be understood that the use of this medication involves several important patient management considerations. For example, the time when the patient should be administered the medication for optimal effect prior to the examination should be considered along with the possibility that there may be an adverse reaction to the drug. Provisions should be available for an area to permit adequate recovery of the patient after the MR procedure. The patient should also have someone available to provide transportation from the MRI facility after receiving medication.

SUMMARY AND CONCLUSIONS

Continued advances in MR technology coupled with more sophisticated MR procedures that aid in the diagnosis and management of an ever increasing number of medical conditions ensures that the number of patients undergoing MR procedures will continue to increase each year.

Thus, the number of patients at risk for experiencing distress during these procedures will likely increase.

Providing a comprehensive explanation of the MR procedure to the patient is perhaps the single most important measure to minimize distress and possible associated adverse outcomes. It is not unreasonable to expect that adherence to the various recommendations developed to alleviate or prevent psychological distress will improve patient comfort and, thereby, greatly reduce the number of incomplete or poor quality studies.

[*Portions of this content were excerpted with permission from Gollub RL and Shellock FG. Claustrophobia, Anxiety, and Emotional Distress in the Magnetic Resonance Environment. In, Magnetic Resonance Procedures: Health Effects and Safety. FG Shellock, Editor, CRC Press, Boca Raton, FL, 2001.]

REFERENCES

Gollub, R.L., and Shellock, F.G., Claustrophobia, Anxiety, and Emotional Distress in the Magnetic Resonance Environment In, Magnetic Resonance Procedures: Health Effects and Safety. FG Shellock, Editor, CRC Press, Boca Raton, FL, 2001.

McGuinness, T.P., Hypnosis in the treatment of phobias: a review of the literature, Am. J. Clin. Hypnosis, 26, 261, 1984.

Murphy, K.J. and Brunberg, J.A., Adult claustrophobia, anxiety and sedation in MRI, Magnetic Resonance Imaging, 15, 51, 1997.

Sarji, S.A., Abdullah, B.J., Kumar, G., et al., Failed magnetic resonance imaging examinations due to claustrophobia., Australas Radiol, 42, 293, 1998.

Shellock, F.G., Claustrophobia, anxiety, and panic disorders associated with MR procedures, in Magnetic Resonance: Bioeffects, Safety, and Patient Management., F.G. Shellock, and Kanal, E., Editor, Lippincott-Raven Press, New York, 1996, pp. 65

Shellock, F.G. and Kanal, E., Policies, guidelines, and recommendations for MR imaging and patient management, J. Mag. Reson. Imaging 1, 97, 1991.

Shellock, F.G., Stone, K.R., Resnick, D., et al., Subjective perceptions of MRI examinations performed using an extremity MR system, Signals, 32, 16, 2000.

Weinreb, J., Maravilla, K.R., Peshock, R., et al., Magnetic resonance imaging: improving patient tolerance and safety, Am. J. Roentgen., 143, 1285, 1984.

EXTREMITY MR SYSTEM

In 1993, a specially designed, low-field-strength (0.2-Tesla MR system, Artoscan, Lunar Corporation Madison, WI and General Electric Medical Systems, Milwaukee, WI/Esaote, Genoa, Italy) MR system became available for MR imaging of extremities. This MR system uses a small-bore permanent magnet to image feet, ankles, knees, hands, wrists, and elbows.

The ergonomic design of the extremity MR system is such that the body part of interest is placed inside the magnet bore, with the patient positioned in a seated or supine position (i.e., depending on the body part that is imaged). The entire extremity MR system weighs approximately 800 kg, has a built-in radiofrequency shield, multiple body-part-specific extremity coils, and 10 mT/m magnetic gradients.

A major advantage of this extremity MR system is that it can be sited in a relatively small space (e.g., approximately 100 square feet) without the need for a special power source, magnetic field shielding, or radiofrequency shielding. Of note is that MR imaging using the extremity MR system has been demonstrated to provide a sensitive, accurate, and reliable assessment of various forms of musculoskeletal pathology.

Because of the unique design features of the extremity MR system (which includes a low-field-strength static magnetic field with a relatively small fringe field) and in consideration of how patients are positioned for MR procedures using this device (i.e., only the body part imaged is placed within the magnet bore while the rest of the body remains outside), it was suggested that it may be possible to safely image patients with aneurysm clips, even if they are made from ferromagnetic materials. Furthermore, it may be possible to perform extremity MR imaging in patients with cardiac pacemakers or implantable cardioverter defibrillators (ICDs). Therefore, investigations were conducted to specifically evaluate these safety issues.

PATIENTS WITH FERROMAGNETIC ANEURYSM CLIPS

A study was performed to assess the magnetic field interaction for a variety of different aneurysm clips exposed to the 0.2-Tesla extremity MR system. Twenty-two different types of aneurysm clips were evaluated including those made from nonferromagnetic, weakly ferromagnetic, and ferromagnetic materials (i.e., a Heifetz aneurysm clip made from 17-7PH and a Yasargil, Model FD aneurysm clip). The results indicated that none of the aneurysm clips tested displayed substantial magnetic field interactions in association with exposure to the 0.2-Tesla extremity MR system.

Due to the design features of the extremity MR system and in consideration of how patients are positioned for MR procedures using this device (i.e., the head does not enter the magnet bore), it is considered safe to perform MR imaging in patients with the specific aneurysm clips that have been evaluated for magnetic field interactions. These findings effectively permit an important diagnostic imaging modality to be used to evaluate the extremities of patients with suspected musculoskeletal abnormalities using the Artoscan MR system.

Notably, various studies have reported that patients with Heifetz (17-7PH) and Yasargil, Model FD aneurysm clips (i.e., two of the clips evaluated in the study using the Artoscan scanner) should not undergo MR imaging using MR systems with conventional designs because of the strong attraction shown by these aneurysm clips, which would pose a substantial hazard to patients.

PATIENTS WITH CARDIAC PACEMAKERS AND IMPLANTABLE CARDIOVERTER DEFIBRILLATORS

In general, patients with cardiac pacemakers and implantable cardioverter defibrillators (ICDs) are not permitted to undergo MR procedures. However, due to the design of the Artoscan extremity MR system it may be possible to safely perform MR examinations in patients with these implanted devices.

Since the magnetic fringe field of the extremity MR system is contained in close proximity to the 0.2-Tesla magnet and this scanner has an integrated Faraday cage, only the patient's extremity is exposed to the MR-related electromagnetic fields when a procedure is performed. Notably, it is not possible for the MR system's gradient or RF electromagnetic fields to induce currents in a pacemaker or ICD because the patient's thorax (i.e., where the pacemaker or ICD is typically implanted) remains outside of the MR system.

Ex vivo experiments were conducted to assess MR safety for seven different cardiac pacemakers and seven different implantable cardioverter defibrillators manufactured by Medtronic, Inc. (Minneapolis, MN). The following devices were tested:

Device	Name	Model
Pacemaker	Elite II	7086
Pacemaker	Thera D	7944
Pacemaker	Thera D	7960I
Pacemaker	Thera DR	7962i
Pacemaker	Thera SR	8940
Pacemaker	Kappa	400
Pacemaker	Kappa	700
ICD	PCD	7217D
ICD	Jewel	7219D
ICD	Jewel Plus	7220C
ICD	Micro Jewel	7221Cx
ICD	Micro Jewel II	7223Cx
ICD	Prototype	7250G
ICD	Prototype	7271

Magnetic field interactions were assessed relative to the 0.2-Tesla static magnetic field of the extremity MR system. Additionally, the cardiac pacemakers and implantable cardioverter defibrillators were operated with various lead systems attached while immersed in a tank containing physiologic saline. This apparatus was used to simulate the thorax. The experimental set-up was oriented in parallel and perpendicular positions relative to the closest part of the MR system to simulate patients undergoing MR procedures using this extremity scanner.

MR studies were performed on a phantom using T1-weighted spin echo and gradient echo sequences. Various functions of the pacemakers and

ICDs were evaluated before, during, and after MR imaging. The results of these tests indicated that magnetic field interactions did not present problems for the devices. The activation of the pacemakers and cardioverter defibrillators did not substantially affect image quality during MR imaging. Importantly, the operation of the extremity MR system produced no alterations in the function of the cardiac pacemakers and implantable cardioverter defibrillators. Therefore, in consideration of these data and in view of how patients are positioned during MR procedures using the Artoscan extremity MR system (i.e., the thorax does not enter the magnet bore), it should be safe to perform examinations in patients with the specific cardiac pacemakers and implantable cardioverter defibrillators evaluated in this study.

REFERENCES

Ahn JM, Sartoris DJ, Kank HS, Resnick D. Gamekeeper thumb: comparison of MR arthrography with conventional arthrography and MR imaging in cadavers. Radiology 1998;206:737-744.

Barile A, Masiocchi C, Mastantuono M, Passariello R, Satragno L. The use of a "dedicated" MRI system in the evaluation of knee joint diseases. Clinical MRI 1995;5:79-82.

Franklin PD, Lemon RA, Barden HS. Accuracy of imaging the menisci on an in-office, dedicated, magnetic resonance imaging extremity system. Am J Sports Med 1998;25:382-388.

Kersting-Sommerhoff B, Hof N, Lenz M, Gerhardt P. MRI of peripheral joints with a low-field dedicated system: a reliable and cost-effective alternative to high-field units? Eur Radiol 1996;6:561-565.

Maschicchi C. Dedicated MR system and acute trauma of the musculo-skeletal system. Eur J Radiol 1996;22:7-10.

Peterfy CG, Roberts T, Genant HK. Dedicated extremity MR imaging: an emerging technology. Radiol Clin North Am 1997;35:1-20.

Shellock FG. Pocket Guide to MR Procedures and Metallic Objects: Update 2001, Seventh Edition, Lippincott Williams & Wilkins Healthcare, Philadelphia, 2001.

Shellock FG. Magnetic Resonance Procedures: Health Effects and Safety. CRC Press, LLC, Boca Raton, FL, 2001.

Shellock FG, Crues JV. Aneurysm clips: Assessment of magnetic field interaction associated with a 0.2-T extremity MR system. Radiology 1998;208:407-409.

Shellock FG, Kanal E. Magnetic Resonance: Bioeffects, Safety, and Patient Management. Second Edition, Lippincott-Raven Press, New York, 1996.

Shellock FG, O'Neil M, Ivans V, Kelly D, O'Connor M, Toay L, Crues JV. Cardiac pacemakers and implantable cardiac defibrillators are unaffected by operation of an extremity MR system. Am J Roentgenol 1999;72:165-17.

Shellock FG, Stone K, Crues JV. Development and clinical applications of kinematic MRI of the patellofemoral joint using an extremity MR system. Med Sci Sport Exerc 1999;31:788-791.

Shellock FG, Mullen M, Stone K, Coleman M, Crues JV. Kinematic MRI evaluation of the effect of bracing on patellar positions: qualitative assessment using an extremity MR system. Journal of Athletic Training 2000;35:44-49.

GUIDELINES FOR THE MANAGEMENT OF THE POST-OPERATIVE PATIENT REFERRED FOR A MAGNETIC RESONANCE PROCEDURE*

There is controversy and confusion regarding the issue of performing a magnetic resonance (MR) procedure during the post-operative period in a patient with a metallic implant or device. Studies in the peer-reviewed literature have supported that, if a metallic object is a "passive implant" (i.e., there is no electronically- or magnetically-activated component associated with the operation of the device) and it is made from a nonferromagnetic material (e.g., Titanium, Titanium alloy, Nitinol, etc.), the patient with the object may undergo an MR procedure immediately after implantation using an MR system operating at 1.5-Tesla or less. In fact, there are several reports that describe placement of vascular stents and other implants using MR-guided procedures that include the use of high-field-strength (1.5-Tesla) MR systems. Additionally, a patient or individual with a nonferromagnetic, passive implant is allowed to enter the MR environment associated with a scanner operating at 1.5-Tesla or less immediately after implantation of such an object. Currently, there is little data to provide guidelines for MR environments using MR systems operating at 3-Tesla or higher.

For an implant or device that exhibits "weakly magnetic" qualities (e.g., certain stents, coils, filters, atrial septal defect occluders, ventricular septal defect occluders, patent ductus arteriosus occluders, etc.), it is typically necessary to wait a period of six to eight weeks after implantation before performing an MR procedure or allowing the individual or patient to enter the MR environment associated with a scanner operating at 1.5-Tesla or less. For example, certain intravascular and intracavitary coils, stents, filters, and cardiac occluders designated as being "weakly" ferromagnetic become firmly incorporated into tissue six to eight weeks following placement. In these cases, retentive or counter-forces provided by tissue ingrowth, scarring, or granulation essentially serve to prevent these objects from presenting risks or hazards to patients or individuals in the MR environment.

However, implants or devices that may be "weakly magnetic" but are rigidly fixed in the body (e.g., bone screws) may be studied immediately after implantation. Specific information pertaining to the recommended post-operative waiting period may be found in the labeling or product insert for a "weakly magnetic" implant or device.

Special Note: If there is any concern regarding the integrity of the tissue with respect to its ability to retain the implant or object in place or the implant cannot be properly identified, the patient or individual should not be exposed to the MR environment.

[*The document, **Guidelines for the Management of the Post-Operative Patient Referred for a Magnetic Resonance Procedure**, was developed by the Medical, Scientific, and Technology Advisory Board and the Corporate Advisory Board of the Institute for Magnetic Resonance Safety, Education, and Research (IMRSER), 2003 and published with permission.]

REFERENCES

Ahmed S, Shellock FG. Magnetic resonance imaging safety: implications for cardiovascular patients. Journal of Cardiovascular Magnetic Resonance 2001;3:171-181.

Bueker A, et al. Real-time MR fluoroscopy for MR-guided iliac artery stent placement. J Magn Reson Imaging 2000;12:616-622.

http://www.IMRSER.org

Manke C, Nitz WR, Djavidani B, et al. MR imaging-guided stent placement in iliac arterial stenoses: A feasibility study. Radiology 2001;219:527-534.

Rutledge JM, Vick GW, Mullins CE, Grifka RG. Safety of magnetic resonance immediately following Palmaz stent implant: a report of three cases. Catheter Cardiovasc Interv 2001;53:519-523.

Sawyer-Glover A, Shellock FG. Pre-MRI procedure screening: recommendations and safety considerations for biomedical implants and devices. J Magn Reson Imaging 2000;12:92-106.

Shellock FG. MRI and post-op patients. Signals, No. 34, Issue 3, p. 8, 2000.

Shellock FG. MR safety update 2002: Implants and devices. J Magn Reson Imaging 2002;16:485-496.

Shellock FG. Magnetic Resonance Procedures: Health Effects and Safety. CRC Press, LLC, Boca Raton, FL, 2001.

Shellock FG. Reference Manual for Magnetic Resonance Safety: 2003 Edition, Amirsys, Inc., Salt Lake City, UT, 2003.

Spuentrup E, et al. Magnetic resonance-guided coronary artery stent placement in a swine model. Circulation 2002;105:874-879.

Teitelbaum GP, Bradley WG, Klein BD. MR imaging artifacts, ferromagnetism, and magnetic torque of intravascular filters, stents, and coils. Radiology 1988;166:657-664.

Teitelbaum GP, Lin MCW, Watanabe AT, et al. Ferromagnetism and MR imaging: safety of cartoid vascular clamps. Am J Neuroradiol 1990;11:267-272.

Teitelbaum GP, Ortega HV, Vinitski S, et al. Low artifact intravascular devices: MR imaging evaluation. Radiology 1988;168:713-719.

Teitelbaum GP, Raney M, Carvlin MJ, et al. Evaluation of ferromagnetism and magnetic resonance imaging artifacts of the Strecker tantalum vascular stent. Cardiovasc Intervent Radiol 1989;12:125-127.

GUIDELINES FOR SCREENING PATIENTS FOR MR PROCEDURES AND INDIVIDUALS FOR THE MR ENVIRONMENT*

The establishment of thorough and effective screening procedures for patients and other individuals is one of the most critical components of a program that guards the safety of all those preparing to undergo magnetic resonance (MR) procedures or to enter the MR environment. An important aspect of protecting patients and individuals from MR system-related accidents and injuries involves an understanding of the risks associated with the various implants, devices, accessories, and other objects that may cause problems in this setting. This requires constant attention and diligence to obtain information and documentation about these objects in order to provide the safest MR setting possible. In addition, because most MR-related incidents have been due to deficiencies in screening methods and/or a lack of properly controlling access to the MR environment (especially with regard to preventing personal items and other potentially problematic objects into the MR system room), it is crucial to set up procedures and guidelines to prevent such incidents from occurring.

Magnetic Resonance (MR) Procedure Screening for Patients

Certain aspects of screening patients for MR procedures may take place during the scheduling process. This should be conducted by a healthcare worker specially trained in MR safety (i.e., this person should be trained to understand the potential hazards and issues associated with the MR environment and MR procedures and be familiar with all of the information contained on the screening forms for patients and individuals). During this time, it may be ascertained if the patient has any implant that may be contraindicated for the MR procedure (e.g., a ferromagnetic aneurysm clip, pacemaker, etc.) or if there is any condition that needs careful consideration (e.g., the patient is pregnant, has a disability, etc.). Preliminary screening helps to prevent scheduling patients that may be inappropriate candidates for MR examinations.

After preliminary screening, every patient must undergo comprehensive screening in preparation for a magnetic resonance (MR) procedure (i.e., MR imaging, MR angiography, functional MRI, MR spectroscopy). Comprehensive patient screening involves the use of a printed form to document the screening procedure, a review of the information on the screening form, and a verbal interview to verify the information on the form and to allow discussion of any question or concern that the patient may have. An MR-safety trained healthcare worker must conduct this aspect of patient screening.

The screening form for patients developed by Sawyer-Glover and Shellock (2000) was revised in consideration of new information in the peer-reviewed literature. This two-page form entitled, *Magnetic Resonance (MR) Procedure Screening Form for Patients*, was also created in conjunction with the Medical, Scientific, and Technology Advisory Board and the Corporate Advisory Board of the Institute for Magnetic Resonance Safety, Education, and Research (IMRSER) (Figure 1). A "downloadable" version of this form may be obtained from the web sites, www.IMRSER.org and www.MRIsafety.com.

Page one of this screening form requests general patient-related information (name, age, sex, height, weight, etc.) as well as information regarding the reason for the MR procedure and/or symptoms that may be present. Pertinent information about the patient is required not only to ensure that the medical records are up-to-date, but also in the event that the MRI facility needs to contact the referring physician for additional information regarding the examination or to verify the patient's medical condition.

The form requests information regarding a prior surgery or operation to help determine if there may be an implant or device present that could create a problem for the patient. Information is also requested pertaining to prior diagnostic imaging studies that may be helpful to review for assessment of the patient's condition.

Next, important questions are posed in an effort to determine if there are possible problems or issues that should be discussed with the patient prior to permitting entry to the MR environment. For example, information is requested regarding any problem with a previous MR examination, an injury to the eye involving a metallic object, or any injury from a metallic object or foreign body. Questions are posed to obtain information about current or recently taken medications as well as the presence of drug allergies. There are also questions asked to assess past and present medical conditions that may affect the MR procedure or the use of an MRI contrast agent in the patient.

At the bottom of page one, there is a section for female patients that poses questions that may impact MR procedures. For example, questions regarding the date of the last menstrual period, pregnancy or late menstrual period are included. A definite or possible pregnancy must be identified prior to permitting the patient into the MR environment so that the risks vs. the benefits of the MR procedure can be considered and discussed with the patient. MR procedures should only be performed in pregnant patients to address important clinical questions. MR facilities should have a clearly defined procedure to follow in the event that the patient has a confirmed or possible pregnancy.

Questions pertaining to the date of the last menstrual period, use of oral contraceptives or hormonal therapy, and fertility medication are necessary for female patients undergoing MR procedures that are performed to evaluate breast disease or for OB/GYN applications, as these may alter the tissue appearance on MR imaging. An inquiry about breastfeeding is included in case the administration of MRI contrast media is being considered for nursing mothers.

The second page of the form has a statement at the top that indicates: "WARNING: Certain implants, devices, or objects may be hazardous to you and/or may interfere with the MR procedure (i.e., MRI, MR angiography, functional MRI, MR spectroscopy). <u>Do not enter</u> the MR system room or MR environment if you have any question or concern regarding an implant, device, or object. Consult the MRI Technologist or Radiologist BEFORE entering the MR system room. The MR system magnet is ALWAYS on."

Next, there is a section that lists various implants, devices, and objects to identify anything that could be hazardous to the patient undergoing the MR procedure or that may produce an artifact that could interfere with the interpretation of the MR procedure. In general, these items are arranged on the checklist in order of the relative safety hazard (e.g., aneurysm clip, cardiac pacemaker, implantable cardioverter defibrillator, electronic implant, etc.), followed by items that may simply produce imaging artifacts that could be problematic for the interpretation of the MR procedure. Additionally, questions are posed to determine if the patient has a breathing problem, movement disorder, or claustrophobia because these are known to present difficulties for MR procedures.

Figures of the human body are included on the second page of the screening form for the patient as a means of showing the location of any object inside of or on the body. This information is particularly useful so that the patient may indicate the approximate position of any object that may be hazardous or that could interfere with the interpretation of the MR procedure as a result of producing an artifact.

Page 2 of the screening form also has an *Important Instructions* section that states: "Before entering the MR environment or MR system room, you must remove all metallic objects including hearing aids, dentures, partial plates, keys, beeper, cell phone, eyeglasses, hair pins, barrettes, jewelry, body piercing jewelry, watch, safety pins, paperclips, money clip, credit cards, bank cards, magnetic strip cards, coins, pens, pocket knife, nail clipper, tools, clothing with metal fasteners, & clothing with metallic threads. Please consult the MRI Technologist or Radiologist if you have any question or concern BEFORE you enter the MR system room."

Finally, there is a statement on the *Magnetic Resonance (MR) Procedure Screening Form for Patients* that indicates hearing protection is "advised or required" to prevent possible problems or hazards related to acoustic noise. In general, this should not be an option for a patient undergoing an MR procedure on a high-field-strength MR system. By comparison, it may not be necessary for the use of hearing protection by patients undergoing MR procedures on low-field-strength MR systems.

It should be noted that undergoing previous MR procedures without incidents does not guarantee a safe subsequent MR examination. Various factors (e.g., the static magnetic field strength of the MR system, the orientation of the patient, the orientation of a metallic implant or object, etc.) can substantially change the scenario. Thus, a written screening form must be completed each time a patient prepares to undergo an MR procedure. This is not an inconsequential matter because a surgical intervention or accident involving a metallic foreign body may have occurred that could impact the safety an MR procedure or of entering the MR environment.

With the use of any type of written questionnaire, limitations exist related to incomplete or incorrect answers provided by the patient. For example, there may be difficulties associated with patients that are impaired with respect to their vision, language fluency, or level of literacy. Therefore, an appropriate accompanying family member or other individual (e.g., referring physician) should be involved in the screening process to verify any information that may impact patient safety. Versions of this form should also be available in other languages, as needed (i.e., specific to the demographics of the MRI facility).

In the event that the patient is comatose or unable to communicate, the written screening form should be completed by the most qualified individual (e.g., physician, family member, etc.) that has knowledge about the patient's medical history and present condition. If the screening information is inadequate, it is advisable to look for surgical scars on the patient and/or to obtain plain films of the skull and/or chest to search for implants that are known to be particularly hazardous in the MR environment (e.g., aneurysm clips, cardiac pacemakers, etc.).

Following completion of the *Magnetic Resonance (MR) Procedure Screening Form for Patients*, an MR-safety trained healthcare worker should review the form's content. Next, a verbal interview should be conducted by the MR-safety trained healthcare worker to verify the information on the form and to allow discussion of any question or concern that the patient may have before undergoing the MR procedure. This allows a mechanism for clarification or confirmation of the answers to the questions posed to the patient so that there is no miscommunication regarding important MR safety issues. In addition, because the patient may not be fully aware of the medical terminology used for a particular implant or device, it is imperative that this particular information on the form be discussed during the verbal interview.

After the comprehensive screening procedure is completed, any patient that is transferred by a stretcher, gurney, or wheelchair to the MR system room should be checked thoroughly and systematically for metal objects under the sheets or blankets such as ferromagnetic oxygen tanks, monitors, or other objects that could pose a hazard.

Magnetic Resonance (MR) Environment Screening for Individuals

Before any "non-patient" individual (e.g., MRI technologist, MR support person, patient's family member, visitor, allied health professional, physician, maintenance worker, custodial worker, fire fighter, security officer, etc.) is allowed into the MR environment, he or she must be screened by an MR-safety trained healthcare worker. Proper screening for individuals involves the use of a printed form to document the screening procedure, a review of the information on the form, and a verbal interview to verify the information on the form and to allow discussion of any question or concern that the individual may have before permitting entry to the MR environment.

In general, magnetic resonance (MR) screening forms were developed with patients in mind and, therefore, pose many questions that are inappropriate or confusing to other individuals that may need to enter the MR environment. Therefore, a screening form was recently created specifically for individuals that need to enter the MR environment and/or MR system room. This form, entitled, *Magnetic Resonance (MR) Environment Screening Form for Individuals* was developed in conjunction with the Medical, Scientific, and Technology Advisory Board and the Corporate Advisory Board of the Institute for Magnetic Resonance Safety, Education, and Research (IMRSER) (Figure 2). A "downloadable" version of this form may be obtained from the web sites, www.IMRSER.org and www.MRIsafety.com.

At the top of this form, the following statement is displayed: "The MR system has a very strong magnetic field that may be hazardous to individuals entering the MR environment or MR system room if they have certain metallic, electronic, magnetic, or mechanical implants, devices, or objects. Therefore, <u>all</u> individuals are required to fill out this form BEFORE entering the MR environment or MR system room. Be advised, the MR system magnet is ALWAYS on."

The Magnetic Resonance (MR) Environment Screening Form for Individuals requests general information (name, age, address, etc.) and poses important questions to determine if there are possible problems or issues that should be discussed with the individual prior to permitting entry to the MR environment. A warning statement is also provided on the form, as follows: "WARNING: Certain implants, devices, or objects may be hazardous to you in the MR environment or MR system room. <u>Do not enter</u> the MR environment or MR system room if you have any question or concern regarding an implant, device, or object." In addition, there is a section that lists various implants, devices, and objects to identify the presence of anything that could be hazardous to an individual in the MR environment (e.g., an aneurysm clip, cardiac pacemaker, implantable cardioverter defibrillator (ICD), electronic or magnetically activated device, metallic foreign body, etc).

Finally, there is an *Important Instructions* section on the form that states: "Remove all metallic objects before entering the MR environment or MR system room including hearing aids, beeper, cell phone, keys, eyeglasses, hair pins, barrettes, jewelry (including body piercing jewelry), watch, safety pins, paperclips, money clip, credit cards, bank cards, magnetic strip cards, coins, pens, pocket knife, nail clipper, steel-toed boots/shoes, and tools. Loose metallic objects are especially prohibited in the MR system room and MR environment. Please consult the MRI Technologist or Radiologist if you have any question or concern BEFORE you enter the MR system room."

The proper use of this written form along with thorough verbal screening of the individual by an MR-safety trained healthcare worker should prevent accidents and injuries in the MR environment.

[*Portions of this text were excerpted with permission from Sawyer-Glover A, Shellock FG. Pre-Magnetic Resonance Procedure Screening, In: Magnetic Resonance Procedures: Health Effects and Safety, FG Shellock, Editor, CRC Press, LLC, Boca Raton, FL, 2001. The screening forms, **Magnetic Resonance (MR) Procedure Screening Form For Patients** and **Magnetic Resonance (MR) Environment Screening Form for Individuals** were developed in conjunction with the Medical, Scientific, and Technology Advisory Board and the Corporate Advisory Board of the Institute for Magnetic Resonance Safety, Education, and Research (IMRSER), 2003 and published with permission]

REFERENCES

http://www.MRIsafety.com

http://www.IMRSER.org

Kanal E, Borgstede JP, Barkovich AJ, Bell C, et al. American College of Radiology, White paper on MR safety. American Journal of Roentgenology 2002;178:1335-1347.

Sawyer-Glover A, Shellock FG. Pre-Magnetic Resonance Procedure Screening, In: Magnetic Resonance Procedures: Health Effects and Safety, FG Shellock, Editor, CRC Press, LLC, Boca Raton, FL, 2001.

Sawyer-Glover A, Shellock FG. Pre-MRI procedure screening: recommendations and safety considerations for biomedical implants and devices. J Magn Reson Imaging 2000;12: 92-106.

Shellock FG. Reference Manual for Magnetic Resonance Safety: 2003 Edition. Amirsys, Salt Lake City, Utah, 2003.

Shellock FG. New recommendations for screening patients for suspected orbital foreign bodies. Signals, No. 36, Issue 4, 2001, pp. 8-9.

Shellock FG. Biomedical implants and devices: assessment of magnetic field interactions with a 3.0-Tesla MR system. J Magn Reson Imaging 2002;16:721-732.

Shellock FG. MR safety update 2002: Implants and devices. Journal of Magnetic Resonance Imaging 2002;16:485-496.

Shellock FG, Crues JV. Commentary. MR safety and the American College of Radiology White Paper. American Journal of Roentgenology 2002;178:1349-1352.

Shellock FG, Kanal E. Policies, guidelines, and recommendations for MR imaging safety and patient management. J Magn Reson Imaging 1991;1:97-101.

Shellock FG, Kanal E. Policies, guidelines, and recommendations for MR imaging safety and patient management. J Magn Reson Imaging 1991;1: 97-101.

Shellock FG, Kanal E. Magnetic Resonance: Bioeffects, Safety, and Patient Management. Second Edition, Lippincott-Raven Press, New York, 1996.

Shellock FG, Kanal E. SMRI Report. Policies, guidelines and recommendations for MR imaging safety and patient management. Questionnaire for screening patients before MR procedures. J Magn Reson Imaging 1994;4:749-751, 1994.

Figure 1. Magnetic Resonance (MR) Procedure Screening Form For Patients (developed in conjunction with the Medical, Scientific, and Technology Advisory Board and the Corporate Advisory Board of the Institute for Magnetic Resonance Safety, Education, and Research, 2002, www.IMRSER.org)

MAGNETIC RESONANCE (MR) PROCEDURE SCREENING FORM FOR PATIENTS

Date ____/____/____   Patient Number _____

Name _____   Age _____ Height _____ Weight _____
Last name   First name   Middle Initial

Date of Birth ____/____/____   Male ☐ Female ☐   Body Part to be Examined _____
month   day   year

Address _____   Telephone (home) (____) ____-_____

City _____   Telephone (work) (____) ____-_____

State _____ Zip Code _____

Reason for MRI and/or Symptoms _____

Referring Physician _____   Telephone (____) ____-_____

1. Have you had prior surgery or an operation (e.g., arthroscopy, endoscopy, etc.) of any kind?   ☐ No   ☐ Yes
   If yes, please indicate the date and type of surgery:
   Date ____/____/____   Type of surgery _____
   Date ____/____/____   Type of surgery _____
2. Have you had a prior diagnostic imaging study or examination (MRI, CT, Ultrasound, X-ray, etc.)?   ☐No   ☐ Yes
   If yes, please list: Body part   Date   Facility

MRI	_____	____/____/____	_____
CT/CAT Scan	_____	____/____/____	_____
X-Ray	_____	____/____/____	_____
Ultrasound	_____	____/____/____	_____
Nuclear Medicine	_____	____/____/____	_____
Other_____	_____	____/____/____	_____

3. Have you experienced any problem related to a previous MRI examination or MR procedure?   ☐ No   ☐ Yes
   If yes, please describe: _____
4. Have you had an injury to the eye involving a metallic object or fragment (e.g., metallic slivers, shavings, foreign body, etc.)?   ☐ No   ☐ Yes
   If yes, please describe: _____
5. Have you ever been injured by a metallic object or foreign body (e.g., BB, bullet, shrapnel, etc.)?   ☐ No   ☐ Yes
   If yes, please describe: _____
6. Are you currently taking or have you recently taken any medication or drug?   ☐ No   ☐ Yes
   If yes, please list: _____
7. Are you allergic to any medication?   ☐ No   ☐ Yes
   If yes, please list: _____
8. Do you have a history of asthma, allergic reaction, respiratory disease, or reaction to a contrast medium or dye used for an MRI, CT, or X-ray examination?   ☐ No   ☐ Yes
9. Do you have anemia or any disease(s) that affects your blood, a history of renal (kidney) disease, or seizures?   ☐ No   ☐ Yes
   If yes, please describe: _____

For female patients:
10. Date of last menstrual period: ____/____/____   Post menopausal?   ☐ No   ☐ Yes
11. Are you pregnant or experiencing a late menstrual period?   ☐ No   ☐ Yes
12. Are you taking oral contraceptives or receiving hormonal treatment?   ☐ No   ☐ Yes
13. Are you taking any type of fertility medication or having fertility treatments?   ☐ No   ☐ Yes
    If yes, please describe: _____
14. Are you currently breastfeeding?   ☐ No   ☐ Yes

WARNING: Certain implants, devices, or objects may be hazardous to you and/or may interfere with the MR procedure (i.e., MRI, MR angiography, functional MRI, MR spectroscopy). <u>Do not enter</u> the MR system room or MR environment if you have any question or concern regarding an implant, device, or object. Consult the MRI Technologist or Radiologist BEFORE entering the MR system room. The MR system magnet is ALWAYS on.

Please indicate if you have any of the following:

☐ Yes ☐ No Aneurysm clip(s)
☐ Yes ☐ No Cardiac pacemaker
☐ Yes ☐ No Implanted cardioverter defibrillator (ICD)
☐ Yes ☐ No Electronic implant or device
☐ Yes ☐ No Magnetically-activated implant or device
☐ Yes ☐ No Neurostimulation system
☐ Yes ☐ No Spinal cord stimulator
☐ Yes ☐ No Internal electrodes or wires
☐ Yes ☐ No Bone growth/bone fusion stimulator
☐ Yes ☐ No Cochlear, otologic, or other ear implant
☐ Yes ☐ No Insulin or other infusion pump
☐ Yes ☐ No Implanted drug infusion device
☐ Yes ☐ No Any type of prosthesis (eye, penile, etc.)
☐ Yes ☐ No Heart valve prosthesis
☐ Yes ☐ No Eyelid spring or wire
☐ Yes ☐ No Artificial or prosthetic limb
☐ Yes ☐ No Metallic stent, filter, or coil
☐ Yes ☐ No Shunt (spinal or intraventricular)
☐ Yes ☐ No Vascular access port and/or catheter
☐ Yes ☐ No Radiation seeds or implants
☐ Yes ☐ No Swan-Ganz or thermodilution catheter
☐ Yes ☐ No Medication patch (Nicotine, Nitroglycerine)
☐ Yes ☐ No Any metallic fragment or foreign body
☐ Yes ☐ No Wire mesh implant
☐ Yes ☐ No Tissue expander (e.g., breast)
☐ Yes ☐ No Surgical staples, clips, or metallic sutures
☐ Yes ☐ No Joint replacement (hip, knee, etc.)
☐ Yes ☐ No Bone/joint pin, screw, nail, wire, plate, etc.
☐ Yes ☐ No IUD, diaphragm, or pessary
☐ Yes ☐ No Dentures or partial plates
☐ Yes ☐ No Tattoo or permanent makeup
☐ Yes ☐ No Body piercing jewelry
☐ Yes ☐ No Hearing aid
           (Remove before entering MR system room)
☐ Yes ☐ No Other implant _____
☐ Yes ☐ No Breathing problem or motion disorder
☐ Yes ☐ No Claustrophobia

Please mark on the figure(s) below the location of any implant or metal inside of or on your body.

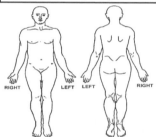

RIGHT   LEFT   LEFT   RIGHT

⚠ **IMPORTANT INSTRUCTIONS**

Before entering the MR environment or MR system room, you must remove **all** metallic objects including hearing aids, dentures, partial plates, keys, beeper, cell phone, eyeglasses, hair pins, barrettes, jewelry, body piercing jewelry, watch, safety pins, paperclips, money clip, credit cards, bank cards, magnetic strip cards, coins, pens, pocket knife, nail clipper, tools, clothing with metal fasteners, & clothing with metallic threads.

Please consult the MRI Technologist or Radiologist if you have any question or concern BEFORE you enter the MR system room.

NOTE: You may be advised or required to wear earplugs or other hearing protection during the MR procedure to prevent possible problems or hazards related to acoustic noise.

I attest that the above information is correct to the best of my knowledge. I read and understand the contents of this form and had the opportunity to ask questions regarding the information on this form and regarding the MR procedure that I am about to undergo.

Signature of Person Completing Form: _____   Date ____ / ____ / ____
                                         Signature

Form Completed By: ☐ Patient ☐ Relative ☐ Nurse _____   _____
                                                    Print name   Relationship to patient

Form Information Reviewed By: _____
                                Print name   Signature

☐ MRI Technologist   ☐ Nurse   ☐ Radiologist   ☐ Other_____

Figure 2. Magnetic Resonance (MR) Environment Screening Form for Individuals (developed in conjunction with the Medical, Scientific, and Technology Advisory Board and the Corporate Advisory Board of the Institute for Magnetic Resonance Safety, Education, and Research, 2003, www.IMRSER.org)

MAGNETIC RESONANCE (MR) ENVIRONMENT SCREENING FORM FOR INDIVIDUALS*

 The MR system has a very strong magnetic field that may be hazardous to individuals entering the MR environment or MR system room if they have certain metallic, electronic, magnetic, or mechanical implants, devices, or objects. Therefore, all individuals are required to fill out this form BEFORE entering the MR environment or MR system room. Be advised, the MR system magnet is ALWAYS on.

*NOTE: If you are a patient preparing to undergo an MR examination, you are required to fill out a different form.

Date ____/____/____ (month day year)   Name _____ (Last Name / First Name / Middle Initial)   Age _____

Address _____   Telephone (home) (____) ____-_____

City _____   Telephone (work) (____) ____-_____

State _____   Zip Code _____

1. Have you had prior surgery or an operation (e.g., arthroscopy, endoscopy, etc.) of any kind?   ❏ No ❏ Yes
   If yes, please indicate date and type of surgery: Date ____/____/____   Type of surgery_____
2. Have you had an injury to the eye involving a metallic object (e.g., metallic slivers, foreign body)?   ❏ No ❏ Yes
   If yes, please describe: _____
3. Have you ever been injured by a metallic object or foreign body (e.g., BB, bullet, shrapnel, etc.)?   ❏ No ❏ Yes
   If yes, please describe: _____
4. Are you pregnant or suspect that you are pregnant?   ❏ No ❏ Yes

⚠ **WARNING:** Certain implants, devices, or objects may be hazardous to you in the MR environment or MR system room. Do not enter the MR environment or MR system room if you have any question or concern regarding an implant, device, or object.

Please indicate if you have any of the following:

❏ Yes ❏ No Aneurysm clip(s)
❏ Yes ❏ No Cardiac pacemaker
❏ Yes ❏ No Implanted cardioverter defibrillator (ICD)
❏ Yes ❏ No Electronic implant or device
❏ Yes ❏ No Magnetically-activated implant or device
❏ Yes ❏ No Neurostimulation system
❏ Yes ❏ No Spinal cord stimulator
❏ Yes ❏ No Cochlear implant or implanted hearing aid
❏ Yes ❏ No Insulin or infusion pump
❏ Yes ❏ No Implanted drug infusion device
❏ Yes ❏ No Any type of prosthesis or implant
❏ Yes ❏ No Artificial or prosthetic limb
❏ Yes ❏ No Any metallic fragment or foreign body
❏ Yes ❏ No Any external or internal metallic object
❏ Yes ❏ No Hearing aid
     (Remove before entering the MR system room)
❏ Yes ❏ No Other implant

⚠ **IMPORTANT INSTRUCTIONS**

Remove all metallic objects before entering the MR environment or MR system room including hearing aids, beeper, cell phone, keys, eyeglasses, hair pins, barrettes, jewelry (including body piercing jewelry), watch, safety pins, paperclips, money clip, credit cards, bank cards, magnetic strip cards, coins, pens, pocket knife, nail clipper, steel-toed boots/shoes, and tools. Loose metallic objects are especially prohibited in the MR system room and MR environment.

Please consult the MRI Technologist or Radiologist if you have any question or concern BEFORE you enter the MR system room.

I attest that the above information is correct to the best of my knowledge. I have read and understand the entire contents of this form and have had the opportunity to ask questions regarding the information on this form.

Signature of Person Completing Form: _____   Date ____/____/____
     Signature

Form Information Reviewed By: _____   _____
     Print name   *Signature*

❏ MRI Technologist   ❏ Radiologist   ❏ Other _____

GUIDELINES TO PREVENT EXCESSIVE HEATING AND BURNS ASSOCIATED WITH MAGNETIC RESONANCE PROCEDURES*

Magnetic resonance (MR) imaging is considered to be a relatively safe diagnostic modality. However, the use of radiofrequency coils, physiologic monitors, electronically-activated devices, and external accessories or objects made from conductive materials has caused excessive heating, resulting in burn injuries to patients undergoing MR procedures. Heating of implants and similar devices may also occur in association with MR procedures, but this tends be problematic primarily for objects made from conductive materials that have elongated shapes such as leads, guidewires, and certain types of catheters (e.g., catheters with thermistors or other conducting components).

Notably, more than 30 incidents of excessive heating have been reported in patients undergoing MR procedures in the United States that were unrelated to equipment problems or the presence of conductive external or internal implants or materials [review of data files from U.S. Food and Drug Administration, Center for Devices and Radiological Health, Manufacturer and User Facility Device Experience Database, MAUDE, http://www.fda.gov/cdrh/maude.html and U.S. Food and Drug Administration, Center for Devices and Radiological Health, Medical Device Report, (http://www.fda.gov/CDRH/mdrfile.html)]. These incidents included first, second, and third degree burns that were experienced by patients. In many of these cases, the reports indicated that the limbs or other body parts of the patients were in direct contact with body radiofre-

45

quency (RF) coils or other RF transmit coils of the MR systems or there were skin-to-skin contact points suspected to be responsible for these injuries.

MR systems require the use of RF pulses to create the MR signal. This RF energy is transmitted readily through free space from the transmit RF coil to the patient. When conducting materials are placed within the RF field, the result may be a concentration of electrical currents sufficient to cause excessive heating and tissue damage. The nature of high frequency electromagnetic fields is such that the energy can be transmitted across open space and through insulators. Therefore, only devices with carefully designed current paths can be made safe for use during MR procedures. Simply insulating conductive material (e.g., wire or lead) or separating it from the patient may not be sufficient to prevent excessive heating or burns from occurring.

Furthermore, certain geometrical shapes exhibit the phenomenon of "resonance" which increases their propensity to concentrate RF currents. At the operating frequencies of present day MR systems, conducting loops of tens of centimeters in size may create problems and, therefore, must be avoided, unless high impedance is used to limit RF current. Importantly, even loops that include small gaps separated by insulation may still conduct current.

To prevent patients from experiencing excessive heating and possible burns in association with MR procedures, the following guidelines are recommended:

(1)   Prepare the patient for the MR procedure by ensuring that there are no unnecessary metallic objects contacting the patient's skin (e.g., metallic drug delivery patches, jewelry, necklaces, bracelets, key chains, etc.).

(2)   Prepare the patient for the MR procedure by using insulation material (i.e., appropriate padding) to prevent skin-to-skin contact points and the formation of "closed-loops" from touching body parts.

(3)   Insulating material (minimum recommended thickness, 1-cm) should be placed between the patient's skin and transmit RF coil that is used for the MR procedure (alternatively, the RF coil itself should be padded). For example, position the patient so that there is no direct contact between the patient's skin and the body RF coil of the MR system. This may be accomplished by having the patient place his/her arms over his/her head or by using elbow pads or foam padding between the patient's

tissue and the body RF coil of the MR system. This is especially important for those MR examinations that use the body coil or other large RF coils for transmission of RF energy.

(4)   Use only electrically conductive devices, equipment, accessories (e.g., ECG leads, electrodes, etc.), and materials that have been thoroughly tested and determined to be safe and compatible for MR procedures.

(5)   Carefully follow specific MR safety criteria and recommendations for implants made from electrically-conductive materials (e.g., bone fusion stimulators, neurostimulation systems, etc.).

(6)   Before using electrical equipment, check the integrity of the insulation and/or housing of all components including surface RF coils, monitoring leads, cables, and wires. Preventive maintenance should be practiced routinely for such equipment.

(7)   Remove all non-essential electrically conductive materials from the MR system (i.e., unused surface RF coils, ECG leads, cables, wires, etc.).

(8)   Keep electrically conductive materials that must remain in the MR system from directly contacting the patient by placing thermal and/or electrical insulation between the conductive material and the patient.

(9)   Keep electrically conductive materials that must remain within the body RF coil or other transmit RF coil of the MR system from forming conductive loops. Note: The patient's tissue is conductive and, therefore, may be involved in the formation of a conductive loop, which can be circular, U-shaped, or S-shaped.

(10)   Position electrically conductive materials to prevent "cross points". For example, a cross point is the point where a cable crosses another cable, where a cable loops across itself, or where a cable touches either the patient or sides of the transmit RF coil more than once. Notably, even the close proximity of conductive materials with each other should be avoided because some cables and RF coils can capacitively-couple (without any contact or crossover) when placed close together.

(11)   Position electrically conductive materials to exit down the center of the MR system (i.e., not along the side of the MR system or close to the body RF coil or other transmit RF coil).

(12)   Do not position electrically conductive materials across an external metallic prosthesis (e.g., external fixation device, cervical fixation device, etc.) or similar device that is in direct contact with the patient.

(13)   Allow only properly trained individuals to operate devices (e.g., monitoring equipment) in the MR environment.

(14)   Follow all manufacturer instructions for the proper operation and maintenance of physiologic monitoring or other similar electronic equipment intended for use during MR procedures.

(15)   Electrical devices that do not appear to be operating properly during the MR procedure should be removed from the patient immediately.

(16)   Closely monitor the patient during the MR procedure. If the patient reports sensations of heating or other unusual sensation, discontinue the MR procedure immediately and perform a thorough assessment of the situation.

(17)   RF surface coil decoupling failures can cause localized RF power deposition levels to reach excessive levels. The MR system operator will recognize such a failure as a set of concentric semicircles in the tissue on the associated MR image or as an unusual amount of image non-uniformity related to the position of the RF coil.

The adoption of these guidelines will help to ensure that patient safety is maintained, especially as more conductive materials and electronically-activated devices are used in association with MR procedures.

[*The document, **Guidelines to Prevent Excessive Heating and Burns Associated with Magnetic Resonance Procedures**, was developed by the Medical, Scientific, and Technology Advisory Board and the Corporate Advisory Board of the Institute for Magnetic Resonance Safety, Education, and Research (IMRSER), 2003 and published with permission.]

REFERENCES

Bashein G, Syrory G. Burns associated with pulse oximetry during magnetic resonance imaging. Anesthesiology 1991;75:382-3.

Brown TR, Goldstein B, Little J. Severe burns resulting from magnetic resonance imaging with cardiopulmonary monitoring. Risks and relevant safety precautions. Am J Phys Med Rehabil 1993;72:166-7.

Chou C-K, McDougall JA, Chan KW. Absence of radiofrequency heating from auditory implants during magnetic resonance imaging. Bioelectromagnetics 1997;44:367-372.

Dempsey MF, Condon B. Thermal injuries associated with MRI. Clin Radiol 2001;56:457-65.

Dempsey MF, Condon B, Hadley DM. Investigation of the factors responsible for burns during MRI. J Magn Reson Imaging 2001;13:627-631.

ECRI, Health Devices Alert. A new MRI complication? Health Devices Alert May 27, pp. 1, 1988.

ECRI. Thermal injuries and patient monitoring during MRI studies. Health Devices Alert 1991;20: 362-363.

Finelli DA, Rezai AR, Ruggieri PM, Tkach JA, Nyenhuis JA, Hrdlicka G, Sharan A, Gonzalez-Martinez J, Stypulkowski PH, Shellock FG. MR imaging-related heating of deep brain stimulation electrodes: In vitro study. Am J Neuroradiol 2002;23:1795-1802.

Hall SC, Stevenson GW, Suresh S. Burn associated with temperature monitoring during magnetic resonance imaging. Anesthesiology 1992;76:152.

Heinz W, Frohlich E, Stork T. Burns following magnetic resonance tomography study. (German) Z Gastroenterol 1999;37:31-2.

http://www.MRIsafety.com

http://www.IMRSER.org

International Electrotechnical Commission (IEC), Medical Electrical Equipment, Particular requirements for the safety of magnetic resonance equipment for medical diagnosis, International Standard IEC 60601-2-33, 2002.

Jones S, Jaffe W, Alvi R. Burns associated with electrocardiographic monitoring during magnetic resonance imaging. Burns 1996;22:420-1.

Kanal E, Shellock FG. Burns associated with clinical MR examinations. Radiology 1990;175: 585.

Kanal E, Shellock FG. Policies, guidelines, and recommendations for MR imaging safety and patient management. J Magn Reson Imaging 1992;2:247-248.

Keens SJ, Laurence AS. Burns caused by ECG monitoring during MRI imaging. Anaesthesia 1996;51:1188-9.

Knopp MV, Essig M, Debus J, Zabel HJ, van Kaick G. Unusual burns of the lower extremities caused by a closed conducting loop in a patient at MR imaging. Radiology 1996;200:572-5.

Knopp MV, Metzner R, Brix G, van Kaick G. Safety considerations to avoid current-induced skin burns in MRI procedures. (German) Radiologe 199838:759-63.

Kugel H, Bremer C, Puschel M, Fischbach R, Lenzen H, Tombach B, Van Aken H, Heindel W. Hazardous situation in the MR bore: induction in ECG leads causes fire. Eur Radiol 2003;13:690-694.

Nakamura T, Fukuda K, Hayakawa K, Aoki I, Matsumoto K, Sekine T, Ueda H, Shimizu Y. Mechanism of burn injury during magnetic resonance imaging (MRI)-simple loops can induce heat injury. Front Med Biol Eng 2001;11:117-29

Nyenhuis JA, Kildishev AV, Foster KS, Graber G, Athey W. Heating near implanted medical devices by the MRI RF-magnetic field. IEEE Trans Magn 1999;35:4133-4135.

Rezai AR, Finelli D, Nyenhuis JA, Hrdlick G, Tkach J, Ruggieri P, Stypulkowski PH, Sharan A, Shellock FG. Neurostimulator for deep brain stimulation: Ex vivo evaluation of MRI-related heating at 1.5-Tesla. Journal of Magnetic Resonance Imaging 2002;15:241-250.

Schaefer DJ. Safety Aspects of radio-frequency power deposition in magnetic resonance. MRI Clinics of North America 1998;6:775-789.

Schaefer DJ, Felmlee JP. Radio-frequency safety in MR examinations, Special Cross-Specialty Categorical Course in Diagnostic Radiology: Practical MR Safety Considerations for Physicians, Physicists, and Technologists, Syllabus, 87th Scientific of the Radiological Society of North America, Chicago, pp 111-123, 2001.

Shellock FG. Magnetic Resonance Procedures: Health Effects and Safety. CRC Press, LLC, Boca Raton, FL, 2001.

Shellock FG. MR safety update 2002: Implants and devices. Journal of Magnetic Resonance Imaging 2002;16:485-496.

Shellock FG. Radiofrequency-induced heating during MR procedures: A review. Journal of Magnetic Resonance Imaging 2000;12: 30-36.

Shellock FG. Reference Manual for Magnetic Resonance Safety: 2003 Edition. Amirsys, Inc., 2003.

Shellock FG, Slimp G. Severe burn of the finger caused by using a pulse oximeter during MRI. American Journal of Roentgenology 1989;153:1105.

Shellock FG, Hatfield M, Simon BJ, Block S, Wamboldt J, Starewicz PM, Punchard WFB. Implantable spinal fusion stimulator: assessment of MRI safety. Journal of Magnetic Resonance Imaging 2000;12:214-223.

Smith CD, Nyenhuis JA, Kildishev AV. Health effects of induced electrical fields: implications for metallic implants. In: Shellock FG, ed. Magnetic resonance procedure: health effects and safety. Boca Raton, FL: CRC Press, 2001; 393-414.

U.S. Food and Drug Administration, Center for Devices and Radiological Health (CDRH), Medical Device Report (MDR) (http://www.fda.gov/CDRH/mdr-file.html). The files contain information from CDRH's device experience reports on devices which may have malfunctioned or caused a death or serious injury. The files contain reports received under both the mandatory Medical Device Reporting Program (MDR) from 1984 - 1996, and the voluntary reports up to June 1993. The database currently contains over 600,000 reports.

U.S. Food and Drug Administration, Center for Devices and Radiological Health (CDRH), Manufacturer and User Facility Device Experience Database, MAUDE, (http://www.fda.gov/cdrh/maude.html). MAUDE data represents reports of adverse events involving medical devices. The data consists of all voluntary reports since June, 1993, user facility reports since 1991, distributor reports since 1993, and manufacturer reports since August, 1996.

METALLIC FOREIGN BODIES AND SCREENING

All patients and individuals with a history of being injured by a metallic foreign body such as a bullet, shrapnel, or other type of metallic object should be thoroughly screened and evaluated prior to admission to the area of the MR system, that is, the MR environment. This is particularly important because serious injury may occur as a result of movement or dislodgment of the metallic foreign body as it is attracted by the magnetic field of the scanner. In addition, excessive heating may occur, although this tends to happen only if the object is made from conductive material and has an elongated shape or forms a loop.

The relative risk of injury is dependent on the ferromagnetic properties of the foreign body, the geometry and dimensions of the object, the strength of the static magnetic field, and the strength of the spatial gradient of the MR system. Additionally, the potential for injury is related to the amount of force with which the object is fixed within the tissue (i.e., counter-force or retention force may prevent movement or dislodgment) and whether or not it is positioned in or adjacent to a particularly sensitive site of the body. These sensitive sites include vital neural, vascular, or soft tissue structures.

The use of plain film radiography is the technique of choice recommended to detect metallic foreign bodies for individuals and patients prior to admission to the MR environment. This includes screening individuals and patients for the presence of metallic orbital foreign bodies (see **Metallic Orbital Foreign Bodies and Screening**). The inherent sensitivity of plain film radiography is considered to be sufficient to identify any metal with a mass large enough to present a hazard to an individual or patient in the MR environment.

REFERENCES

Boutin, R. D., Briggs, J. E., and Williamson, M. R., Injuries associated with MR imaging: survey of safety records and methods used to screen patients for metallic foreign bodies before imaging. Am J Roentgenol 162, 189, 1994.

Dempsey MF, Condon B, Hadley DM. Investigation of the factors responsible for burns during MRI. J Magn Reson Imaging 2001;13:627-631.

Elster, A. D., Link, K. M., and Carr, J. J., Patient screening prior to MR imaging: a practical approach synthesized from protocols at 15 U.S. medical centers. Am J Roentgenol 162, 195, 1994.

Jarvik JG, Ramsey JG. Radiographic screening for orbital foreign bodies prior to MR imaging: Is it worth it? Am J Neuroradiol 21, 245, 2000.

Mani, R. L., In search of an effective screening system for intraocular metallic foreign bodies prior to MR - An important issue of patient safety. Am J Neuroradiol 9, 1032, 1988.

Murphy KJ, Burnberg JA. Orbital plain film as a prerequisite for MR imaging: is a known history of injury sufficient screening criteria? Am J Roentgenol 167, 1053, 1996.

Otto PM, et al. Screening test for detection of metallic foreign objects in the orbit before magnetic resonance imaging. Invest Radiol 1992;27:308-311.

Seidenwurm DJ, McDonnell CH, Raghavan N, Breslau J. Cost utility analysis of radiographic screening for an orbital foreign body before MR imaging. Am J Neuroradiol 21, 426, 2000.

Shellock FG, Kanal E. Magnetic resonance: Bioeffects, Safety and Patient Management, 2nd edition. Lippincott-Raven, New York, 1996.

Shellock FG. Magnetic Resonance Procedures: Health Effects and Safety. CRC Press, LLC, Boca Raton, FL, 2001.

Shellock FG. Pocket Guide to MR Procedures and Metallic Objects: Update 2001, Seventh Edition, Lippincott Williams & Wilkins Healthcare, Philadelphia, 2001.

Shellock FG, Kanal E, SMRI Safety Committee. Policies, guidelines, and recommendations for MR imaging safety and patient management. J Magn Reson Imaging 1991;1:97-101.

Williams S, Char D H, Dillon WP, LincoffN, MoseleyM. Ferrous intraocular foreign bodies and magnetic resonance imaging. Am J Ophthalmology 105, 398, 1988.

METALLIC ORBITAL FOREIGN BODIES AND SCREENING

In the past, any individual with a suspected orbital foreign body was required to have plane film radiographs of the orbits acquired to determine the presence of a metallic fragment prior to entering the MR environment. Thus, screening plain films of the orbits were deemed necessary for every individual with a history exposure to intraocular or periorbital foreign body or when a patient had a history of exposure to potential metallic ocular injury (e.g., welders, grinders, metal workers, sculptors, etc.). This was considered the standard of care to prevent serious injuries to the eye associated with the MR environment. However, based on an investigation by Seidenwurm et al., new guidelines for radiographic screening of patients with suspected metallic foreign bodies have been proposed and implemented in the clinical MR setting.

NEW GUIDELINES FOR ORBITAL FOREIGN BODY SCREENING

The single case report in 1986 by Kelly et al. regarding a patient that sustained an ocular injury from a retained metallic foreign body has led to great controversy regarding the procedure required to screen patients prior to MR procedures. Notably, this incident is the only serious eye-related injury that has occurred in the MR environment (i.e., based on a recent review of the peer-reviewed literature). In consideration of this, the standard policy of performing radiographic screening for orbital foreign bodies in patients or individuals simply because of a history of occupational or other similar exposure to metal fragments needs to be carefully reconsidered.

54

A study by Seidenwurm et al. evaluated the cost-effectiveness of using a clinical versus radiographic technique to screen patients for orbital foreign bodies before MR procedures. The costs of screening were determined on the basis of published reports, disability rating guides, and a practice survey. A sensitivity analysis was performed for each variable. For this analysis, the benefits of screening were avoidance of immediate, permanent, nonameliorable, or unilateral blindness.

The findings of Seidenwurm et al. support the fact that the use of clinical screening before radiography increases the cost-effectiveness of foreign body screening by an order of magnitude (i.e., assuming base case ocular foreign body removal rates). From a clinical screening standpoint for a metallic foreign body located in the orbit, asking the patient "Did a doctor get it all out?" serves this purpose.

Seidenwurm et al. implemented the following policy with regard to screening patients with suspected metallic foreign bodies, "If a patient reports injury from an ocular foreign body that was subsequently removed by a doctor or that resulted in negative findings on any examination, we perform MR imaging...Those persons with a history of injury and no subsequent negative eye examination are screened radiographically." Of note is that Seidenwurm et al. has performed approximately 100,000 MRI procedures under this protocol without incident.

Thus, an occupational history of exposure to metallic fragments, by itself, is not sufficient to mandate radiographic orbital screening. Therefore, guidelines for foreign body screening should be altered in consideration of this new information and because radiographic screening before MR procedures on the basis of occupational exposure alone is not cost effective. Furthermore, it is not clinically necessary.

Clinical Screening Protocol. The procedure to follow with regard to patients with suspected foreign bodies involves an initial clinical screening protocol, as recommended by Seidenwurm et al. This involves asking patients whether they have a high-risk occupation and whether they have had an ocular injury. If they sustained an ocular injury from a metallic object, they are asked whether they had a medical examination at the time of the injury, and whether they were told by the doctor, "It's all out." If they did not have an injury, if they were told their ophthalmologic examination was normal, and/or if the foreign body was removed at the time of the injury, then they proceed to MR imaging.

Radiographic Screening Protocol. Based on the results of the clinical screening protocol, patients are screened radiographically if they sustained an ocular injury related to a metallic foreign object and they were not told their post-injury eye examination was normal. In these cases, the MR examination is postponed and the patient is scheduled for screening radiography.

SCREENING ADOLESCENTS FOR METALLIC ORBITAL FOREIGN BODIES

A published case report by Elmquist and Shellock illustrates that special precautions are needed for screening adolescent patients prior to MR procedures. This article described an incident in which a 12-year-old patient accompanied by his parent completed all routine screening procedures prior to preparation for MR imaging of the lumbar spine. The patient and parent provided negative answers to all questions regarding prior injuries by metallic objects and the presence of metallic foreign bodies.

While entering the MR system room, the adolescent patient appeared to be anxious about the examination. He was placed in a feet-first, supine position on the MR system table and prepared for the procedure. During this time, the patient became more anxious and restless, shifting his position several times on the table. As the patient was moved slowly toward the opening of the bore of a 1.5-Tesla MR system, he complained of a pressure sensation in his left eye. The MRI technologist immediately removed the patient from the MR environment.

Once again, the patient was questioned regarding a previous eye injury. The patient denied sustaining such an injury. Despite that patient's response, a metallic foreign body in the orbit was suspected. Therefore, plain film radiographs of the orbits were obtained. The plain films revealed a metallic foreign body in the left orbit, curvilinear in shape and approximately 5-mm in size. The patient and parent were counseled regarding the implications of future MR procedures with respect to the possibility of significant eye injury related to movement or dislodgment of the metallic foreign body.

This case demonstrates that routine guidelines and safety protocols may not always be sufficient for evaluation of potential hazardous situations that may be present particularly in adolescents referred for MR procedures. There are possible additional risks involved whenever a parent or guardian fills out the pre-MR procedure screening form because a child may not be willing to disclose a previous injury or accident.

In consideration of this incident and to avoid unfortunate accidents related to the electromagnetic fields used for MR procedures, it is recommended that adolescents be provided additional screening that includes private counseling about the hazards associated with the MR environment.

REFERENCES

Boutin, R. D., Briggs, J. E., and Williamson, M. R., Injuries associated with MR imaging: survey of safety records and methods used to screen patients for metallic foreign bodies before imaging, Am. J. Roentgenol., 162, 189, 1994.

Elmquist, C., Shellock, F.G., and Stoller, D., Screening adolescents for metallic foreign bodies before MR procedures, J. Magn. Reson. Imaging, 5, 784, 1996.

Elster, A. D., Link, K. M., and Carr, J. J., Patient screening prior to MR imaging: a practical approach synthesized from protocols at 15 U.S. medical centers, Am. J. Roentgenol., 162, 195, 1994.

Jarvik, J. G., and Ramsey, J.G., Radiographic screening for orbital foreign bodies prior to MR imaging: Is it worth it?, Am J Neuroradiol 21, 245, 2000.

Kelly, W. M., Pagle, P. G., Pearson, A., San Diego, A. G., and Soloman, M. A., Ferromagnetism of intraocular foreign body causes unilateral blindness after MR study, Am J Neuroradiol 7, 243, 1986.

Mani, R. L., In search of an effective screening system for intraocular metallic foreign bodies prior to MR - An important issue of patient safety. Am J Neuroradiol 9, 1032, 1988.

Murphy, K. J., and Brunberg, J. A., Orbital plain film as a prerequisite for MR imaging: is a known history of injury sufficient screening criteria? Am. J. Roentgenol., 167, 1053, 1996.

Seidenwurm, D. J., McDonnell, C. H., Raghavan, N., and Breslau, J., Cost utility analysis of radiographic screening for an orbital foreign body before MR imaging, Am J Neuroradiol 21, 426, 2000.

Shellock FG. Pocket Guide to MR Procedures and Metallic Objects: Update 2001, Seventh Edition, Lippincott Williams & Wilkins Healthcare, Philadelphia, 2001.

Shellock FG. Magnetic Resonance Procedures: Health Effects and Safety. CRC Press, LLC, Boca Raton, FL, 2001.

Shellock FG. New recommendations for screening patients for suspected orbital foreign bodies. Signals, No. 36, Issue 4, pp. 8-9, 2001.

Shellock FG, Kanal E, SMRI Safety Committee. Policies, guidelines, and recommendations for MR imaging safety and patient management. J Magn Reson Imaging 1991;1:97-101.

Shellock FG, Kanal E. Magnetic Resonance: Bioeffects, Safety, and Patient Management. Second Edition, Lippincott-Raven Press, New York, 1996.

Williams, S., Char, D. H., Dillon, W. P., Lincoff, N., and Moseley, M., Ferrous intraocular foreign bodies and magnetic resonance imaging, Am J Opthalmol 105, 398, 1988.

MONITORING PATIENTS IN THE MR ENVIRONMENT

Conventional monitoring equipment and accessories were not designed to operate in the harsh magnetic resonance (MR) environment where static, gradient, and radiofrequency (RF) electromagnetic fields can adversely effect or alter the operation of these devices. Fortunately, various monitors and other patient support devices have been developed or specially-modified to perform properly during MR procedures.

MR healthcare workers must carefully consider the ethical and medico-legal ramifications of providing proper patient care that includes identifying patients who require monitoring in the MR environment and following a proper protocol to ensure their safety by using appropriate equipment, devices, accessories. The early detection and treatment of complications that may occur in high-risk, critically-ill, sedated, or anesthetized patients undergoing MR procedures can prevent relatively minor problems from becoming life-threatening situations.

GENERAL POLICIES AND PROCEDURES

In general, monitoring during an MR examination is indicated whenever a patient requires observations of vital physiologic parameters due to an underlying health problem or whenever a patient is unable to respond or alert the MRI technologist or other healthcare worker regarding pain, respiratory problem, cardiac distress, or other difficulty that might arise during the examination. In addition, a patient should be monitored if there is a greater potential for a change in physiologic status during the MR procedure. Besides patient monitoring, various support devices and accessories may be needed for use in the high-risk patient to ensure safety.

Because of the wide-spread use of MRI contrast agents and the potential for adverse effects or idiosyncratic reactions to occur, it is prudent to have

59

MR-compatible monitoring equipment and accessories readily available for the proper management and support of patients who may experience side-effects. This is emphasized because adverse events, while extremely rare, may be serious or even fatal.

In 1992, the Safety Committee of the Society for Magnetic Resonance Imaging published guidelines and recommendations concerning the monitoring of patients during MR procedures. This information indicates that all patients undergoing MR procedures should, at the very least, be visually (e.g., using a camera system) and/or verbally (e.g., intercom system) monitored, and that patients who are sedated, anesthetized, or are unable to communicate should be physiologically monitored and supported by the appropriate means.

Severe injuries and fatalities have occurred in association with MR procedures. These may have been prevented with the proper use of monitoring equipment and devices. Of note is that guidelines issued by the Joint Commission on Accreditation of Healthcare Organizations (JCAHO) indicate that patients that receive sedatives or anesthetics require monitoring during administration and recovery from these medications. Other professional organizations similarly recommend the need to monitor certain patients using proper equipment and techniques. Table 1 summarizes the patients that may require monitoring and support during MR procedures.

Table 1. Patients that may require monitoring and support during MR procedures.

- Patients that are physically or mentally unstable.
- Patients that have compromised physiologic functions.
- Patients that are unable to communicate.
- Neonatal and pediatric patients.
- Sedated or anesthetized patients.
- Patients undergoing MR-guided interventional procedures.
- Patients undergoing MR procedures using experimental MR systems.
- Patients that may have a reaction to an MRI contrast agent.
- Critically ill or high-risk patients.

SELECTION OF PARAMETERS TO MONITOR

The proper selection of the specific physiologic parameter(s) that should be monitored during the MR procedure is crucial for patient safety. Various factors must be considered including the patient's medical history, present condition, the use of medication and possible side effects, as well as the aspects of the MR procedure to be performed. For example, if the patient is to receive a sedative, it is mandatory to monitor respiratory rate, apnea, and/or oxygen saturation. If the patient requires general anesthesia during the MR procedure, monitoring multiple physiologic parameters is required.

Policies and procedures for the management of the patient in the MR environment should be comparable to those used in the operating room or critical care setting, especially with respect to monitoring and support requirements. Specific recommendations for physiologic monitoring of patients during MR procedures should be developed in consideration of "standard of care" issues as well as in consultation with anesthesiologists and other similar specialists.

PERSONNEL INVOLVED IN PATIENT MONITORING

Only healthcare professionals with appropriate training and experience should be permitted to be responsible for monitoring patients during MR procedures. This includes several facets of training and experience. The healthcare professional must be well-acquainted with the operation of the monitoring equipment and accessories used in the MR environment and should be able to recognize equipment malfunctions, device problems, and recording artifacts. Furthermore, the person responsible for monitoring the patient should be well-versed in screening patients for conditions that may complicate the procedure. For example, patients with asthma, congestive heart failure, obesity, obstructive sleep apnea, and other conditions are at increased risk for having problems during sedation. Also, the healthcare professional must be able to identify and manage adverse events using appropriate equipment and procedures in the MR environment.

Additionally, there must be policies and procedures implemented to continue appropriate physiologic monitoring and management of the patient by trained personnel after the MR procedure is performed. This is especially needed for a patient recovering from the effects of a sedative or general anesthesia.

The monitoring of physiologic parameters and management of the patient during an MR procedure may be the responsibility of one or more individuals depending on the level of training for the healthcare worker and in consideration of the condition, medical history, and procedure that is to be performed for the patient. These individuals include anesthesiologists, nurse anesthetists, nurses, MR technologists, or radiologists.

EMERGENCY PLAN

The development, implementation, and regular practice of an emergency plan that addresses and defines the activities, use of equipment, and other pertinent issues pertaining to a medical emergency are important for patient safety in the MR environment. For example, a plan needs to be developed for handling patients if there is the need to remove them from the MR system room to perform cardiopulmonary resuscitation in the event of a cardiac or respiratory arrest. Obviously, taking necessary equipment such as a cardiac defibrillator, intubation instruments, or other similar devices near the MR system could pose a substantial hazard to the patients and healthcare workers if these are not safe for use in the MR environment. Appropriate healthcare professionals that are in charge of the Code Blue team, "running" the code, maintaining the patient's airway, administering drugs, recording events, and conducting other emergency-related duties must be trained and continuously practiced in the performance of these critical activities.

For out-patient or mobile MR facilities, it is necessary to educate outside emergency personnel regarding the potential hazards associated with the MR environment. Typically, MR facilities not affiliated with or in close proximity to a hospital must contact paramedics to handle medical emergencies and to transport patients to the hospital for additional care. Therefore, personnel responsible for summoning the paramedics, notifying the hospital, and performing other integral activities must be designated beforehand to avoid problems and confusion during an actual emergent event.

TECHNIQUES AND EQUIPMENT USED TO MONITOR AND SUPPORT PATIENTS

Physiologic monitoring and support of patients is not a trivial task in the MR environment. A variety of potential problems and hazards exist. Furthermore, the types of equipment for patient monitoring and support must be considered carefully and implemented properly to ensure the safety of both patients and MR healthcare workers.

Several potential problems and hazards are associated with the performance of patient monitoring and support in the MR environment. Physiologic monitors and accessories that contain ferromagnetic components (e.g., transformers, outer casings, etc.) can be strongly attracted by the static magnetic field used of the MR system, posing a serious "missile" or projectile hazard to patients and MR healthcare workers. Additionally, the MR system may be damaged as a result of being struck by a "flying" monitor or similar device.

If possible, necessary or critical devices that have ferromagnetic components should be permanently fixed to the floor and properly labeled with warning information to prevent them from being moved too close to the MR system. All personnel involved with MR procedures should be aware of the importance of the placement and use of the equipment, especially with regard to the hazards of moving portable equipment too close to the MR system.

Electromagnetic fields associated with the MR system can significantly effect the operation of monitoring equipment, especially those with displays that involve electron beams (i.e., CRTs) or video display screens (with the exception of those with liquid crystal displays or LCDs). In addition, the monitoring equipment, itself, may emit spurious noise that, in turn, produces distortion or artifacts on the MR images

Physiologic monitors that contain microprocessors or other similar components may leak RF, producing electromagnetic interference that can substantially alter MR images. To prevent adverse radiofrequency-related interactions between the MR system and physiologic monitors, RF-shielded cables, RF filters, special outer RF-shielded enclosures, or fiber-optic techniques can be utilized to prevent image-related or problems in the MR environment.

During the operation of MR systems, electrical currents may be generated in the conductive materials of monitoring equipment that are used as part of the interface to the patient. These currents may be of sufficient magnitude to cause excessive heating and thermal injury to the patient. Numerous first, second, and third degree burns have occurred in association with MR procedures that were directly attributed to the use of monitoring devices. These thermal injuries have been associated with the use of electrocardiographic lead wires, plethysmographic gating systems, pulse oximeters, and other types of monitoring equipment and accessories comprised of wires, cables, or similar components made from conductive materials. Therefore, in consideration of the various problems and hazards associated with the use of monitoring equipment and accessories, it is important to follow closely the instructions and recommendations from the manufacturers with regard to the use of the devices in the MR environment.

MONITORING EQUIPMENT AND SUPPORT DEVICES

This section describes the physiologic parameters that may be assessed in patients during MR procedures using MR-compatible monitoring equipment. In addition, various devices and accessories that are useful for the support and management of patients are presented.

Electrocardiogram and Heart Rate

Monitoring the patient's electrocardiogram (ECG) in the MR environment is particularly challenging because of the inherent distortion of the ECG waveform that occurs using MR systems operating at high field strengths. This effect is observed as blood, a conductive fluid, flows through the large vascular structures in the presence of the static magnetic field of the MR system. The resulting induced biopotential is seen primarily as an augmented T-wave amplitude, although other non-specific waveform-changes are also apparent on the ECG. Since altered T-waves or ST segments may be associated with cardiac disorders, static magnetic field-induced ECG-distortions may be problematic for certain patients. For this reason, it may be necessary to obtain a baseline recording of the ECG prior to placing the patient inside the MR system along with a recording obtained immediately after the MR procedure to determine the cardiac status of the patient.

Additional artifacts caused by the static, gradient, and RF electromagnetic fields can severely distort the ECG, making observation of morphologic changes and detection of arrhythmias quite difficult. ECG artifacts that occur in the MR environment may be decreased substantially by implementing several simple techniques that include, the following:

- use ECG electrodes that have minimal metal,

- select electrodes and cables that contain nonferromagnetic metals,

- place the limb electrodes in close proximity to one another,

- position the line between the limb electrodes and leg electrodes parallel to the magnetic field flux lines,

- maintain a small area between the limb and leg electrodes,

- position the area of the electrodes near or in the center of the MR system, and

- twist or braid the ECG leads.

The use of MR-safe/MR-compatible ECG electrodes is strongly recommended to ensure patient safety and proper recording of the electrocardiogram in the MR environment. Accordingly, ECG electrodes have been specially developed for use during MR procedures to protect the patient from potentially hazardous conditions. These ECG electrodes were also designed to reduce MRI-related artifacts.

It is well known that the use of standard ECG electrodes, leads, and cables may cause excessive heating that could burn the patient during an MR procedure. This occurs as electrical current is generated in the ECG cable or leads during the operation of the MR system. Accordingly, monitoring equipment has been modified to record the ECG while ensuring patient safety in the MR environment.

Special fiber-optic ECG recording techniques may be used to prevent burns during MR procedures. For example, one such fiber-optic system acquires the ECG waveform using an MR-safe transceiver that resides in the MR system bore along with the patient and is located very near the ECG electrodes. A module digitizes and optically encodes the patient's ECG waveform and transmits it out from the MR system to the monitor using a fiber-optic cable. The use of this fiber-optic ECG technique eliminates the potential for burns associated with hard-wired ECG systems by removing the conductive patient cable and its antenna effect that are typical responsible for excessive heating.

Besides using an ECG monitor, the patient's heart rate may be determined continuously during the MR procedure using various types of MR-compatible devices including a photoplethysmograph and a pulse oximeter. A noninvasive, heart rate and blood pressure monitor (see section below) can also be utilized to obtain intermittent or semi-continuous recordings of heart rate during the MR procedure.

Blood Pressure

Conventional, manual sphygmomanometers may be adapted for use during MR procedures. This is typically accomplished by lengthening the tubing from the cuff to the device so that the mercury column and other primary components may be positioned an acceptable distance (e.g., 8 to 10 feet from the bore of a 1.5-Tesla MR system) from the fringe field of the MR system.

Blood pressure measuring devices that incorporate a spring-gauge instead of a mercury column may be adversely affected by magnetic fields, causing them to work erroneously in the MR setting. Therefore, spring-gauge blood pressure devices should undergo pre-clinical testing before being used to monitor patients undergoing MR procedures.

Blood pressure monitors that use other noninvasive techniques, such as the oscillometric method, may be used to obtain semi-continuous recordings of systolic, diastolic, and mean blood pressures as well as pulse rate. These devices can be utilized to record systemic blood pressure in adult, pediatric, and neonate patients, selecting the appropriate blood pressure cuff size for a given patient.

It should be noted that the intermittent inflation of the blood pressure cuff by an automated, noninvasive blood pressure monitor may disturb lightly-sedated patients, especially infants or neonates, causing them to move and disrupt the MR procedure. For this reason, the use of a noninvasive blood pressure monitor may not be the best instrument to conduct physiologic monitoring in every type of patient.

Respiratory Rate and Apnea

Because respiratory depression and upper airway obstruction are frequent complications associated with the use of sedatives and anesthetics, monitoring techniques that detect a decrease in respiratory rate, hypoxemia, or airway obstruction should be used during the administration of these drugs. This is particularly important in the MR environment because visual observation of the patient's respiratory efforts is often difficult to accomplish, especially for the patients in conventional, closed-configured MR systems.

Respiratory rate monitoring can be performed during MR procedures by various techniques. The impedance method that utilizes chest leads and electrodes (similar to those used to record the ECG) can be used to monitor respiratory rate. This technique of recording respiratory rate measures a difference in electrical impedance induced between the leads that correspond to changes in respiratory movements. Unfortunately, the electrical impedance method of assessing respiratory rate may be inaccurate in pediatric patients because of the small volumes and associated motions of the relatively small thorax area.

Respiratory rate may also be monitored during MR procedures using a rubber bellows placed around the patient's thorax or abdomen (i.e., for "chest" or "belly" breathers). The bellows is attached to a remote pressure transducer that records body movement changes associated with inspiration and expiration. However, the bellows monitoring technique, like the electrical impedance method, is only capable of recording body movements associated with respiratory efforts. Therefore, these respiratory rate monitoring techniques do not detect apneic episodes related to upper airway obstruction (i.e., absent airflow despite respiratory effort) and may not provide sufficient sensitivity for assessing patients during MR procedures. For this reason, assessment of respiratory rate and identification of apnea should be accomplished using other, more appropriate monitoring devices.

Respiratory rate and apnea may be monitored during MR procedures using an end-tidal carbon dioxide monitor or a capnometer. These devices measure the level of carbon dioxide during the end of the respiratory cycle (i.e., end-tidal carbon dioxide), when carbon dioxide is at its maximum level. Additionally, capnometers provide quantitative data with respect to end-tidal carbon dioxide that is important for determining certain aspects of gas exchange in patients. The waveform provided on the end-tidal carbon dioxide monitors is also useful for assessing whether the patient is having any difficulties with breathing. The interface between the patient for the end-tidal carbon dioxide monitor and capnometer is a nasal or oronasal cannula that is made out of plastic. This type of interface prevents any potential adverse interaction between the monitor and the patient during an MR procedure.

Oxygen Saturation

Oxygen saturation is a crucial variable to measure in sedated and anesthetized patients. This physiologic parameter is measured using pulse oximetry, a monitoring technique that assesses the oxygenation of tissue. A pulse oximeter is utilized to record oxygen saturation. Because oxygen saturated blood absorbs differing quantities of light compared with unsaturated blood, the amount of light that is absorbed by the blood can be readily used to calculate the ratio of oxygenated hemoglobin to total hemoglobin and displayed as the oxygen saturation. Additionally, the patient's heart rate may be calculated by measuring the frequency that pulsations occur as the blood moves through the vascular bed. Thus, the pulse oximeter determines oxygen saturation and pulse rate on a continuous basis by measuring the transmission of light through a vascular measuring site such as the ear lobe, finger tip, or toe. Notably, the use of pulse oximetry is considered by anesthesiologists as the standard practice for monitoring sedated or anesthetized patients.

Commercially-available, specially-modified pulse oximeters that have hard-wire cables have been used to monitor sedated patients during MR procedures with moderate success. Unfortunately, these pulse oximeters tend to work intermittently during the operation of the MR system, primarily due to interference from the gradient and/or radio frequency electromagnetic fields. Of greater concern is the fact that many patients have been burned using pulse oximeters with hard-wire cables, presumably as a result of excessive current being induced in inappropriately looped conductive cables attached to the patient probes of the pulse oximeters.

Fortunately, pulse oximeters have been developed that use fiber-optic technology to obtain and transmit the physiologic signals from the patient. These devices operate without interference by the electromagnetic fields used for MR procedures. It is physically impossible for a patient to be burned by the use of a fiber-optic pulse oximeter during an MR procedure because there are no conductive pathways formed by any metallic materials that connect to the patient. There are several different MR-compatible, fiber-optic pulse oximeters that are commercially available for use in the MR environment.

Temperature

There are several reasons to monitor skin and/or body temperatures during MR procedures. These include recording temperatures in neonates with inherent problems retaining body heat (a tendency that is augmented during sedation), in patients during MR procedures that require high levels of RF power, and in patients with underlying conditions that impair their ability to dissipate heat.

Skin and body temperatures may be monitored during MR procedures using a variety of techniques. However, it should be noted that the use of hard-wire, thermistor or thermocouple-based techniques to record temperatures in the MR environment may cause artifacts or erroneous measurements due to direct heating of the temperature probes. Nevertheless, if properly modified, temperature recordings may be obtained in the MR setting using specially-modified hard-wire leads and thermistors.

A more effective and easier technique of recording temperatures during MR procedures is with the use of a fluoroptic thermometry system. Experiments and clinical studies have shown that this method is both safe and reliable and can be used with MR systems that have static magnetic field strengths up to 8-Tesla.

The fluoroptic monitoring system has several important features that make it particularly useful for temperature monitoring during MR procedures. For example, the device incorporates fiber-optic probes that are small but efficient in carrying optical signals over long paths, it provide noise-free applications in electromagnetically hostile environments, and has fiber-optic components that will not pose a risk to patients.

Multi-Parameter Physiologic Monitoring Systems

In certain cases, it may be necessary to monitor several different physiologic parameters simultaneously in patients undergoing MR procedures. While several different stand-alone units may be used to accomplish this task, the most efficient means of recording multiple parameters is by utilizing a monitoring system that permits the measurement of different physiologic functions such as heart rate, respiratory rate, blood pressure, and oxygen saturation.

Currently, there are a number of multi-parameter patient monitoring systems that are MR-compatible (Table 2). Typically, these devices are designed with components positioned within the MR system room and incorporate special circuitry to substantially reduce the artifacts that effect the recording of ECG and other physiologic variables, making them also useful for the performance of "gated" MR procedures.

Ventilators

Devices used for mechanical ventilation of patients typically contain mechanical switches, microprocessors, and ferromagnetic components that may be adversely effected by the electromagnetic fields used by MR systems. Ventilators that are activated by high pressure oxygen and controlled by the use of fluidics (i.e., no requirements for electricity) may still have ferromagnetic parts that can malfunction as a result of interference from MR systems.

MR-compatible ventilators have been modified or specially designed for use during MR procedures performed in adult as well as neonatal patients. These devices tend to be constructed from non-ferromagnetic materials and have undergone pre-clinical evaluations to ensure that they operate properly in the MR environment, without producing artifacts on MR images.

Additional devices and Accessories

A variety of devices and accessories are often necessary for support and management of high-risk or sedated patients in the MR environment. MR-compatible gurneys, oxygen tanks, stethoscopes, suction devices, infusion pumps, power injectors, and other similar devices and accessories may be obtained from various manufacturers and distributors listed in Table 2. Additionally, there are MR-compatible gas anesthesia systems available that have been designed for use in patients undergoing MR procedures (Table 2).

Table 2. List of manufacturers and suppliers of monitors and support devices for use in the MR environment.

Company and Location	Device(s)
Biochem International Waukesha, WI	Monitoring equipment
Datex-Engstrom Instrumentarium Corp. Helsinki, Finland	Monitoring equipment

Datex-Ohmeda Inc. Madison, WI	Anesthesia equipment
InVivo Research, Inc. Orlando, FL	Monitoring equipment
Luxtron Santa Clara, CA	Temperature monitor
Magmedix Gardner, MA	Monitoring equipment Patient support devices
MedPacific Seattle, WA	Monitoring equipment
Medrad Indianola, PA	Monitoring equipment
Nonin Medical, Inc. Plymouth, MN	Monitoring equipment
North American Draeger Telford, PA	Monitoring equipment Anesthesia equipment
Ohmeda, Inc. Madison, WI	Anesthesia equipment
Omnivent Topeka, KS	Ventilator
Vasomed, Inc. St. Paul, MN	Blood flow monitor

(*Note that these monitors and devices may require modifications to make them MR-compatible. Consult manufacturers to determine additional information related to compatibility with specific MR systems.)

SEDATION

Whenever sedatives are used, it is imperative to perform physiologic monitoring to ensure patient safety. In addition, it is important to have the necessary equipment readily available in the event of an emergency. These requirements should also be followed for patients undergoing sedation in the MR environment.

There is controversy regarding who should be responsible for performing sedation of patients in the MR environment. (For the sake of discussion, the terms "sedation" and "anesthesia" are used interchangeably since they are actually part of the same continuum. Thus, when a patient is sedated, he/she may actually be anesthetized and all of the associated risks are present.) Obviously, there are medical, regulatory, administrative, and financial issues to be considered.

In general, for patients that do not have conditions that may complicate sedation procedures, a nurse under the direction of a radiologist may be responsible for preparing, sedating, monitoring, and recovering these cases. However, for patients that have serious medical or other unusual problems, it is advisable to utilize anesthesia consultation to properly manage these individuals before, during, and after MR procedures.

In addition, with regard to the use of sedation in the MR environment, the MRI facility should establish policies and guidelines for patient preparation, monitoring, sedation, and management during the post-sedation recovery period. These policies and guidelines should be based on standards set forth by the American Society of Anesthesiologists (ASA), the American College of Radiology (ACR), the American Academy of Pediatrics Committee on Drugs (AAP-COD) and the Joint Commission on Accreditation of Healthcare Organizations (JCAHO).

For example, Practice Guidelines for Sedation and Anesthesia from the American Society of Anesthesiologists indicate that a person must be present that is responsible for monitoring the patient if sedative or anesthetic medications are used. Furthermore, the following aspects of patient monitoring must be performed:

(1) visual monitoring,

(2) assessment of the level of consciousness,

(3) evaluation of ventilatory status,

(4) oxygen status assessed via the use of pulse oximetry, and

(5) determination of hemodynamic status via the use of blood pressure monitoring and electrocardiography if significant cardiovascular disease is present in the patient.

The healthcare professional must be able to recognize complications of sedation such as hypoventilation and airway obstruction as well as be able to establish a patent airway for postive-pressure ventilation.

Patient Preparation

Special patient screening must be conducted to identify conditions that may complicate sedation in order to properly prepare the patient for the administration of a sedative. This screening procedure should request important information from the patient that includes the following: major organ system disease (e.g., diabetes, pulmonary, cardiac, or hepatic disease), prior experience or adverse reactions to sedatives or anesthetics, current medications, allergies to drugs, and a history of alcohol, substance, or tobacco abuse.

In addition, the nothing by mouth (NPO) interval for the patient must be determined to reduce the risk of aspiration during the procedure. The ASA "Practice Guidelines Regarding Preoperative Fasting" recommend a minimum NPO periods of two hours for clear liquids, four hours for breast milk, six hours for infant formula, and six hours for a "light meal". The NPO period is extremely important because sedatives depress the patient's gag reflex.

Administration of Sedation

A thorough discussion of sedation techniques especially with regard to the use of various pharmacologic agents is outside the scope of this monograph. Therefore, interested readers are referred to the excellent, comprehensive review of this topic written by Reinking Rothshild (2000), a board certified anesthesiologist with extensive experience sedating patients in the MR environment.

Documentation

During the use of sedation, written records should be maintained that indicate the patient's vital signs as well as the name, dosage, and time of administration for all drugs that are given to the patient. The use of a time-based anesthesia-type record, such as that recommended by Reinking Rothschild (2000), is the best means of maintaining written documentation for sedation of patients in the MR environment.

Post-Sedation Recovery

After sedation, medical care of the patient that underwent sedation must continue. This is especially important for pediatric patients because certain medications have relatively long half-lives (e.g., chloral hydrate, pentobarbitol, etc.). Therefore, an appropriate room with monitoring and emergency equipment must be available to properly manage these patients.

Prior to allowing the patient to leave the MRI facility, the patient should be alert, oriented, and have stable vital signs. In addition, a responsible adult should accompany the patient home. Written instructions that include an emergency telephone number should be provided to the patient.

SUMMARY AND CONCLUSIONS

The care and management of high-risk, critically-ill, sedated or anesthetized patients in the MR environment presents special challenges. These challenges are related to requirements for MR-safe and MR-compatible equipment and devices as well as the need for MRI facilities to implement proper policies and procedures to handle these patients.

MR-compatible monitoring equipment and support accessories are commercially available from several manufacturers and distributors. Importantly, MRI facilities need to carefully consider the implementation of policies, procedures, and guidelines that have been developed and recommended by well-established professional organizations.

REFERENCES

American College of Radiology, ACR standard for the use of intravenous conscious sedation, and ACR standard for pediatric sedation/analgesia. In, 1998 ACR Standards, Reston, VA, American College of Radiology, 1998, p. 123.

American Academy of Pediatrics Committee on Drugs, Guidelines for monitoring and management of pediatric patients during and after sedation for diagnostic and therapeutics procedures, Pediatrics, 89, 1100, 1992.

A Report by the ASA Task Force on Sedation and Analgesia by Non-Anesthesiologists. Practice guidelines for sedation and analgesia by non-anesthesiologists, Anesthesiology, 84, 459, 1996.

Holshouser, B., Hinshaw, D. B., and Shellock, F. G., Sedation, anesthesia, and physiologic monitoring during MRI, J. Magn. Reson. Imaging, 3: 553-558, 1993.

Kanal, E., and Shellock, F. G., Policies, guidelines, and recommendations for MR imaging safety and patient management. Patient monitoring during MR examinations, J. Magn. Reson. Imaging, 2, 247, 1992.

Kanal, E., and Shellock, F. G., Patient monitoring during clinical MR imaging, Radiology, 185, 623, 1992.

McArdle, C., Nicholas, D., Richardson, C., and Amparo, E., Monitoring of the neonate undergoing MR imaging: Technical considerations, Radiology, 159, 223, 1986.

Reinking Rothschild, D., Chapter 5, Sedation for open magnetic resonance imaging, In, Open MRI, P. A. Rothschild and D. Reinking Rothschild, Editors, Lippincott, Williams and Wilkins, Philadelphia, 2000, pp. 39.

Shellock, F. G., and Kanal, E., Magnetic Resonance: Bioeffects, Safety, and Patient Management. Second Edition, Lippincott-Raven Press, New York, 1996.

Shellock FG. Chapter 11, Patient monitoring in the MR environment. In: Magnetic Resonance Procedures: Health Effects and Safety. CRC Press, Boca Raton, FL, 2001, pp. 217-241.

PREGNANT PATIENTS AND MR PROCEDURES*

Magnetic resonance (MR) imaging has been used to evaluate obstetrical, placental, and fetal abnormalities in pregnant patients for more than 17 years. MR imaging is recognized as a beneficial diagnostic tool and is utilized to assess a wide range of diseases and conditions that affect the pregnant patient as well as the fetus.

Initially, there were substantial technical problems with the use of MR imaging primarily due to the presence of image degradation from fetal motion. However, several technological improvements, including the development of high-performance gradient systems and rapid pulse sequences, have provided major advances especially useful for imaging pregnant patients. Thus, high quality MR imaging examinations for obstetrical and fetal applications may now be accomplished routinely in the clinical setting.

Because of the importance and prevalence of MR procedures in pregnant patients, it is crucial to understand the safety aspects of this technology and the possible bioeffects associated with the presence of the electromagnetic fields used for MR examinations.

PREGNANCY AND MR SAFETY

The use of diagnostic imaging is often required in pregnant patients. Thus, it is not surprising, that the question of whether or not a patient should undergo an MR procedure during pregnancy will often arise. Unfortunately, there have been few studies directed toward determining the relative safety of using MR procedures in pregnant patients. The main safety issues are related to possible bioeffects of the static magnetic field of the MR system, the risks associated with exposure to the gradient magnetic fields, the potential adverse effects of the radiofrequency (RF) electromagnetic fields, and possible adverse effects related to the combination of these three different electromagnetic fields.

MR environment-related risks are difficult to assess for pregnant patients due to the number of possible permutations of the various factors that are present in this setting (e.g., differences in field strengths, pulse sequences, etc.). This becomes even more complicated since new hardware and software is continually developed for clinical MR systems.

There have been a number of laboratory and clinical research investigations conducted to determine the effects of the use of MR imaging during pregnancy. Most of the laboratory studies showed no evidence of injury or harm to the fetus, while a few studies reported adverse outcomes for the laboratory animals. However, whether or not these findings can be extrapolated to human subjects is debatable.

By comparison, there have been relatively few studies performed in humans exposed to MR imaging or the MR environment *in-utero*. Each investigation reported no adverse outcomes for human subjects. For example, Baker et al. reported no demonstrable increase in disease, disability, or hearing loss in 20 children examined in-utero using echo-planar MRI for suspected fetal compromise. Myers et al. reported no significant reduction in fetal growth vs. matched controls in 74 volunteer subjects exposed *in-utero* to echo-planar MRI at 0.5-Tesla. A longer-term study is ongoing. A survey of reproductive health among 280 pregnant MR healthcare workers performed by Kanal et al. showed no substantial increase in common adverse reproductive outcomes.

In consideration of the literature published to date, there apparently are discrepancies with respect to the experimental findings of the effects of electromagnetic fields used for MR procedures and the pertinent safety aspects of pregnancy. These discrepancies may be explained by a variety of factors, including the differences in the scientific methodology used for the experiment, the type of organism examined, and the variance in exposure duration as well as the conditions of the exposure to the electromagnetic fields. Additional investigations are warranted before the risks associated with exposure to MR procedures can be absolutely known and properly characterized.

RECOMMENDED GUIDELINES FOR THE USE OF MR PROCEDURES IN PREGNANT PATIENTS

As stated in the Policies, Guidelines, and Recommendations for MR Imaging Safety and Patient Management issued by the Safety Committee of the Society for Magnetic Resonance Imaging in 1991, "MR imaging may be used in pregnant women if other nonionizing forms of diagnostic imaging are inadequate or if the examination provides important information that would otherwise require exposure to ionizing radiation (e.g., fluoroscopy, CT, etc.). Pregnant patients should be informed that, to date, there has been no indication that the use of clinical MR imaging during pregnancy has produced deleterious effects." This policy has been adopted by the American College of Radiology and is considered to be the "standard of care" with respect to the use of MR procedures in pregnant patients.

Thus, MR procedures may be used in pregnant patients to address important clinical problems or to manage potential complications for the patient or fetus. The overall decision to utilize an MR procedure in a pregnant patient involves answering a series of important questions including, the following:

- Is sonography satisfactory for diagnosis?

- Is the MR procedure appropriate to address the clinical question?

- Is obstetrical intervention prior to the MR procedure a possibility? That is, is termination of pregnancy a consideration? Is early delivery a consideration?

With regard to the use of MR procedures in pregnant patients, this diagnostic technique should not be withheld for the following cases:

- Patients with active brain or spine signs and symptoms requiring imaging.

- Patients with cancer requiring imaging.

- Patients with chest, abdomen, and pelvic signs and symptoms of active disease when sonography is non-diagnostic.

- In specific cases of suspected fetal anomaly or complex fetal disorder.

SUMMARY AND CONCLUSIONS

The information pertaining to the safety aspects of MR procedures for pregnant patients has been presented along with a discussion of the various important clinical applications for this diagnostic imaging technique. In cases where the referring physician and radiologist can defend that the findings of the MR examination has the potential to change or alter the care or management of the mother or fetus, the MR procedure (i.e., MR imaging, angiography, functional MRI, or spectroscopy) may be performed with verbal and written informed consent

[*Portions of the content on this topic were excerpted with permission from Colletti P M. Magnetic Resonance Procedures and Pregnancy, In: Magnetic Resonance Procedures: Health Effects and Safety, FG Shellock, Editor, CRC Press, Boca Raton, FL, 2001.]

REFERENCES

Baker, P.N., Johnson IR, Harvey PR, et al, A three-year follow-up of children imaged in utero with echo-planar magnetic resonance. Am J Obstet Gynecol 170, 32-33, 1994.

Benson, R.C., Colletti, P.M., Platt, L.D., et al, MR imaging of fetal anomalies, Am. J. Roentgenol., 156, 1205, 1991.

Brown, C.E.L., and Weinreb, J.C., Magnetic resonance imaging appearance of growth retardation in a twin pregnancy, Obstet. Gynecol. 71, 987, 1988.

Carnes, K.I., and Magin, R.L., Effects of in utero exposure to 4.7T MR imaging conditions on fetal growth and testicular development in the mouse, Magn. Reson. Imaging, 14, 263, 1996.

Carswell, H., Fast MRI of fetus yields considerable anatomic detail, Diag. Imaging, Nov, 11-12, 1988.

Colletti, P.M., Computer-assisted imaging of the fetus with magnetic resonance imaging, Comput. Med. Imaging Graph., 20, 491, 1996.

Colletti, P.M., and Platt, L.D., When to use MRI in obstetrics, Diag. Imaging, 11, 84, 1989.

Colletti, P.M., and Sylvestre, P.B., Magnetic resonance imaging in pregnancy, MRI Clin. N. Am., 2, 291, 1994.

Dinh, D.H., Wright, R.M., and Hanigan, W.C., The use of magnetic resonance imaging for the diagnosis of fetal intracranial anomalies, Child Nerv. Syst., 6, 212, 1990.

Dunn, R.S., and Weiner, S.N., Antenatal diagnosis of sacrococcygeal teratoma facilitated by combined use of Doppler sonography and MR imaging, Am. J. Roentgenol., 156,1115, 1991.

Fitamorris-Glass, R., Mattrey, R.F., Cantrell, C.J., Magnetic resonance imaging as an adjunct to ultrasound in oligohydramnios, J Ultrasound Med., 8,159, 1989.

Fraser, R., Magnetic resonance imaging of the fetus, Initial experience [letter], Gynecol. Obstet. Invest., 29, 255, 1990.

Gardens AS, Weindling AM, Griffiths RD, et al, Fast-scan magnetic resonance imagin of fetal anomalies. Br J Obstet Gynecol 98,1217-1222, 1991

Hill MC, Lande IM, Larsen JW Jr, Prenatal diagnosis of fetal anomalies using ultrasound and MRI. Radiol Clin North Am 26,287-307, 1988

Horvath L, Seeds JW, Temporary arrest of fetal movement with pancuronium bromide to enable antenatal magnetic resonance imagin of holosencephaly. Am. J. Roentgenol., 6,418-420, 1989

Heinrichs, W.L., Fong, P., Flannery, M., Meinrichs, S.C., et al., Midgestational exposure of pregnant BALB/c mice to magnetic resonance imaging conditions, Magn. Reson. Imaging, 6,305, 1988.

Kanal, E., Gillen, J., Evans, J.A., et al, Survey of reproductive health among female MR workers, Radiology, 187,395, 1993.

Kanal, E, Shellock, F.G., and Sonnenblick, D., MRI clinical site safety: Phase I results and preliminary data, Magn. Reson. Imaging, 7[Suppl] 1, 106, 1988.

Kay, H.H., Herfkens, R.J., and Kay, B.K., Effect of magnetic resonance imaging on Xenopus laevis embryogenesis, Magn. Reson. Imaging, 6, 501-6, 1988.

Lenke RR, Persutte WH, Nemes JM, Use of pancuronium bromide to inhibit fetal movement during magnetic resonance imaging. J. Reprod. Med. 34,315-317, 1989

Malko, J.A., Constatinidis, I, Dillehay, D., et al., Search for influence of 1.5 T magnetic field on growth of yeast cells, Bioelectromagnetics, 15, 495, 1987.

Mansfield P, Stehling MK, Ordidge RJ, et al, Study of internal structure of the human fetus in utero at 0.5 T. Br J Radiol 13,314-318, 1990

McCarthy, S.M., Stark, D.D., Filly, R.A., et al, Uterine neoplasms, MR imaging, Radiology, 170, 125, 1989.

McCarthy SM, Filly RA, Stark DD, et al, Magnetic resonance imaging of fetal anomalies

Smith FW, Magnetic resonance tomography of the pelvis. Cadiovasc Intervent Radiol 8,367-376, 1986

McRobbie, D., and Foster, M.A., Pulsed magnetic field exposure during pregnancy and implications for NMR foetal imaging, A study with mice. Magn. Reson. Imaging, 3, 231, 1985.

Myers, C., Duncan, K.R., Gowland, P.A., et al., Failure to detect intrauterine growth restriction following in utero exposure to MRI, Br J Radiol, 71, 549, 1998.

Murakami, J., Toril, Y., and Masuda, K., Fetal developmet of mice following intrauterine exposure to a static magnetic field of 6.3 T, Magn. Reson. Imaging, 10, 433, 1992.

Nara, V.R., Howell, R.W., Goddu, S.M., et al., Effects of a 1.5T static magnetic field on spermatogenesis and embryogenesis in mice, Invest. Radiol, 31, 586, 1996.

Shellock, F.G., and Kanal, E., Policies, guidelines, and recommendations for MR imaging safety and patient management, J. Magn. Reson. Imaging, 1, 97, 1991.

Shellock, F.G., and Kanal, E., Magnetic resonance procedures and pregnancy, Magnetic Resonance Bioeffects, Safety, and Patient Management, Second Edition, Lippincott-Ravin, Philadelphia, New York, 1996, Chapter 4, pp. 49.

Smith FW, Kent C, Abramovich DR, et al, Nuclear magnetic resonance imaging - a new look at the fetus. Br J Gynaecol 92,1024-1033, 1985

Smith FW, Sutherland HW, Magnetic resonance imaging: The use of the inversion recovery sequence to display fetal morphology. Br J Radiol 61,338-341, 1988.

Stark, D.D., McCarthy, S.M., Filly, R.A., et al, Pelvimetry by magnetic resonance imaging. Am. J. Roentgenol., 144, 947, 1985.

Tesky, G.C., Ossenkopp, K.P., Prato, F.S., and Sestini, E., Survivability and long-term stress reactivity levels following repeated exposure to nuclear magnetic resonance imaging procedures in rats, Physiol. Chem. Phys. Med. NMR, 19, 43, 1987.

Tyndall, D.A., MRI effects on the teratogenicity of x-irradiation in the C57BL/6J mouse, Magn. Reson. Imaging, 8, 423, 1990.

Tyndall, R.J., and Sulik, K.K., Effects of magnetic resonance imaging on eye development in the C57BL/6J mouse, Teratology, 43, 263, 1991.

Tyndall, D.A., MRI effects on craniofacial size and crown-rump length in C57BL/6J mice in 1.5T fields, Oral Surg. Oral Med. Oral Pathol., 76, 655, 1993.

Yip, Y.P.,Capriotti, C., Talagala ,S.L., and Yip, J.W., Effects of MR exposure at 1.5 T on early embryonic development of the chick, J. Magn. Reson. Imaging, 4, 742, 1994.

Yip, Y.P., Capriotti, C., and Yip, J.W., Effects of MR exposure on axonal outgrowth in the sympathetic nervous system of the chick, J. Magn. Reson. Imaging, 4, 457, 1995.

Yip, Y.P., Capriotti, C., Norbash, S.G., Talagala, S.L., and Yip, J.W., Effects of MR exposure on cell proliferation and migration of chick motor neurons, J. Magn. Reson. Imaging, 4,799, 1994.

Vadeyar, S.H., Moore, R.J., Strachan, B.K., et al., Effect of fetal magnetic resonance imaging on fetal heart rate patterns, Am. J. Obstet. Gynecol., 182, 666, 2000.

Weinreb, J.C., Brown, C.E., Lowe, T.W., et al., Pelvic masses in pregnant patients, MR and US imaging, Radiology, 159, 717, 1986.

Weinreb JC, Lowe T, Santos-Ramos R, et al, Magnetic resonance imaging in obstetric diagnosis. Radiology 154,157-161, 1985.

Wenstrom KD, Williamson RA, Weiner CP, et al, Magnetic resonance imaging of fetuses with intracranial defects. Obstet Gynecol., 77,529-532, 1991.

Wilbur, A.C., Langer, B.G., and Spigos, D.G., Diagnosis of sacroiliac joint infection in pregnancy by magnetic resonance imaging, Magn. Reson. Imaging, I 6, 341, 1988.

Wilcox, A., Weinberg, C., O'Connor J, et al., Incidence of early loss of pregnancy, New Engl. J. Med., 319,189-194, 1988.

Williamson, RA, Weiner CP, Yuh WTC, et al, Magnetic resonance imaging of anomalous fetuses. Obstet Gynecol 71,952, 1988.

PREGNANT TECHNOLOGISTS AND OTHER HEALTHCARE WORKERS IN THE MR ENVIRONMENT

Due to the concern with regard to pregnant technologists and other health-care workers in the MR environment, a survey of reproductive health among female MR system operators was conducted in 1990 by Kanal et al. Questionnaires were sent to all female MR technologists and nurses at the majority of clinical MR facilities in the United States. The question-naire addressed menstrual and reproductive experiences as well as work activities. This study attempted to account for known potential confound-ing variables (e.g., age, smoking, alcohol use) for this type of data.

Of the 1915 completed questionnaires analyzed, there were 1421 preg-nancies: 280 occurred while working as an MR employee (technologist or nurse), 894 while employed at another job, 54 as a student, and 193 as a homemaker. Five categories were analyzed that included spontaneous abortion rate, pre-term delivery (less than 39 weeks), low birth weight (less than 5.5 pounds), infertility (taking more than eleven months to con-ceive), and gender of the offspring.

The data indicated that there were no statistically significant alterations in the five areas studied for MR healthcare workers relative to the same group studied when they were employed elsewhere, prior to becoming MR healthcare employees. Additionally, adjustment for maternal age, smok-ing, and alcohol use failed to markedly change any of the associations.

Menstrual regularity, menstrual cycle, and related topics were also exam-ined in this study. These included inquiries regarding the number of days of menstrual bleeding, the heaviness of the bleeding, and the time between menstrual cycles. Admittedly, this is a very difficult area to objectively examine, because it entirely depends upon both subjective memory and the memory of the respondent for a topic, where subjective memory is notoriously inadequate. Nevertheless, the data suggested that there were no clear correlations between MR workers and specific modifications of the menstrual cycle.

The data from this extensive epidemiological investigation were reassuring insofar as that there did not appear to be any deleterious effects from exposure to the static magnetic field component of the MR system. Therefore, a policy is recommended that permits pregnant technologists and healthcare workers to perform MR procedures, as well as to enter the MR system room, and to attend to the patient during pregnancy, regardless of the trimester. Importantly, the technologists or healthcare worker should not remain within the MR system room or magnet bore during the actual operation of the scanner.

This later recommendation is especially important for those MR healthcare workers involved in interventional MR-guided examinations and procedures, since it may be necessary for them to be directly exposed to the MR system's electromagnetic fields at levels similar to those used for patients. These recommendations are not based on indications of adverse effects, but rather, from a conservative point of view and the feeling that there are insufficient data pertaining to the effects of the other electromagnetic fields of the MR system to support or allow unnecessary exposures.

REFERENCES

Kanal E, Gillen J, Evans J, Savitz D, Shellock FG. Survey of reproductive health among female MR workers. Radiology 1993;187: 395-399.

Shellock FG. Magnetic Resonance Procedures: Health Effects and Safety. CRC Press, LLC, Boca Raton, FL, 2001.

Shellock FG, Kanal E. Policies, guidelines, and recommendations for MR imaging safety and patient management. J Magn Reson Imaging 1991;1: 97-101.

Shellock FG, Kanal E. Magnetic Resonance: Bioeffects, Safety, and Patient Management. Second Edition, Lippincott-Raven Press, New York, 1996.

PREVENTION OF MISSILE EFFECT ACCIDENTS

The "missile effect" refers to the capability of the fringe field component of the static magnetic field of an MR system to attract a ferromagnetic object, drawing it rapidly into the scanner by considerable force. The missile effect can pose a significant risk to the patient inside the MR system and/or anyone who is in the path of the projectile. Therefore, a strict policy should be established by the MRI facility to detect metallic objects prior to allowing individuals or patients to enter the MR environment in order to avoid accidents and potential injuries related to the missile effect. In addition, to guard against accidents from metallic projectiles, the immediate area around the MR system should be clearly demarcated, labeled with appropriate danger signs, and secured by trained staff aware of proper MR safety procedures.

For patients preparing to undergo MR procedures, all metallic or other potentially problematic personal belongings (i.e., hearing aids, analogue watches, jewelry, etc.) and devices must be removed as well as clothing items that have metallic fasteners or other metallic components (e.g., threads). The most effective means of preventing a ferromagnetic object from inadvertently becoming a missile is to require the patient to where a gown.

Non-ambulatory patients must only be allowed to enter the area of the MR system using a nonferromagnetic wheelchair or nonferromagnetic gurney. Wheelchairs and gurneys should also be inspected for the presence of a ferromagnetic oxygen tank or other similar components or accessories before allowing the patient into the MR setting. Fortunately, there are several commercially available, MR-safe or MR-compatible devices that may be used to transport and support patients to and from the MR system room.

Any individual accompanying the patient must be required to remove all metallic objects before entering the MR area and should undergo a careful and thorough screening procedure. All hospital and outside personnel that may need to enter the MR environment periodically or in response to an emergency (e.g., custodial staff, maintenance workers, housekeeping staff, bioengineers, nurses, security officers, fire fighters, etc.) should be educated about the potential hazards associated with the magnetic fringe field of the MR system. These individuals should, likewise, be instructed to remove metal objects before entering the MR environment to prevent missile-related accidents.

Many serious incidents have occurred when individuals, who were unaware of the powers of the fringe field, entered the MR environment with items such as oxygen tanks, wheel chairs, monitors, and other similar ferromagnetic objects. In July 2002, a fatal accident occurred which illustrated the importance of careful attention to ferromagnetic objects that may pose hazards in the MR environment. In this widely publicized incident, a young patient suffered a blow to the head from a ferromagnetic oxygen tank that became a projectile in the presence of a 1.5-Tesla MR system.

While MR safety guidelines and procedures are well known, accidents related to the missile effect continue to occur. Guidelines and recommendations for preventing these hazards are presented in Table 1.

Table 1. Guidelines and recommendations for preventing hazards related to "missile effects" in the MR environment.*

(1) Appoint a safety officer or other person responsible for ensuring that proper procedures are in effect, enforced, and updated to ensure safety in the MR environment.

(2) Establish and routinely review MR safety policies and procedures, and assess the level of compliance by all staff members.

(3) Provide all MR staff, along with other personnel who have an opportunity or need to enter the MR environment (e.g., transport personnel, security officers, housekeeping staff, maintenance workers, fire department personnel, etc.) with formal training on MR safety.

(4) Understand and emphasize to all personnel that the MR system's static magnetic field is always "on" and treat the MR environment, accordingly.

(5) Don't allow equipment and devices containing ferromagnetic components into the MR environment, unless they have been tested and labeled "MR-safe".

(6) Adhere to any restrictions provided by suppliers regarding the use of MR-safe and/or MR-compatible equipment and devices in your MR environment. A label of MR-safe means that, "the device, when used in the MR environment, has been demonstrated to present no additional risk to the patient or other individuals, but may affect the quality of diagnostic information" (CDRH Magnetic Resonance Working Group, 1997). MR-compatible equipment, on the other hand, is not only MR-safe, but also can be used in the MR environment with no significant effect on its operation or on the quality of diagnostic information.

(7) Maintain a list of MR-safe and MR-compatible equipment, including restrictions for use. This list should be kept and updated in every MRI facility by the MR safety officer (see www.MRIsafety.com for comprehensive information).

(8) Bring non-ambulatory patients into the MR environment using a non-magnetic wheelchair or gurney. Ensure that no oxygen tanks, sandbags with metal shot, or other ferromagnetic objects are concealed under blankets or sheets or stowed away on the transport equipment.

(9) Ensure that IV poles accompanying patients into the MR environment are nonferromagnetic.

(10) Carefully screen all individuals entering the MR environment for magnetic objects in their bodies (e.g., implants, bullets, shrapnel, etc.), on their bodies (e.g., hair pins, brassieres, buttons, zippers, jewelry), or attached to their bodies (e.g., body piercing jewelry, body modification implants, halo vests, cervical fixation devices). Magnetic objects on or attached to the bodies of patients, family members, or staff members should be removed, if feasible, before the individuals enter the MR environment.

(11) Have patients wear hospital gowns without metallic fasteners for MR procedures. The patient's clothing may contain metallic objects or threads that may pose a hazard in the MR environment.

[*Adapted and published permission from ECRI Report, 2001 and Shellock FG, 2001).

REFERENCES

Chaljub G, Kramer LA, Johnson RF III, Johnson RF Jr., Singh H, Crow WN. Projectile cylinder accidents resulting from the presence of ferromagnetic nitrous oxide or oxygen tanks in the MR suite. American Journal of Roentgenology 2001;177:27-30.

ECRI. Patient Death Illustrates the Importance of Adhering to Safety Precautions in Magnetic Resonance Environments. ECRI, Plymouth Meeting, PA, Aug. 6, 2001.

http://www.MRIsafety.com

Shellock FG. Magnetic Resonance Procedures: Health Effects and Safety. CRC Press, LLC, Boca Raton, FL, 2001.

Shellock FG. Pocket Guide to MR Procedures and Metallic Objects: Update 2001, Seventh Edition, Lippincott Williams & Wilkins Healthcare, Philadelphia, 2001.

Shellock FG, Kanal E. Magnetic Resonance: Bioeffects, Safety, and Patient Management. Second Edition, Lippincott-Raven Press, New York, 1996.

SIGNS TO HELP CONTROL ACCESS TO THE MR ENVIRONMENT

To guard against accidents and injuries to patients and other individuals as well as damage to magnetic resonance (MR) systems, the general and immediate areas associated with the scanner (also referred to as the MR environment) must have supervised and controlled access. Supervised and controlled access involves having MR safety-trained personnel present at all times during the operation of the facility to ensure that no unaccompanied or unauthorized individuals are allowed to enter the MR environment. Importantly, MR safety-trained personnel should be responsible for performing comprehensive screening of patients and other individuals before allowing them to enter the MR system room.

Additionally, it is necessary to educate everyone who needs to enter the MR environment on a regular or intermittent basis (e.g., custodial workers, transporters, security personnel, firefighters, nurses, anesthesiologists, etc.) regarding the potential hazards related to the powerful magnetic field of the MR system. Unfortunately, even with proper MR safety procedures in place, many individuals and patients have inadvertently "wandered" unattended into the MR environment, and these situations have resulted in problematic or disastrous consequences.

As one means of helping to control access to the MR environment, the area must be clearly demarcated and labeled with prominently displayed signs to make all individuals and patients aware of the risks associated with the MR system. The content of these signs is particularly important. However, the information shown on some signs currently in use is out-of-date, erroneous, or not displayed in a prominent enough manner. Therefore, new signs with revised content and new information were designed to promote a safe MR environment.

Old "Warning" Sign. One sign commonly found in MRI centers has information that states:

WARNING

STRONG MAGNETIC FIELD

NO PACEMAKERS

NO METALLIC IMPLANTS

NO NEUROSTIMULATION SYSTEMS

NO LOOSE OBJECTS

Obviously, given the present state of knowledge pertaining to MR safety, much of this information is outdated or simply incorrect. In fact, according to the Food and Drug Administration document entitled, Guidance for the Submission Of Premarket Notifications for Magnetic Resonance Diagnostic Devices (issued November 14, 1998), Attachment B, states: "The controlled access area should be labeled "Danger - High Magnetic Field" at all entries." Also, this FDA document indicates, "Operators should be warned by appropriate signs about the presence of magnetic fields and their force and torque on magnetic materials, and that loose ferrous objects should be excluded. "

New "DANGER" Sign. In consideration of the above, the old "WARNING" sign was revised to include the guidance from the FDA as well as the most current findings for MR safety, especially with regard to implanted objects. Because the term "warning" does not adequately convey the importance of a situation that is potentially hazardous and that has been responsible for serious injuries and deaths, the revised sign states:

DANGER!

Additionally, to inform everyone about the powerful static magnetic field associated with the MR system, especially individuals unacquainted with MR technology, the following information is prominently shown on this sign:

RESTRICTED ACCESS

STRONG MAGNETIC FIELD

THE MAGNET IS ALWAYS ON!

With respect to the information for implants and devices, in addition to cardiac pacemakers, implantable cardioverter defibrillators (or ICDs) are also potentially hazardous for patients and individuals in the MR environment. Therefore, this information is included on the new sign. Furthermore, because recently published reports have indicated that certain neurostimulation systems are safe for patients undergoing MR procedures if highly specific guidelines are followed, the statement regarding neurostimulation systems was deleted to avoid undue confusion.

Notably, recent articles in the peer-reviewed literature have reported that many types of metallic implants are actually safe for patients undergoing MR procedures. Accordingly, this information is now clarified on the revised sign, with individuals and patients informed to consult MRI professionals if there are any questions regarding this matter, as follows:

"Persons with certain metallic, electronic, magnetic, or mechanically-activated implants, devices, or objects may not enter this area. Serious injury may result.

Do not enter this area if you have any question regarding an implant, device, or object. Consult the MRI Technologist or Radiologist."

Finally, the statement, "No Loose Objects" on the old "WARNING" sign is rather simplistic and does not address other aspects of concern with respect to bringing potentially problematic items into the MR environment. Accordingly, the new sign states:

"Objects made from ferrous materials must not be taken into this area. Serious injury or property damage may result. Electronic objects such as hearing aids, cell phones, and beepers may also be damaged."

Thus, this new sign is more prominent, the term "DANGER" rather that "WARNING" is used (which, hopefully, will make individuals and patients readily take notice), and the overall content is more accurate with respect to current MR safety information. A Spanish language version of this sign has also been created*.

Interestingly, many individuals fail to realize that the MR system's static magnetic field is always on. In fact, investigations of accidents that involved relatively large ferromagnetic objects like oxygen cylinders, chairs, IV poles, and wheelchairs revealed that the offending hospital personnel thought that the powerful magnetic field was activated *only* during the MR procedure. Therefore, a smaller sign or decal that states:

DANGER!

THE MAGNET IS ALWAYS ON!

may be used to further emphasize the potentially hazardous nature of the MR environment.

Sign Placement. The strategic placement of signs in and around the MR environment is crucial to ensure that all individuals and patients see them before entering this area. In general, a sign should be placed on the door or entrance to MR system room and near the doorframe to be viewed by individuals and patients, especially if the door to the MR system room is open.

[*To obtain the signs designed to help control access to the MR environment, please visit www.Magmedix.com]

REFERENCES

Finelli DA, Rezai AR, Ruggieri P, Tkach J, Nyenhuis J, Hridlicka G, Sharan A, Gonzalez-Martinez J, Stypulkowski PH, Shellock FG. MR-related heating of deep brain stimulation electrodes: an in vitro study of clinical imaging sequences. American Journal of Neuroradiology 2002;23:1795-1802.

Rezai AR, Finelli D, Nyenhuis JA, Hrdlick G, Tkach J, Ruggieri P, Stypulkowski PH, Sharan A, Shellock FG. Neurostimulator for deep brain stimulation: Ex vivo evaluation of MRI-related heating at 1.5-Tesla. Journal of Magnetic Resonance Imaging 2002;15:241-250.

Shellock FG. MR safety update 2002: Implants and devices. Journal of Magnetic Resonance Imaging 2002;16:485-496.

Shellock FG. Biomedical implants and devices: assessment of magnetic field interactions with a 3.0-Tesla MR system. Journal of Magnetic Resonance Imaging 2002;16:721-732.

Shellock FG. Reference Manual For Magnetic Resonance Safety: 2003 Edition. Amirsys, Inc., Salt Lake City, UT.

U. S. Department Of Health and Human Services, Center for Devices and Radiological Health Food and Drug Administration, the document entitled, Guidance for the, Submission Of Premarket Notifications for Magnetic Resonance Diagnostic Devices, Issued November 14, 1998.

TATTOOS, PERMANENT COSMETICS, AND EYE MAKEUP

Traditional (i.e., decorative) and cosmetic tattoo procedures have been performed for thousands of years. In the United States, cosmetic tattoos or "permanent cosmetics" are used to reshape, recolor, recreate, or modify eye shadow, eyeliner, eyebrows, lips, beauty marks, and cheek blush. Additionally, permanent cosmetics are often used aesthetically to enhance nipple-areola reconstruction procedures.

Magnetic resonance (MR) imaging is a frequently used imaging modality, particularly for evaluating the brain, head and neck, and other anatomic regions where cosmetic tattoos are typically applied. Unfortunately, there is much confusion regarding the overall MR safety aspects of permanent cosmetics. For example, based on a few reports of symptoms localized to the tattooed area during MR imaging, many radiologists have refused to perform MR procedures on individuals with permanent cosmetics, particularly tattooed eyeliner (Unpublished Observations. F. G. Shellock, 2000). This undue concern for possible adverse events prevents patients with cosmetic tattoos from access to an extremely important diagnostic imaging technique.

While it is well-known that permanent cosmetics and tattoos may cause MR imaging artifacts and that both cosmetic and decorative tattoos may cause relatively minor, short-term cutaneous reactions, the frequency and severity of soft tissue reactions or other related problems associated with MR imaging and cosmetic tattoos is unknown. Therefore, a study was conducted by Tope and Shellock (2002) to determine the frequency and severity of adverse events associated with MR imaging in a population of subjects with permanent cosmetics. A questionnaire was distributed to clients of cosmetic tattoo technicians. This survey asked study subjects for demographic data, information about their tattoos, and for their experiences during MR imaging procedures. Results from 1,032 surveys were tabulated. One hundred thirty-five (13.1%) study subjects underwent MR imaging after having permanent cosmetics applied. Of these, only two individuals (1.5%) experienced problems associated with MR imaging. One subject reported a sensation of "slight tingling" and the other subject reported a sensation of "burning". Both of these incidents were transient in nature and did not prevent the MR procedures from being performed.

Based on these findings and additional information in the peer-reviewed literature, it appears that MR imaging may be performed in patients with permanent cosmetics without any serious soft tissue reactions or adverse events. Therefore, the presence of permanent cosmetics should not prevent a patient from undergoing MR imaging. Furthermore, when one considers the many millions of clinical MR procedures that have been conducted in patients over the past 18 years and that only a very small percentage of these individuals have had minor, short-term problems related to the presence of permanent cosmetics, it is apparent that this MR safety concern has an extremely low rate of occurrence and relatively insignificant consequences.

Before undergoing an MR procedure, the patient should be asked if he or she has ever had any type of permanent coloring technique (i.e., tattooing) applied to any part of the body. This includes cosmetic applications such as eyeliner, lip-liner, lip coloring, as well as decorative designs. This question is necessary because of the associated imaging artifacts and, more importantly, because a small number of patients (fewer than 10 documented cases) have experienced transient skin irritation, cutaneous swelling, or heating sensations at the site of the permanent colorings in association with MR procedures (review of Medical Device Reports, 1985 to 2003).

Interestingly, decorative tattoos tend to cause worse problems (including first- and second-degree burns) for patients undergoing MR imaging compared to those that have been reported for cosmetic tattoos. With regard to decorative tattoos, a letter to the editor described a second-degree burn that occurred on the skin of the deltoid from a decorative tattoo. The authors suggested that "the heating could have come either from oscillations of the gradients or, more likely from the RF-induced electrical currents". However, the exact mechanism(s) responsible for complications or adverse events in the various cases that have occurred related to decorative tattoos is unknown.

Additionally, Kreidstein et al. reported that a patient experienced a sudden burning pain at the site of a decorative tattoo while undergoing MR imaging of the lumbar spine using a 1.5-Tesla MR system. Swelling and erythema was resolved within 12 hours, without evidence of permanent sequelae. The tattoo pigment used in this case was ferromagnetic, which possibly explains the symptoms experienced by the patient. Surprisingly, in order to permit completion of the MR examination, an excision of the tattooed skin with primary closure of the site was performed.

The authors of this report stated, "Theoretically, the application of a pressure dressing of the tattoo may prevent any tissue distortion due to ferromagnetic pull". However, this simple, relatively benign procedure was not attempted for this patient. They also indicated that, "In some cases, removal of the tattoo may be the most practical means of allowing MRI".

Kanal and Shellock commented on this report in a letter to the editor, suggesting that the response to this situation was "rather aggressive". Clearly the trauma, expense, and morbidity associated with excision of a tattoo far exceed those that may be associated with ferromagnetic tattoo interactions. A firmly applied pressure bandage may be used if there is any concern related to "movement" of the ferromagnetic particles in the tattoo pigment. Additionally, direct application of a cold compress to the site of a tattoo would likely mitigate any heating sensation that may occur in association with MR imaging.

Artifacts. Imaging artifacts associated with permanent cosmetics and certain types of eye makeup have been reported. These artifacts are predominantly associated with the presence of pigments that use iron oxide or other type of metal and occur in the immediate area of the applied pigment or material. As such, tattoo- and makeup-related MR imaging artifacts should not prevent a diagnostically adequate MR imaging procedure from being performed, especially in consideration that careful selection of imaging parameters may easily minimize artifacts related to metallic materials.

The only possible exception to this is if the anatomy of interest is in the exact same position of where the tattoo was applied using an iron oxide-based pigment. For example, Weiss et al. reported that heavy metal particles used in the pigment base of mascara and eyeliner tattoos, have a paramagnetic effect that causes alteration of the local magnetic field in adjacent tissues. Changes in the typical MR signal pattern may result in distortion of the globes. In some cases, the distortion may mimic actual ocular disease, such as a ciliary body melanoma or cyst.

GUIDELINES AND RECOMMENDATIONS

In consideration of the available literature and experience pertaining to MR procedures and patients with permanent cosmetics and tattoos, guidelines to manage these individuals include, the following:

(1) The screening form used for patients should include a question pertaining to the presence of permanent cosmetics or decorative tattoos.

(2) Before undergoing an MR procedure, the patient should be asked if he or she has a permanent coloring technique (i.e., tattooing) applied to any part of the body. This includes cosmetic applications such as eyeliner, lip-liner, lip coloring, as well as decorative designs.

(3) The patient should be informed of the relatively minor risk associated with the site of the tattoo.

(4) The patient should be advised to immediately inform the MRI technologist regarding any unusual sensation felt at the site of the tattoo in association with the MR procedure.

(5) The patient should be closely monitored using visual and auditory means throughout the entire operation of the MR system to ensure safety.

(6) As a precautionary measure, a cold compress (e.g., wet washcloth) may be applied to the tattoo site during the MR procedure.

In addition to the above, information and recommendations have been provided for patients by the United States Food and Drug Administration, Center for Food Safety and Applied Nutrition, Office of Cosmetics and Colors Fact Sheet, as follows: "... the risks of avoiding an MRI when your doctor has recommended one are likely to be much greater than the risks of complications from an interaction between the MRI and tattoo or permanent makeup. Instead of avoiding an MRI, individuals who have tattoos or permanent makeup should inform the radiologist or technician of this fact in order to take appropriate precautions, avoid complications, and assure the best results."

SUMMARY AND CONCLUSIONS

Because of the relatively remote possibility of having an incident occur in a patient with permanent cosmetic application or tattoo and due to the relatively minor, short-term complication that may develop (i.e., transient cutaneous redness and swelling), the patient should be permitted to undergo an procedure without reservation as long as proper precautions are followed (e.g., application of cold compress). Any problem associated with using an MR procedure in a patient with a cosmetic or decorative tattoo should not prevent the performance of the examination, since the information obtained by this diagnostic modality is typically critical to the care and management of the patient.

REFERENCES

Becker H. The use of intradermal tattoo to enhance the final result of nipple-areola reconstruction. Plast Reconstr Surg 1986;77:673.

Carr JJ. Danger in performing MR imaging on women who have tattooed eyeliner or similar types of permanent cosmetic injections. AJR Am J Roentgenol 1995; 165:1546-1547.

U. S. Food and Drug Administration, Center for Food Safety and Applied Nutrition, Office of Cosmetics and Colors Fact Sheet, Tattoos and permanent makeup. November 29, 2000.

Gomey M. Tattoo pigments, patient clothing, and magnetic resonance imaging. Risk Management Bulletin #12748-8/95, The Doctors' Company: Napa, CA. August, 1995.

Halder RM, Pham HN , Hreadon JY, Johnson HA. Micropigmentation for the treatment of vitiligo. J Dermatol Surg Oncol 1989;15:1092-1098.

Jackson JG, Acker H. Permanent eyeliner and MR imaging. (letter) AJR Am J Roentgenol 1987;149:1080.

Kanal E, Borgstede JP, Barkovich AJ, Bell C, et al. American College of Radiology, White paper on MR safety. American Journal of Roentgenology 2002;178:1335-1347.

Kanal E, Shellock FG. MRI interaction with tattoo pigments. (letter) Plast Reconstr Surg 1998;101:1150-1151.

Kreidstein ML, Giguere D, Freiberg A. MRI interaction with tattoo pigments: case report, pathophysiology, and management. Plast Reconstr Surg 1997;99:1717-1720.

Lund A, Nelson ID, Wirtschafter ID, Williams PA. Tattooing of eyelids: magnetic resonance imaging artifacts. Ophthalmic Surg 1986;17:550-553.

Sacco D, et al. Artifacts caused by cosmetics in MR imaging of the head. Am J Roentgenol 1987;148:1001-1004.

Shellock FG. Magnetic Resonance Procedures: Health Effects and Safety, CRC Press, LLC, Boca Raton, FL, 2001.

Shellock FG. Guide to MR Procedures and Metallic Objects: Update 2001. Seventh Edition, Lippincott Williams & Wilkins Healthcare, Philadelphia, 2001.

Tattoos. FDA Medical Bulletin 1994;24:8.

Tope WD, Shellock FG. Magnetic resonance imaging and permanent cosmetics (tattoos): survey of complications and adverse events. J Magn Reson Imaging 2002;15:180-184.

Vahlensieck M. Tattoo-related cutaneous inflammation (burn grade I) in a mid-field MR scanner. (letter) Eur Radiol 2000;10:97.

Wagle WA, Smith M. Tattoo-induced skin burn during MR imaging. (letter) Am J Roentgenol 2000: 174:1795.

Weiss RA, Saint-Louis LA, Haik BG, McCord CD, Taveras JL. Mascara and eye-lining tattoos: MRI artifacts. Ann Ophthalmology 1989;21:129-131.

SECTION II

MR PROCEDURES
AND
IMPLANTS, DEVICES,
AND MATERIALS

GENERAL INFORMATION

Magnetic resonance (MR) procedures may be contraindicated for patients primarily because of risks associated with movement or dislodgment of a ferromagnetic biomedical implant, material, device, or object. There are other possible hazards and problems related to the presence of a metallic object that include induction of currents (e.g., in materials that are conductors), excessive heating, and the misinterpretation of an imaging artifact as an abnormality.

Induced Electrical Currents. The potential for MR procedures to injure patients by inducing electrical currents in conductive materials or devices such as gating leads, indwelling catheters with metallic components (e.g., thermodilution catheters), guide wires, disconnected or broken surface coils, certain cervical fixation devices, cochlear implants, infusion pumps, or improperly used physiologic monitors has been previously reported. Recommendations concerning techniques to protect patients from injuries related to induced currents that may develop during an MR procedure, especially those associated with the use of monitoring equipment, have been presented.

Heating. Increases in temperature produced in association with MR procedures have been studied using *ex vivo* testing techniques to evaluate various metallic implants, materials, devices, and objects of a variety of sizes, shapes, and metallic compositions. In general, reports have indicated that only minor temperature changes occur in association with MR procedures involving relatively small metallic objects that are "passive" implants (i.e., those that are not electronically-activated). Therefore, heat generated during an MR procedure involving a patient with a metallic "passive" implant does not appear to be a substantial hazard. To date, there has been no report of a patient being seriously injured as a result of excessive heat developing in a "passive" metallic biomedical implant or device, with the exception of objects that have an elongated shape or those that form a conducting loop. In fact, according to Dempsey et al., heating tends to be

100

more likely for objects that form resonant conducting loops or for extended wires that form resonant antennae. Additionally, first-, second-, or third-degree burns have occurred in association with conductive devices that were not used according to the manufacturer's recommendations.

Artifacts. The type and extent of artifacts caused by the presence of metallic implants, materials, and devices have been described and tend to be easily recognized on MR images. Artifacts and image distortion associated with metallic objects are predominantly caused by a disruption of the local magnetic field that perturbs the relationship between position and frequency, which is crucial for proper image reconstruction. For this reason, implants, materials, devices, or objects that incorporate magnets can produce artifacts on MR images that are especially profound because of the accentuated effect of altering the local magnetic field.

The relative amount of artifact seen on an MR image is dependent on the magnetic susceptibility, quantity, shape, orientation, and position of the object in the body as well as the technique used for imaging (i.e., the specific pulse sequence parameters) and the image processing method. An artifact caused by the presence of a metallic object in a patient during MR imaging is seen typically as a local or regional distortion of the image and/or as a signal void. In some cases, there may be areas of high signal intensity seen along the edges of the signal void.

Nonferromagnetic objects tend to produce artifacts that are less severe than ferromagnetic objects for a given set of MR imaging parameters. Artifacts associated with nonferromagnetic implants result from eddy currents that can be generated in the objects by gradient magnetic fields used for MR imaging that, in turn, disrupt the local magnetic field.

Magnetic Field Interactions. Numerous studies have assessed magnetic field interactions for biomedical implants, materials, devices, and objects by measuring deflection forces, translational attraction, torque or other interactions associated with magnetic fields generated by MR systems. These investigations have demonstrated that, for certain implants, MR procedures may be performed safely in patients with metallic objects that are nonferromagnetic or "weakly" ferromagnetic (i.e., only minimally attracted by the magnetic field in relation to its *in vivo* application), such that the associated magnetic field interactions are insufficient to move or dislodge them, *in situ*. Furthermore, the "intended *in vivo* use" of the implant or device must be taken into consideration, because this can impact MR-safety for a given object.

Each implant, material, device, or object (particularly those made from unknown materials) should be evaluated using *ex vivo* techniques before allowing an individual or patient with the object to enter the MR environment and/or before performing an MR procedure. By following this guideline, the relative magnetic susceptibility for an object may be determined so that a competent decision can be made concerning possible risks associated with exposure to the MR system. Because movement or dislodgment of an implanted metallic object in a patient undergoing an MR procedure is the primary mechanism responsible for an injury, this aspect of testing is considered to be of the utmost importance.

Various factors influence the risk of performing an MR procedure in a patient with a metallic object including the strength of the magnetic field, the magnetic susceptibility of the object, the mass of the object, the geometry of the object, the location and orientation of the object *in situ*, the presence of retentive mechanisms (i.e., fibrotic tissue, bone, sutures, etc.) and the length of time the object has been in place. These factors should be carefully considered before subjecting a patient or individual with a ferromagnetic object to an MR procedure or allowing entrance to the MR environment. This is particularly important if the object is located in a potentially dangerous area of the body such as near a vital neural, vascular, of soft tissue structure where movement or dislodgment could injure the patient. Furthermore, in certain cases, there is a possibility of changing the operational or functional aspects of the implant, material, or device as a result of exposure to the electromagnetic fields of the MR system.

Notably, patients with certain implants or devices that have relatively strong ferromagnetic qualities may be safely scanned using MR procedures because the objects are held in place by sufficient retentive forces that prevent them from being moved or dislodged. Again, this is with reference to the "intended *in vivo* use" of an implant or device. For example, there is an interference screw (i.e., the Perfix Interference Screw) used for reconstruction of the anterior cruciate ligament that is screwed into the patient's bone, preventing it from being moved, even if the patient is exposed to a 1.5-Tesla MR system.

MR systems with very low (0.2-Tesla or less) or very high (8.0-Tesla) static magnetic fields are currently used for clinical and research MR applications. Considering that most metallic objects evaluated for magnetic field interactions were assessed at 1.5-Tesla, an appropriate variance or modification of the information provided regarding the safety of performing an MR procedure in a patient with a metallic object may exist when an MR system with a lower or higher static magnetic field strength is used.

That is, it may be acceptable to adjust safety recommendations depending on the static magnetic field strength and other aspects of a given MR system. Obviously, performing an MR procedure using a 0.2 -Tesla MR system has different risk implications for a patient with a ferromagnetic object compared with using an 8.0-Tesla MR system.

Information Pertaining to Implants, Materials, Devices and Objects. The information contained in this textbook is a compilation of the current data available for assessment of magnetic field interactions and other safety tests for implants, materials, devices, and objects and is based primarily on published reports in the peer-reviewed literature. This compilation also includes unpublished data acquired from *ex vivo* tests of objects that were conducted using commonly performed, standardized, and well-accepted techniques to assess MR-safety and/or MR-compatibility. Furthermore, information that was obtained from manufacturers (e.g., product insert information) for various implants and devices is provided.

Although every attempt was made to provide comprehensive and accurate information, there are many other implants, materials, devices, and objects in existence that remain to be evaluated with regard to MR safety. Therefore, to ensure the safety of an individual or patient in the MR environment, MR healthcare professionals should follow the guideline whereby MR procedures should only be performed in a patient with a metallic object that has been previously tested and demonstrated to be safe. A similar guideline should be followed with regard to whether or not to allow an individual with an implant or device to enter the MR environment.

REFERENCES

Arena L, Morehouse HT, Safir J. MR imaging artifacts that simulate disease: how to recognize and eliminate them. Radiographics 1995;15:1373-1394.

American Society for Testing and Materials (ASTM) Designation: F 2052. Standard test method for measurement of magnetically induced displacement force on passive implants in the magnetic resonance environment. In: Annual Book of ASTM Standards, Section 13, Medical Devices and Services, Volume 13.01 Medical Devices; Emergency Medical Services. West Conshohocken, PA, 2002; pp. 1576-1580.

Davis P L, Crooks L, Arakawa M, et al. Potential hazards in NMR imaging: heating effects of changing magnetic fields and RF fields on small metallic implants. AJR Am J Roentgenol 1981;137:857-860.

Dempsey MF, Condon B, Hadley DM. Investigation of the factors responsible for burns during MRI. J Magn Reson Imaging 2001;13:627-631.

Nyenhuis JA, Kildishev AV, Foster KS, Graber G, Athey W. Heating near implanted medical devices by the MRI RF-magnetic field. IEEE Trans Magn 1999;35:4133-4135.

Schenck JF. Chapter 1, Health Effects and Safety of Static Magnetic Fields. In: Magnetic Resonance Procedures: Health Effects and Safety. CRC Press, LLC, Boca Raton, FL, 2001; pp. 1-31.

Shellock FG. Pocket Guide to MR Procedures and Metallic Objects: Update 2001, Seventh Edition, Lippincott Williams & Wilkins Healthcare, Philadelphia, 2001.

Shellock FG. Magnetic Resonance Procedures: Health Effects and Safety. CRC Press, LLC, Boca Raton, FL, 2001.

Shellock FG. Biomedical implants and devices: assessment of magnetic field interactions with a 3.0-Tesla MR system. J Magn Reson Imaging 2002;16:721-732.

Shellock FG. MR safety update 2002: Implants and devices. J Magn Reson Imaging 2002;16:485-496.

Shellock FG. Surgical instruments for interventional MRI procedures: assessment of MR safety. J Magn Reson Imaging 2001;13:152-157.

Shellock FG, Crues JV. High-field strength MR imaging and metallic biomedical implants: an ex vivo evaluation of deflection forces. AJR Am J Roentgenol 1988;151:389-392.

Shellock FG, Kanal E. Magnetic Resonance: Bioeffects, Safety, and Patient Management. Second Edition, Lippincott-Raven Press, New York, 1996.

Shellock FG, Mink JH, Curtin S, et al. MRI and orthopedic implants used for anterior cruciate ligament reconstruction: assessment of ferromagnetism and artifacts. J Magn Reson Imaging 1992;2:225-228.

Shellock FG, Tkach JA, Ruggieri PM, Masaryk TJ, Rasmussen P. Aneurysm clips: evaluation of magnetic field interactions and translational attraction using "long-bore" and "short-bore" 3.0-Tesla MR systems. AJNR Am J Neuroradiology 2003;24:463-471.

Shellock FG, Tkach JA, Ruggieri PM, Masaryk TJ. Cardiac pacemakers, ICDs, and loop recorder: Evaluation of translational attraction using conventional ("long-bore") and "short-bore" 1.5- and 3.0-Tesla MR systems. Journal of Cardiovascular Magnetic Resonance 2003;5:387-397.

Smith CD, Kildishev AV, Nyenhuis JA, Foster KS, Bourland JD, Interactions of MRI magnetic fields with elongated medical implants. J Appl Physics 2000; 87:6188-6190.

Smith CD, Nyenhuis JA, Kildishev AV. Chapter 16. Health effects of induced electrical currents: Implications for implants. In: Magnetic resonance: Health

Effects and Safety, FG Shellock, Editor, CRC Press, Boca Raton, FL, 2001; pp. 393-413.

3.0-TESLA MR SAFETY INFORMATION FOR IMPLANTS AND DEVICES

The clinical use of 3.0-Tesla MR systems for brain, musculoskeletal, body, cardiovascular, and other applications is increasing worldwide. Because previous investigations performed to determine MR safety for implants and devices used mostly MR systems with static magnetic fields of 1.5-Tesla or less, it is crucial to perform *ex vivo* testing at 3.0-Tesla to characterize MR safety for these objects, especially with regard to magnetic field interactions. Importantly, a metallic object that displayed "weakly" ferromagnetic qualities in association with a 1.5-Tesla MR system may exhibit substantial magnetic field interactions during exposure to a 3.0-Tesla MR system. Therefore, recently it has been necessary to conduct *ex vivo* testing to identify potentially hazardous implants and devices prior to subjecting patients or individuals with these objects to the 3.0-Tesla MR environment. This is especially crucial because most 3.0-Tesla MRI facilities currently do not perform MR procedures on patients with metallic objects due to the lack of safety information.

Magnetic Field Interactions at 3.0-Tesla. From a magnetic field interaction consideration, translational attraction and/or torque may cause movement or dislodgment of a ferromagnetic implant resulting in an uncomfortable sensation or injury to a patient or individual. Translational attraction is proportional to the strength of the static magnetic field, the strength of the spatial gradient, the mass of the object, the shape of the object, and the magnetic susceptibility of the object. The effects of translational attraction on external and implanted ferromagnetic objects are predominantly responsible for possible hazards in the immediate area around the MR system. That is, as one moves in close proximity to the MR system or is moved into the MR system for an examination. An evaluation of torque is also important for a metallic object, especially if it has an elongated configuration. Qualitative and quantitative techniques have been used to determine magnetic field-related torque for implants and devices.

Translational attraction is typically assessed for metallic implants and devices using the deflection angle test recommended by the American Society for Testing and Materials (ASTM, 2002). According to ASTM guidelines, the deflection angle for an implant is generally measured at the point of the "highest spatial gradient" for the specific MR system used for testing. The deflection angle test is commonly performed as an integral part of MR safety testing for metallic implants and devices.

The ASTM guideline for deflection angle testing of implants and devices states that, "...if the implant deflects less than 45°, then the magnetically induced deflection force is less than the force on the implant due to gravity (its weight)". For this condition, it is assumed that any risk imposed by the application of the magnetically induced force is no greater than any risk imposed by normal daily activity in the Earth's gravitational field. Basically, findings from the deflection angle test permit implants and devices made from nonferromagnetic or weakly ferromagnetic materials that display deflection angles between 0 and 44° to be present in patients or individuals undergoing MR procedures without concerns for movement of dislodgment. Additionally, a torque value for an implant or device that is less than that produced by normal daily activities (which might include rapidly accelerating vehicles or amusement park rides) is assumed to be safe.

From a practical consideration, in addition to the findings for translational attraction and torque, the "intended *in vivo* use" of the implant or device must be considered as well as mechanisms that may provide retention of the object *in situ* (e.g., implants or devices held in place by sutures, granulation or ingrowth of tissue, or by other means) with regard to MR safety for a given metallic object.

Long-Bore vs. Short-Bore 3.0-Tesla MR Systems. Various types of magnets exist for commercially available 3.0-Tesla MR systems, including magnet configurations that are conventional "long-bore" scanners and newer "short-bore" systems. Because of physical differences in the position and magnitude of the highest spatial gradient for different magnets, measurements of deflection angles for implants using long-bore vs. short-bore MR systems can produce substantially different results for deflection angle measurements, as reported by Shellock et al. In fact, studies conducted with 3.0-Tesla MR systems indicated that, in general, there were significantly (p<0.01) higher deflection angles measured for implants and devices in association with exposure to short-bore vs. the long-bore MR systems. Basically, the differences in deflection angle measurements for metallic objects were related to differences in the highest spatial gradients for long-bore vs. short-bore scanners.

These MR safety implications are primarily for magnetic field-related translational attraction with regard to long-bore vs. short-bore 3.0-Tesla MR systems. For example, the deflection angle measured for an implant on a short-bore can be substantially higher (and thus, potentially unsafe from a magnetic field interaction consideration) compared to the deflection angle measured on a long-bore MR system. Therefore, MR safety information for measurements of magnetic field interactions for metallic objects must be considered with regard to the specific type of MR system used for the evaluation (e.g., long-bore vs. short-bore 3.0-Tesla MR system).

Heating of Implants and Devices at 3.0-Tesla. Because the radiofrequency exposure guidelines indicated by the United States Food and Drug Administration for 3.0- and 1.5-Tesla MR systems are the same, temperature elevations or heating produced in association with MR procedures conducted at 3.0-Tesla for passive implants are likely to be comparable. *Ex vivo* testing techniques have been used to evaluate MRI-related heating for various metallic implants, materials, devices, and objects of a variety of sizes, shapes, and metallic compositions. In general, reports have indicated that only minor temperature changes occur in association with MR procedures involving metallic objects that are passive implants (e.g., those that are not electronically-activated), with the exception of those implants that have an elongated shape or form a conducting loop. Therefore, heat generated during an MR procedure performed at 3.0-Tesla involving a patient with passive metallic implant does not appear to be a substantial hazard.

However, because excessive heating and burns have occurred in association with implants and devices that have an elongated configuration, that form conducting loops, or that are electronically-activated, patients with these objects should not undergo MR procedures at 3.0-Tesla until *ex vivo* heating assessments are performed to determine the relative risks.

SUMMARY OF 3.0-TESLA MR SAFETY INFORMATION FOR IMPLANTS AND DEVICES

In consideration of the importance of 3.0-Tesla MR safety information for implants and devices, the findings for various objects that have been tested are summarized here and presented in other pertinent sections of this book.

Aneurysm Clips. Various aneurysm clips have been tested for magnetic field interactions in association with 3.0-Tesla MR systems (refer to **The List** for information pertaining to the aneurysm clips tested at 3.0-Tesla). Findings for these specific aneurysm clips indicated that they either exhibited no magnetic field interactions or relatively minor or "weak" magnetic field interactions. Accordingly, these specific aneurysm clips are considered safe for patients undergoing MR procedures using MR systems operating at 3.0-Tesla.

Cardiac Pacemakers, ICDs, and the Reveal Loop Recorder. For cardiac pacemakers and ICDs evaluated using 3.0-Tesla MR systems, seven implants exhibited deflection angles greater than 45 degrees on the long-bore, 3.0-Tesla MR system, while 13 exhibited deflection angles greater than 45 degrees on the short-bore, 3.0-T MR system (refer to **The List** for information on these implants). Notably, other factors exist that may impact MR safety for cardiac pacemakers and ICDs. Therefore, regardless of the fact that magnetic field interactions may not present a risk for some of these cardiovascular implants tested at 3.0-Tesla, the other potentially hazardous mechanisms should be considered carefully.

For the Reveal 9525, Insertable Loop Recorder (Medtronic, Minneapolis, MN), findings at 3.0-Tesla indicated that problems may exist related to movement of this device in association with this high field strength environment. Again, other factors that may impact the MR-safety for this device must also be considered.

Coils and Stents. Several different coils and stents have been evaluated at 3.0-Tesla. Of these implants, two displayed magnetic field interactions that exceeded the ASTM guideline for MR safety (i.e., the deflection angles were greater than 45°). However, similar to other coils and stents, tissue ingrowth may be sufficient to prevent these implants from posing a substantial risk to a patient or individual in the 3.0-Tesla MR environment. Furthermore, certain stents have hooks or barbs to prevent migration after placement that may also help to retain these implants in place. Thus, these issues warrant further study or analysis.

Heart Valve Prostheses and Annuloplasty Rings. Findings obtained at 3.0-Tesla for various heart valve prostheses and annuloplasty rings that underwent testing indicated that one annuloplasty ring (Carpentier-Edwards Physio Annuloplasty Ring, Mitral Model 4450, Edwards Lifesciences, Irvine, CA) showed relatively minor magnetic field interactions, while the others exhibited no magnetic field interactions. Therefore, similar to heart valves prostheses and annuloplasty rings tested at 1.5-Tesla, because the actual magnetic field interactions exerted on this particular implant are deemed minor compared to the force exerted by the beating heart (i.e., approximately 7.2 N), MR procedures at 3.0-Tesla are not considered to be hazardous.

Additional heart valves and annuloplasty rings from the Medtronic Heart Valve Division have undergone MR safety testing at 3.0-Tesla (work conducted by E. Kanal for Medtronic Heart Valve Division, Medtronic, Inc., Minneapolis, MN). These implants were tested for magnetic field interactions and artifacts using a shielded, 3.0-Tesla MR system. According to information provided by the manufacturer, (Medtronic Heart Valves, Technical Service, Medtronic, Inc., Minneapolis, MN; Personal Communication, Kathryn M. Bayer, Senior Technical Consultant), these specific products are safe for patients undergoing MR procedures using scanners operating up to 3.0-Tesla. No adverse effects have been experienced with MR imaging (even when imaging close to the devices) using an MR system operating up to 3.0-Tesla.

Hemostatic Clips, Other Clips, Fasteners, and Staples. At 3.0-Tesla, a variety of hemostatic clips, other types of clips, fasteners, and staples have been evaluated for MR safety. The Surgiclip spring made from carbon steel (United States Surgical, North Haven, CT) showed a deflection angle of 90° and a qualitative torque of +4. However, considering the "intended *in vivo* use" of this device, the closing force may provide substantial counterforce to prevent it from being moved or dislodged. However, this remains to be determined by further experimental findings. In lieu of this information, this implant is currently categorized as "unsafe" at 3.0-Tesla.

Miscellaneous Implants and Devices. Several implants and devices (refer to **The List** to determine specific implants and devices) have been tested for MR safety in association with 3.0-Tesla MR systems. Of these, several exhibited measurable magnetic field interactions but none were at a level considered to present a hazard to a patient undergoing an MR procedure at this static magnetic field strength. For example, different cranial or burr hole fixation implants and devices made from titanium have been evaluated at 3.0-Tesla. These were found to be safe for patients undergoing MR procedures using MR systems operating at this field strength. The Essure Device was evaluated for MR safety at 3.0-Tesla and found to be safe for patients.

Orthopedic Implants, Materials, and Devices. A variety of orthopedic implants have been evaluated for magnetic field interactions at 3.0-Tesla (refer to **The List**). All of these are considered to be safe based on findings for deflection angles, qualitative torque measurements, and the intended *in vivo* uses of these devices.

MR Safety at 3.0-Tesla and Penile Implants. Several different penile implants have been tested for MR safety in association with 3.0-Tesla MR systems. Findings for these specific penile implants indicated that they either exhibited no magnetic field interactions or relatively minor or "weak" magnetic field interactions. Accordingly, these specific penile implants are considered safe for patients undergoing MR procedures using MR systems operating at 3.0-Tesla.

Sutures. At 3.0-Tesla, most sutures with needles removed that underwent evaluation displayed no magnetic field interactions, while two (Flexon suture and Steel suture, United States Surgical, North Haven, CT) showed minor deflection angles and torque. For these two sutures, the *in situ* application of these materials is likely to provide sufficient counter-forces to prevent movement or dislodgment. Therefore, in consideration of the intended *in vivo* use of these materials, all of the sutures with the needles removed that have been tested are regarded to be safe at 3.0-Tesla.

Vascular Access Ports, Infusion Pumps, and Catheters. None of the various vascular access ports and infusion pumps assessed for magnetic field interactions at 3.0-Tesla exhibited magnetic field interactions. Therefore, these implants and device will not move or dislodge in this MR environment. For the accessories, the infusion set and needles showed measurable ferromagnetism, with the PORT-A-CATH Needle (Deltec, Inc., St. Paul, MN) exceeding the recommended ASTM deflection angle safety guideline (i.e., greater than 45 degrees). However, during the actual use of this accessory, it is unlikely that it will present a problem in the 3.0-Tesla MR environment considering that the simple application of a small amount of adhesive tape effectively counterbalances the relatively minor ferromagnetism determined for this device (Unpublished 0bservations, F.G. Shellock, 2002).

REFERENCES

American Society for Testing and Materials (ASTM) Designation: F 2052. Standard test method for measurement of magnetically induced displacement force on passive implants in the magnetic resonance environment. In: Annual Book of ASTM Standards, Section 13, Medical Devices and Services, Volume 13.01 Medical Devices; Emergency Medical Services. West Conshohocken, PA, 2002; pp. 1576-1580.

Davis P L, Crooks L, Arakawa M, et al. Potential hazards in NMR imaging: heating effects of changing magnetic fields and RF fields on small metallic implants. AJR Am J Roentgenol 1981;137:857-860.

Dempsey MF, Condon B, Hadley DM. Investigation of the factors responsible for burns during MRI. J Magn Reson Imaging 2001;13:627-631.

Hennemeyer CT, Wicklow K, Feinberg DA, Derdeyn CP. In vitro evaluation of platinum Guglielmi detachable coils at 3-T with a porcine model: safety issues and artifacts. Radiology 2001;219:732-737.

Medtronic Heart Valves, Medtronic, Inc., Minneapolis, MN, Permission to publish 3-Tesla MR testing information for Medtronic Heart Valves provided by Kathryn M. Bayer, Senior Technical Consultant, Medtronic Heart Valves, Technical Service.

Nyehnuis JA, Kildishev AV, Foster KS, Graber G, Athey W. Heating near implanted medical devices by the MRI RF-magnetic field. IEEE Trans Magn 1999;35:4133-4135.

Schenck JF. Chapter 1, Health Effects and Safety of Static Magnetic Fields. In: Magnetic Resonance Procedures: Health Effects and Safety. CRC Press, LLC, Boca Raton, FL, 2001; pp. 1-31.

Shellock FG Radiofrequency-induced heating during MR procedures: A review. J Magn Reson Imaging 2000;12:30-36.

Shellock FG. MR safety update 2002: Implants and devices. J Magn Reson Imaging 2002;16:485-496.

Shellock FG. Biomedical implants and devices: assessment of magnetic field interactions with a 3.0-Tesla MR system. J Magn Reson Imaging 2002;16:721-732.

Shellock FG, Tkach JA, Ruggieri PM, Masaryk TJ. Cardiac pacemakers, ICDs, and loop recorder: Evaluation of translational attraction using conventional ("long-bore") and "short-bore" 1.5- and 3.0-Tesla MR systems. Journal of Cardiovascular Magnetic Resonance 2003;5:387-397.

Shellock FG, Tkach JA, Ruggieri PM, Masaryk T, Rasmussen P. Aneurysm clips: evaluation of magnetic field interactions and translational attraction using "long-bore" and "short-bore" 3.0-Tesla MR systems. American Journal of Neuroradiology 2003;24:463-471.

Smith CD, Kildishev AV, Nyenhuis JA, Foster KS, Bourland JD, Interactions of MRI magnetic fields with elongated medical implants. J Appl Physics 2000; 87:6188-6190.

Smith CD, Nyenhuis JA, Kildishev AV. Chapter 16. Health effects of induced electrical currents: Implications for implants. In: Magnetic resonance: health effects and safety, FG Shellock, Editor, CRC Press, Boca Raton, FL, 2001; pp. 393-413.

ACCURX CONSTANT FLOW IMPLANTABLE PUMP AND DURACATH INTRASPINAL CATHETER

AccuRx Constant Flow Implantable Pump. The AccuRx Constant Flow Implantable Pump (Advanced Neuromodulation Systems, Plano, TX) is an implantable device that stores and dispenses medication at a constant factory-set flow rate to a specific site. The pumps are available in a range of flow rates to allow physicians to tailor therapy to the needs of the patient.

The AccuRx Constant Flow Implantable Pump consists of a single sealed chamber formed between a rigid titanium shell and a polymeric diaphragm. The chamber holds the drug to be infused. The diaphragm exerts pressure on the medication causing it to flow out of the chamber through a series of filters and a flow restrictor, and then out the catheter (see DuraCath Intraspinal Catheter information below) to the delivery site. The AccuRx Constant Flow Implantable Pump is refilled through a raised refill port in the center of the titanium shell. Refilling the pump expands the diaphragm outward and starts the pump on the next cycle of drug infusion.

DuraCath Intraspinal Catheter. The DuraCath Intraspinal Catheter (Advanced Neuromodulation Systems, Plano, TX) is designed for long-term intraspinal (epidural or intrathecal) implantation. The trimmable, flexible, elastic radiopaque catheter incorporates insertion depth markings and a removable guide wire to facilitate implantation. This catheter has a closed tip and multiple side-exit channels to facilitate dispersion of the drug to reduce the probability of catheter tip complications. There are one- and two-piece versions of the DuraCath Intraspinal Catheter to suite a

114

variety of clinical needs. The DuraCath Intraspinal Catheter is made from silicone and has a small connector made from 316L stainless steel.

MAGNETIC RESONANCE IMAGING (MRI) INFORMATION

Exposure of the AccuRx Constant Flow Implantable Pump and DuraCath Intraspinal Catheter to magnetic resonance imaging (MRI) fields of 1.5-Tesla has demonstrated no impact on pump or catheter performance and a limited effect on the quality of the diagnostic information. Testing performed on the AccuRx Constant Flow Implantable Pump and DuraCath Intraspinal Catheter has established the following with regard to MRI safety and diagnostic issues:

Static Magnetic Field. Testing demonstrated that a 1.5-Tesla magnetic resonance environment produced no measurable magnetic field interactions (i.e., translational attraction and torque) for the AccuRx Constant Flow Implantable Pump and DuraCath Intraspinal Catheter.

Heating During MRI. Testing demonstrated that there was very little heating for the AccuRx Constant Flow Implantable Pump in association with MR imaging conducted using an excessive amount of RF energy (i.e., a whole-body averaged specific absorption rate of 1.5 W/kg; spatial peak SAR of 2.9 W/kg). Furthermore, this amount of heating is considered to be physiologically inconsequential and will not impose an additional risk to the patient undergoing an MR procedure under the conditions used for the evaluation. In the unlikely event that the patient experiences warmth near the pump, the MR procedure should be stopped immediately.

Peripheral Nerve Stimulation. The presence of the AccuRx Constant Flow Implantable Pump may cause a one-fold increase (doubling) of the induced gradient current in tissues near the device. This increase in the induced gradient current is similar to that already present elsewhere in the body (e.g., at bone-tissue interfaces) and, as such, is unlikely to enhance the propensity for nerve stimulation. In the unlikely event that the patient reports stimulation during the scan, the MR procedure should be stopped immediately.

REFERENCES

AccuRX Constant Flow Implantable Pumps, Technical Manual, Advanced Neuromodulation Systems, Inc., Plano, TX.

ANEURYSM CLIPS

The surgical management of intracranial aneurysms and arteriovenous malformations (AVMs) by the application of aneurysm clips is a well-established procedure. The presence of an aneurysm clip in a patient referred for an MR procedure represents a situation that requires the utmost consideration because of the associated risks.

Certain types of intracranial aneurysm clips (e.g., those made from martensitic stainless steels such as 17-7PH or 405 stainless steel) are an absolute contraindication to the use of MR procedures because excessive, magnetically induced forces can displace these implants and cause serious injury or death. By comparison, aneurysm clips classified as "nonferromagnetic" or "weakly ferromagnetic" (e.g., those made from Phynox, Elgiloy, austentitic stainless steels, titanium alloy, or commercially pure titanium) are safe for patients undergoing MR procedures (additional information on performing MR procedures in patients with aneurysm clips is provided in the section entitled, **Extremity MR System**).

[For the sake of discussion, the term "weakly ferromagnetic" refers to metal that may demonstrate some extremely low ferromagnetic qualities using highly sensitive measurements techniques (e.g., vibrating sample magnetometer, superconducting quantum interference device or SQUID magnetometer, etc.) and as such, may not be technically referred to as being "nonferromagnetic." All metals possess some degree of magnetism, such that no metal is considered to be totally "nonmagnetic" or "nonferromagnetic."]

MR procedures have been used to evaluate patients with certain types of aneurysm clips. Becker et al. using MR systems that ranged from 0.35 to 0.6-Tesla, studied three patients with nonferromagnetic aneurysm clips (one patient, Yasargil, 316 LVM stainless steel; two patients, Vari-Angle McFadden, MP35N; 316 LVM) and one patient with a ferromagnetic aneurysm clip (Heifetz aneurysm clip, 17-7PH) without incident. Dujovny et al. similarly reported no adverse effects in patients with nonferromagnetic aneurysm clips that underwent procedures using 1.5-Tesla MR systems.

Pride et al. performed a study in patients with nonferromagnetic aneurysm clips that underwent MR imaging. There were no objective adverse outcomes for the patients, confirming that MR procedures can be performed safely in patients with nonferromagnetic clips. Brothers et al. also demonstrated that MR imaging at 1.5-Tesla may be performed safely in patients with nonmagnetic aneurysm clips. This report was particularly important because, according to Brothers et al., MR imaging was found to be better than CT in the postoperative assessment of aneurysm patients, especially in showing small zones of ischemia.

Of note is that only one ferromagnetic aneurysm clip-related fatality has been reported in the peer-reviewed literature, to date. According to this report, the patient became symptomatic at a distance of approximately 1.2-meters from the bore of the MR system, suggesting that translational attraction of the aneurysm clip was partially responsible for dislodgment of this implant.

This unfortunate incident was the result of erroneous information pertaining to the type of aneurysm clip that was present in the patient. That is, the clip was thought to be a nonferromagnetic Yasargil aneurysm clip (Aesculap Inc., Central Valley, PA) and turned out to be a ferromagnetic Vari-Angle clip (Codman & Shurtleff, Randolf, MA).

There has never been a report of an injury to a patient or individual in the MR environment related to the presence of an aneurysm clip made from a nonferromagnetic or "weakly" ferromagnetic material. In fact, there have been cases in which patients with ferromagnetic aneurysm clips (based on the extent of the artifact seen during MR imaging or other information) have undergone MR procedures without sustaining injuries (Personal communications, D. Kroker, 1995; E. Kanal, 1996; A. Osborne, 2002).

In these cases, the aneurysm clips were exposed to magnetic-induced translational attraction and torque associated with MR systems that had static magnetic fields of up to 1.5-Tesla. Although these cases do not prove or suggest safety, they do demonstrate the difficulty of predicting the outcome for patients with ferromagnetic aneurysm clips that undergo MR procedures.

Unfortunately, there is much controversy and confusion regarding the amount of ferromagnetism that needs to be present in an aneurysm clip to constitute a hazard for a patient in the MR environment. Consequently, this issue has not only created problems for MR healthcare workers but for manufacturers of aneurysm clips, as well.

For example, MR healthcare workers performing test procedures on aneurysm clips similar to that described in the report by Kanal et al. (1996) presumably identified the presence of magnetic field interactions and returned several clips made from Phynox to the manufacturer (Personal Communication, Aesculap, Inc., South San Francisco, CA, 1997). However, the testing method used by Kanal et al. (1996) was admittedly crude and developed to primarily obtain rapid, qualitative screening data for large numbers of aneurysm clips to determine if quantitative assessments were necessary. Notably, the test technique used by Kanal et al. (1996) may be problematic and yield spurious results, especially if the aneurysm clip has a shape or configuration that is somewhat "unstable" (Unpublished Observations, F.G. Shellock, 1997). For example, aneurysm clips with blades that are bayonet, curved, or angled shapes are less stable on a piece of plate glass (i.e., using the testing method described by Kanal et al.) when placed in certain orientations compared with aneurysm clips with blades that have a straight-shape.

A variety of more appropriate testing techniques have been developed and utilized over the years to evaluate the relative amount of ferromagnetism present for implants and devices prior to allowing patients with these objects to enter the MR environment. In 2002, the American Society for Testing and Materials (ASTM) provided recommendations for testing passive implants that involves the use of the deflection angle test, originally described by New et al., to assess translational attraction. Additionally, the Food and Drug Administration recommends that an evaluation of torque should be performed on aneurysm clips, as well. Thus, procedures such as the deflection angle test and some form of evaluation of torque are the most appropriate means of determining which specific aneurysm clip may present a hazard to a patient or individual in the MR environment.

Aneurysm Clips and MR Safety at 1.5-Tesla and Less. In consideration of the current knowledge pertaining to aneurysm clips, the following guidelines are recommended with regard to performing an MR procedure in a patient with an aneurysm clip or before allowing an individual with an aneurysm clip into the MR environment:

1. Specific information (i.e., manufacturer, type or model, material, lot and serial numbers) about the aneurysm clip must be known, especially with respect to the material used to make the aneurysm clip, so that only patients or individuals with nonferromagnetic or weakly fer-

romagnetic clips are allowed into the MR environment. The manufacturer typically provides this information in the labeling of the aneurysm clip. The implanting surgeon is responsible for properly communicating this information in the patient's or individual's records.

2. An aneurysm clip that is in its original package and made from Phynox, Elgiloy, MP35N, titanium alloy, commercially pure titanium or other material known to be nonferromagnetic or weakly ferromagnetic does not need to be evaluated for ferromagnetism. Aneurysm clips made from nonferrromagnetic or "weakly" ferromagnetic materials in original packages do not require testing of ferromagnetism because the manufacturers ensure the pertinent MR safety aspects of these clips and, therefore, are responsible for the accuracy of the labeling.

3. If the aneurysm clip is not in its original package and/or properly labeled, it should undergo testing for magnetic field interactions following appropriate testing procedures to determine if it is safe or unsafe for the MR environment.

4. The radiologist and implanting surgeon should be responsible for evaluating the available information pertaining to the aneurysm clip, verifying its accuracy, obtaining written documentation, and deciding to perform the MR procedure after considering the risk vs. benefit aspects for a given patient or individual.

MR Safety at 3.0-Tesla and Aneurysm Clips. Various aneurysm clips have been tested for magnetic field interactions in association with 3.0-Tesla MR systems (refer to **The List** for information for aneurysm clips tested at 3.0-Tesla). Findings for these specific aneurysm clips indicated that they either exhibited no magnetic field interactions or relatively minor or "weak" magnetic field interactions. Accordingly, these particular aneurysm clips are considered safe for patients undergoing MR procedures using MR systems operating at 3.0-Tesla.

MR Safety at 8.0-Tesla and Aneurysm Clips. Currently, the most powerful, clinical MR system in existence operates at a static magnetic field strength of 8.0-Tesla. *Ex vivo* testing has been conducted to identify potentially hazardous for various implants and devices using this ultra-high-field-strength scanner. The first investigation to determine magnetic field interactions for aneurysm clips exposed to an 8.0-Tesla MR system was conducted by Kangarlu and Shellock.

Twenty-six different aneurysm clips were tested for magnetic field inter-actions using previously-described techniques. These aneurysm clips were specifically selected for this investigation because they represent various types of clips made from nonferromagnetic or weakly ferromagnetic materials used for temporary or permanent treatment of aneurysms or arteriovenous malformations. Additionally, these aneurysm clips were reported previously to be safe for patients undergoing MR procedures using MR systems with static magnetic field strengths of 1.5-Tesla or less.

According to the test results, six aneurysm clips (i.e., type, model, blade length) made from stainless steel alloy (Perneczky) and Phynox (Yasargil, Models FE 748 and FE 750) displayed deflection angles above 45° (i.e., referring to the ASTM guideline) and relatively high qualitative torque values. These findings indicated that these specific aneurysm clips may be unsafe for individuals or patients in an 8.0-Tesla MR environment.

Aneurysm clips made from commercially pure titanium (Spetzler), Elgiloy (Sugita), titanium alloy (Yasargil, Model FE 750T), and MP35N (Sundt) displayed deflection angles less than 45° (i.e., referring to the ASTM guideline) and qualitative torque values that were relatively minor. Accordingly, these aneurysm clips are considered to be safe for patients or individuals exposed to an 8.0-Tesla MR system.

As previously indicated, at 1.5-Tesla, aneurysm clips that are considered to be acceptable for patients or others in the MR environment include those made from commercially pure titanium, titanium alloy, Elgiloy, Phynox, and austentic stainless steel. By comparison, findings from the 8.0-Tesla study indicated that deflection angles for the aneurysm clips made from commercially pure titanium and titanium alloy ranged from 5 to 6°, suggesting that these aneurysm clips would be safe for patients or individuals in the 8.0-Tesla MR environment. However, deflection angles for aneurysm clips made from Elgiloy ranged from 36 to 42°, such that further consideration must be given to the specific type of Elgiloy clip that is present. For example, an Elgiloy clip that has a greater mass than those tested in this study may exceed a deflection angle of 45° (i.e., referring to the ASTM guideline) in association with an 8.0-Tesla MR system.

Depending on the actual dimensions and mass, an aneurysm clip made from Elgiloy may or may not be acceptable for a patient or individual in the 8.0-Tesla MR environment. It should be noted that the results of this investigation are highly specific to the types of intracranial aneurysm clips that underwent testing (i.e., with regard model, shape, size, blade length, material, etc.) by Kangarlu and Shellock.

Effects of Long-Term and Multiple Exposures to the MR System. MR testing procedures used for aneurysm clips over the past few years would result in the potential for reintroduction of aneurysm clips into strong MR magnetic fields several times prior to implantation into the patient. Furthermore, there are patients with implanted aneurysm clips previously tested and designated as "MR-safe" or "MR-compatible" that have undergone repeated exposures strong magnetic fields during follow-up MR examinations.

A concern that has emerged is that there are potential alterations in magnetic properties of pre-or post-implanted aneurysm clips resulting from long-term or multiple exposures to strong magnetic fields. Long-term or multiple exposures to strong magnetic fields (such as those associated with MR imaging systems) have been suggested to grossly "magnetize" aneurysm clips, even if they are made from nonferromagnetic or weakly ferromagnetic materials. This could present a substantial hazard to an individual in the MR environment. Therefore, an investigation was conducted to study intracranial aneurysm clips *in vitro* prior to and following long-term and multiple exposures to 1.5-Tesla MR systems. This was done to quantify possible alterations in the magnetic properties of aneurysm clips.

Aneurysm clips made from Elgiloy, Phynox, titanium alloy, commercially pure titanium, and austenitic stainless steel were tested in association with long-term and multiple exposures to 1.5-Tesla MR systems. The findings of this investigation indicated that there was a lack of response to the magnetic field exposure conditions that were used, such that long-term or multiple exposures to 1.5-Tesla MR systems should not result in significant changes in their magnetic properties.

Artifacts Associated with Aneurysm Clips. An additional problem related to aneurysm clips is that artifacts produced by these metallic implants may substantially detract from the diagnostic aspects of MR procedures. MR imaging, MR angiography, and functional MRI are frequently used to evaluate the brain or cerebral vasculature of patients with aneurysm clips. For example, to reduce morbidity and mortality after subarachnoid hemorrhage, it is imperative to assess the results of the surgical treatment of cerebral aneurysms.

The extent of the artifact produced by a given aneurysm clip will have a direct effect on the diagnostic aspects of the MR procedure. Therefore, an investigation was conducted to characterize artifacts associated with aneurysm clips made from nonferromagnetic or weakly ferromagnetic materials. Five different aneurysm clips made from five different materials were evaluated in this investigation, as follows:

(1) Yasargil, Phynox (Aesculap, Inc., Central Valley,PA),

(2) Yasargil, titanium alloy (Aesculap, Inc., Central Valley, PA),

(3) Sugita, Elgiloy (Mizuho American, Inc., Beverly, MA),

(4) Spetzler Titanium Aneurysm Clip, commercially pure titanium (Elekta Instruments, Inc., Atlanta, GA), and

(5) Perneczky, cobalt alloy (Zepplin Chirurgishe Instrumente, Pullach, Germany).

These aneurysm clips were selected for testing because they are made from nonferromagnetic or weakly ferromagnetic materials. Furthermore, they represent the most frequently used, commercially available aneurysm clips in the United States. These aneurysm clips have been previously reported to be safe for patients in the 1.5-Tesla MR environment and, as such, are often found in patients referred for MR procedures.

MR imaging artifact testing revealed that the size of the signal voids were directly related to the type of material (i.e., the magnetic susceptibility) used to make the particular clip. Arranged in decreasing order of artifact size, the materials responsible for the artifacts associated with the aneurysm clips were, as follows: Elgiloy (Sugita), cobalt alloy (Perneczky), Phynox (Yasargil), titanium alloy (Yasargil), and commercially pure titanium (Spetzler). These results have implications when one considers the various critical factors that are responsible for the decision to use a particular type of aneurysm clip (e.g., size, shape, closing force, biocompatibility, corrosion resistance, material-related effects on diagnostic imaging examinations, etc.).

An aneurysm clip that causes a relatively large artifact is less desirable because it can reduce the diagnostic capabilities of the MR procedure if the area of interest is in the immediate location of where the aneurysm clip was implanted. Fortunately, aneurysm clips exist that are made from materials (i.e., commercially pure titanium and titanium alloy) that minimize such artifacts.

Additional artifact research has been conducted by Burtscher et al. with the intent of determining the extent to which titanium aneurysm clips could improve the quality of MR imaging compared to stainless steel aneurysm clips and to determine whether the associated artifacts could be reduced by controlling MR imaging parameters. The results of this investigation indicated that the use of titanium aneurysm clips reduced MR artifacts by approximately 60% compared to stainless steel aneurysm clips. MR imaging artifacts were further reduced by using spin echo-based pulse sequences with high bandwidths or, if necessary, gradient echo pulse sequences with a low echo times (TE).

REFERENCES

American Society for Testing and Materials (ASTM) Designation: F 2052. Standard test method for measurement of magnetically induced displacement force on passive implants in the magnetic resonance environment. In: Annual Book of ASTM Standards, Section 13, Medical Devices and Services, Volume 13.01 Medical Devices; Emergency Medical Services. West Conshohocken, PA, 2002, pp. 1576-1580.

Becker RL, Norfray JF, Teitelbaum GP, et al. MR imaging in patients with intracranial aneurysm clips. Am J Roentgenol 1988;9:885-889.

Brothers MF, Fox AJ, Lee DH, Pelz DM, Deveikis JP. MR imaging after surgery for vertebrobasilar aneurysm. Am J Neuroradiol 1990;11:149-161.

Brown MA, Carden JA, Coleman RE, et al. Magnetic field effects on surgical ligation clips. Magn Reson Imaging 1987;5:443-453.

Burtscher IM, Owman T, Romner B, Stahlberg F, Holtas S. Aneurysm clip MR artifacts. Titanium versus stainless steel and influence of imaging parameters. Acta Radiology 1998;39:70-76.

Dujovny M, Kossovsky N, Kossowsky R, et al. Aneurysm clip motion during magnetic resonance imaging: in vivo experimental study with metallurgical factor analysis. Neurosurgery 1985;17:543-548.

FDA stresses the need for caution during MR scanning of patients with aneurysm clips. In: Medical Devices Bulletin, Center for Devices and Radiological Health. March, 1993;11:1-2.

Johnson GC. Need for caution during MR imaging of patients with aneurysm clips [Letter]. Radiology 1993;188:287.

Kanal E, Shellock FG. MR imaging of patients with intracranial aneurysm clips. Radiology 1993;187:612-614.

Kanal E, Shellock FG. Aneurysm clips: effects of long-term and multiple exposures to a 1.5-Tesla MR system. Radiology 1999;210:563-565.

Kanal E, Shellock FG, Lewin JS. Aneurysm clip testing for ferromagnetic properties: clip variability issues. Radiology 1996;200:576-578.

Kangarlu A, Shellock FG. Aneurysm clips: evaluation of magnetic field interactions with an 8.0-T MR system. J Magn Reson Imaging 2000;12:107-111.

Klucznik RP, Carrier DA, Pyka R, Haid RW. Placement of a ferromagnetic intracerebral aneurysm clip in a magnetic field with a fatal outcome. Radiology 1993;187:855-856.

New PFJ, Rosen BR, Brady TJ, et al. Potential hazards and artifacts of ferromagnetic and nonferromagnetic surgical and dental materials and devices in nuclear magnetic resonance imaging. Radiology 1983;147:139-148.

Pride GL, Kowal J, Mendelsohn DB, Chason DP, Fleckenstein JL. Safety of MR scanning in patients with nonferromagnetic aneurysm clips. J Magn Reson Imaging 2000;12:198-200.

Shellock FG. Magnetic Resonance Procedures: Health Effects and Safety. CRC Press, LLC, Boca Raton, FL, 2001.

Shellock FG. Magnetic resonance procedures and aneurysm clips: A review. Signals, No. 33, Issue 2, pp. 17-20, 2000.

Shellock FG. Biomedical implants and devices: assessment of magnetic field interactions with a 3.0-Tesla MR system. J Magn Reson Imaging 2002;16:721-732.

Shellock FG, Tkach JA, Ruggieri PM, Masaryk T, Rasmussen P. Aneurysm clips: evaluation of magnetic field interactions and translational attraction using "long-bore" and "short-bore" 3.0-Tesla MR systems. American Journal of Neuroradiology 2003;24:463-471.

Shellock FG. Pocket Guide to MR Procedures and Metallic Objects: Update 2001, Seventh Edition, Lippincott Williams & Wilkins Healthcare, Philadelphia, 2001.

Shellock FG, Crues JV. High-field strength MR imaging and metallic biomedical implants: an ex vivo evaluation of deflection forces. Am J Roentgenol 1988;151:389-392.

Shellock FG, Crues JV. Aneurysm clips: Assessment of magnetic field interaction associated with a 0.2-T extremity MR system. Radiology 1998;208:407-409.

Shellock FG, Kanal E. Aneurysm clips: Evaluation of MR imaging artifacts at 1.5-Tesla. Radiology 1998;209:563-566.

Shellock FG, Kanal E. Magnetic Resonance: Bioeffects, Safety, and Patient Management. Second Edition, Lippincott-Raven Press, New York, 1996.

Shellock FG, Kanal E. Yasargil aneurysm clips: evaluation of interactions with a 1.5-Tesla MR system. Radiology 1998;207:587-591.

Shellock FG, Shellock VJ. MR-compatibility evaluation of the Spetzler titanium aneurysm clip. Radiology 1998;206:838-841.

BIOPSY NEEDLES, MARKERS, AND DEVICES

MR imaging has been used to guide tissue biopsy and apply markers with encouraging results. These specialized procedures require tools that are compatible with MR systems. Many commercially available biopsy needles, markers, and devices (i.e., guide wires, stylets, marking wires, marking clips, biopsy guns, etc.) have been evaluated with respect to compatibility with MR procedures, not only to determine magnetic susceptibility but also to characterize imaging artifacts. The results have indicated that most of the commercially available biopsy needles, markers, and devices are not useful for MR-guided biopsy procedures due to the presence of excessive ferromagnetism and the associated imaging artifacts that may limit or obscure the area of interest.

For many of the commercially available devices, studies have reported that the presence of ferromagnetic biopsy needles and lesion marking wires in a tissue phantom used for testing produced such substantial artifacts that they would not be useful for MR-guided procedures. Needles or devices containing any type of ferromagnetic material tend to have too much magnetic susceptibility to allow effective use for MR-guided procedures.

Fortunately, several needles, markers, and devices have been constructed out of nonferromagnetic materials specifically for use in MR-guided procedures. Of note is that certain of these nonferromagnetic materials, such as titanium, do not appear to have the same properties, which should be carefully considered whenever selecting an MR-compatible biopsy needle for an MR-guided procedure. For example, Faber et al. reported that titanium alloy needles (Somatex, Germany) were weaker and bent easily during insertion compared to the Inconell (Cook, Germany) or high nickel alloy (E-Z-Em, Westbury, NY) biopsy needles.

Although most of the biopsy guns tested for magnetic field interactions and artifacts were found to be ferromagnetic, since they are not used in the immediate area of the target tissue, artifacts associated with these devices are unlikely to affect the resulting images during MR-guided biopsy procedures. Nevertheless, the presence of ferromagnetism is likely to preclude the optimal use of most biopsy guns in the MR environment. Currently, there is at least one commercially available biopsy gun developed specifically for use in MR-guided procedures that does not have ferromagnetic components.

A metallic marking clip, the MicroMark Clip, made from 316L stainless steel by Biopsys Medical (Irvine, CA), has been developed for percutaneous placement after stereotactic breast biopsy. The placement of a marking clip is of obvious use, especially in cases where mammographic findings are not apparent or visible. The use of a marking clip enables the accurate localization of the surgical excision site and is a useful surrogate target, even when the entire lesion is removed and there is a subsequent need for wire localization prior to surgery.

A current limitation of MR-guided needle localization procedures is that there is an inability to document lesion retrieval, because it is not possible to perform contrast enhancement of the resected specimen. A MicroMark Clip placed during MR-guided biopsy or localization can permit radiography to be performed on the surgical specimen to confirm retrieval of the clip and, thus, document retrieval of the lesion.

Tests conducted to assess magnetic field interactions, heating, and artifacts indicated that the presence of the MicroMark Clip presents no risk to a patient undergoing an MR procedure using an MR system operating with a static magnetic field of 1.5-Tesla or less. Unfortunately the probe used with the MicroMark is strongly attracted by a 1.5-Tesla MR system, preventing its optimal use in the MR environment. However, the marking clip could be placed outside of the influence of the magnetic field after placement of an MR-compatible introducer.

The MicroMark II Clip (316LVM stainless steel, Ethicon Endosurgery, Cincinnati, Ohio) has been tested for MR safety using 1.5-Tesla and 3.0-Tesla MR systems. The findings indicated that there were no magnetic field interactions associated with exposure to 1.5-Tesla and 3.0-Tesla MR systems, there was no MRI-related heating, and the artifacts were shown to be relatively minor.

REFERENCES

Lewin JS, et al. Needle localization in MR-guided biopsy and aspiration: Effect of field strength, sequence design, and magnetic field orientation. AJR Am J Roentgenol 1996;166:1337-1341.

Lufkin R, Layfield L. Coaxial needle system of MR- and CT-guided aspiration cytology. J Computer Assist Tomogr 1989;13:1105-1107.

Lufkin R, Teresi L, Hanafee W. New needle for MR-guided aspiration cytology of the head and neck. AJR Am J Roentgenol 1987;149:380-382.

Moscatel M, Shellock FG, Morisoli S. Biopsy needles and devices: assessment of ferromagnetism and artifacts during exposure to a 1.5-Tesla MR system. J Magn Reson Imaging 1995;5:369-372.

Shellock FG. Magnetic Resonance Procedures: Health Effects and Safety. CRC Press, LLC, Boca Raton, FL, 2001.

Shellock FG. Pocket Guide to MR Procedures and Metallic Objects: Update 2001, Seventh Edition, Lippincott Williams & Wilkins Healthcare, Philadelphia, 2001.

Shellock FG, Kanal E. Magnetic Resonance: Bioeffects, Safety, and Patient Management. Second Edition, Lippincott-Raven Press, New York, 1996.

Shellock FG, Shellock VJ. Additional information pertaining to the MR-compatibility of biopsy needles and devices. J Magn Reson Imaging 1996;6:411.

Shellock FG, Shellock VJ. Metallic marking clips used after stereotactic breast biopsy: *ex vivo* testing of ferromagnetism, heating, and artifacts associated with MRI. AJR Am J Roentgenol 1999;172:1417-1419.

BONE FUSION STIMULATOR/SPINAL FUSION STIMULATOR

The implantable bone fusion or spinal fusion stimulator (Electro-Biology, Inc., Parsippany, NJ) is designed for use as an adjunct therapy to a spinal fusion procedure. The implantable spinal fusion stimulator consists of a direct current generator with a lithium iodine battery and solid-state electronics encased in a titanium shell, partially-coated with platinum that acts as an anode.

The generator weighs 10 grams and has the following dimensions: 45-mm x 22-mm x 6-mm. Two nonmagnetic silver/stainless steel leads insulated with silastic provide a connection to two titanium electrodes that serve as the cathodes. This device produces a continuous 20 microamp current. The cathodes are comprised of insulated wire leads that terminate as bare wire leads, which are embedded in pieces of bone grafted onto the lateral aspects of fusion sites. The generator is implanted beneath the skin and muscle near the vertebral column and provides the full-rated current for approximately 24 to 26 weeks. The use of this electronic implant provides a faster consolidation of the bone grafts, leading to higher fusion rates and improved surgical outcomes, along with a reduced need for orthopedic instrumentation.

Studies using excessively-high electromagnetic fields under highly-specific experimental conditions and modeling scenarios for the lumbar/torso area (i.e., 1.5-Tesla MR system, excessive exposures to RF fields, excessive exposures to gradient magnetic fields, etc.) have demonstrated that the implantable spinal fusion stimulator will not present a hazard to a patient undergoing MR imaging with respect to movement, heating, or induced electric fields during the use of conventional MR techniques.

Additionally, there was no evidence of malfunction of the implantable spinal fusion stimulator based on *in vitro* and *in vivo* experimental findings. These studies addressed the use of conventional pulse sequences and parameters with an acknowledgement that echo planar techniques or imaging parameters that require excessive RF power will have different implications and consequences for the patient with an implantable spinal fusion stimulator.

To date, MR examinations have been performed in more than 120 patients (conceivably, using MR imaging conditions that involved a wide-variety of imaging parameters and conditions) with implantable spinal fusion stimulators with no reports of substantial adverse events (based on a review of data obtained through the Freedom of Information Act and Unpublished Observations, Simon BJ, Electro-Biology, Inc., Parsippany, NJ, 1999). Furthermore, the manufacturer of this implant and the Food and Drug Administration have not received complaints of injuries associated with the presence of this device in patients undergoing MR procedures.

In an *in vivo* study, there were no reports of immediate or delayed (minimum of one month follow up) adverse events from patients with implantable spinal fusion stimulators who underwent MR imaging at 1.5-Tesla. Each patient was visually inspected following the MRI study and there was no evidence of excessive heating (i.e., change in skin color or other similar response).

One patient indicated a sensation of "warming" felt at the site of the stimulator, however, this feeling was described as minor and the MR examination was completed without further indication of unusual sensations or problems. Of further note is that there were no reports of excessive heating or neuromuscular stimulation in association with the presence of the implantable spinal fusion stimulators in patients that underwent MR procedures.

Chou et al. conducted a thorough investigation of the effect of heating of the implantable spinal fusion stimulator associated with MR imaging. This work was performed using a human phantom during MR procedures involving a relatively high exposure to RF energy (i.e., at whole body averaged specific absorption rates of approximately 1.0 W/kg). Fluoroptic thermometry probes were placed at various positions on and near the cathodes, leads, and the stimulator for each experiment to record temperature changes.

The phantom used by Chou et al. did not include the effects of blood flow, which obviously would help dissipate heating that may occur during MR imaging and, therefore, it further represents an excessive RF exposure condition. With the implantable spinal fusion stimulator in place and the leads intact, the maximum temperature rise after 25 minutes of scanning occurred at the center of the stimulator and was less than 2.0°C.

The temperature rise at the cathodes was less than 1.0°C. When the stimulator and leads were removed, the maximum temperature rise was less than 1.5°C, recorded at the tip of the electrode with insignificant temperature changes occurring at the cathode. These temperature changes are within physiologically acceptable ranges for the tissues where the implantable spinal fusion stimulator is implanted, especially considering that the temperatures for muscle and subcutaneous tissues are at levels that are known to be several degrees below the normal core temperature of 37°C.

Chou et al. also investigated heating of the tips of broken leads of the implantable spinal fusion stimulator (this device was the same as that which underwent testing in the present study). Temperature changes occurred in localized regions that were within a few millimeters of the cut ends of the leads, with maximum temperature increases that ranged from 11 to 14°C.

If these levels of temperatures occurred during MR imaging, the amount of possible tissue damage would be comparable in characteristics and clinical significance to a small electrosurgical lesion and would likely occur in the scar tissue that typically forms around the implanted leads. Additionally, the potential for tissue damage is only theoretical and a brief temperature elevation around a broken lead, over an approximated volume of 2 to 3-mm radius may not be clinically worse than the scar tissue that forms over the leads during implantation. Fortunately, broken leads are rare, occurring in approximately 10 out of the 70,000 devices implanted over the last ten years (Personal Communication, Simon BJ, Electro-Biology, Inc.).

Based on the available findings from the various investigations that have been conducted, RF energy-induced heating during MR imaging does not appear to present a major problem for a patient with the implantable spinal fusion stimulator, as long as there is no broken lead. The integrity of the leads should be assessed using a radiograph prior to the MR procedure.

In general, the implantable spinal fusion stimulator is considered to be safe for patients undergoing MR procedures if specific guidelines are followed. Guidelines recommended for conducting an MR examination in a patient with the implantable spinal fusion stimulator are, as follows:

(1) The cathodes of the implantable spinal fusion stimulator should be positioned a minimum of 1-cm from nerve roots to reduce the possibility of nerve excitation during an MR procedure.

(2) Plain films should be obtained prior to MR imaging to verify that there are no broken leads present for the implantable spinal fusion stimulator. If this cannot be reliably determined, then the potential risks and benefits to the patient requiring MR imaging must be carefully assessed in consideration of the possibility of the potential for excessive heating to develop in the leads of the stimulator.

(3) MR imaging should be performed using MR systems with static magnetic fields of 1.5-Tesla or less and conventional techniques including spin-echo, fast spin-echo, and gradient echo pulse sequences should be used. Pulse sequences (e.g., echo planar techniques) or conditions that produce exposures to high levels of RF energy (i.e., exceeding a whole body averaged specific absorption rate of 1.0 W/kg) or exposure to gradient fields that exceed 20-T/second, or any other unconventional MR technique should be avoided.

(4) Patients should be continuously observed during MR imaging and instructed to report any unusual sensations including any feelings of warming, burning, or neuromuscular excitation or stimulation.

(5) The implantable spinal fusion stimulator should be placed as far as possible from the spinal canal and bone graft since this will decrease the likelihood that artifacts will affect the area of interest on MR images.

(6) Special consideration should be given to selecting an imaging strategy that minimizes artifacts if the area of interest for MR imaging is in close proximity to the implantable spinal fusion stimulator. The use of fast spin-echo pulse sequences will minimize the amount of artifact associated with the presence of the implantable spinal fusion stimulator.

REFERENCES

Chou C-K, McDougall JA, Chan KW. RF heating of implanted spinal fusion stimulator during magnetic resonance imaging. IEEE Trans Biomed Engineering 1997;44:357-373.

Shellock FG. Pocket Guide to MR Procedures and Metallic Objects: Update 2001, Seventh Edition, Lippincott Williams & Wilkins Healthcare, Philadelphia, 2001.

Shellock FG. Magnetic Resonance Procedures: Health Effects and Safety. CRC Press, LLC, Boca Raton, FL, 2001.

Shellock FG, Hatfield M, Simon BJ, Block S, Wamboldt J, Starewicz PM, Punchard WFB. Implantable spinal fusion stimulator: assessment of MRI safety. J Magn Reson Imaging 2000;12:214-223.

Shellock FG, Kanal E. Magnetic Resonance: Bioeffects, Safety, and Patient Management. Second Edition, Lippincott-Raven Press, New York, 1996.

BREAST TISSUE EXPANDERS AND IMPLANTS

Adjustable breast tissue expanders and mammary implants are utilized for breast reconstruction following mastectomy, for the correction of breast and chest-wall deformities and underdevelopment, for tissue defect procedures, and for cosmetic augmentation. These devices are typically equipped with either an integral injection site or a remote injection dome that is utilized to accept a needle for placement of saline for expansion of the prosthesis intra-operatively and/or postoperatively.

The Becker and the Siltex prostheses are additionally equipped with a choice of a standard injection dome or a micro-injection dome. The Radovan expander is indicated for temporary implantation only. The injection port contains 316L stainless steel to guard against piercing the injection port by the needle used to fill the implant.

There are various breast tissue expanders that are constructed with magnetic ports to allow for a more accurate detection of the injection site. These devices are substantially attracted to the static magnetic fields of MR systems and, therefore, may be uncomfortable, injurious, or contraindicated for patients undergoing MR procedures. One such device is the Contour Profile Tissue Expander (Mentor, Santa Barbara, CA), which contains a magnetic injection dome and is considered to be unsafe for an MR procedure.

The relative amount of image artifacts and distortion caused by the metallic components of these devices should not greatly affect the diagnostic quality of an MR examination, unless the imaging area of interest is at the same location as the metallic portion of the breast tissue expander. Breast tissue expanders with magnetic ports produce relatively large artifacts on

MR images and, as such, assessment of the breast using MR imaging tends to be particularly problematic. Notably, there may be a situation during which a patient is referred for MR imaging for the determination of breast cancer or a breast implant rupture, such that the presence of the metallic artifact could obscure the precise location of the abnormality. In view of this possibility, it is recommended that patients be identified with breast tissue expanders that have metallic components so that radiologists interpreting the MR images are aware of the potential problems related to the generation of artifacts.

McGhan Medical Breast Tissue Expanders and MR Safety. McGhan Medical Breast Tissue Expanders are intended for temporary subcutaneous implantation to develop surgical flaps and additional tissue coverage (Product Information, Style 133 Family of Breast Tissue Expanders with MAGNA-SITE Injection Sites, McGhan Medical/INAMED Aesthetics, Santa Barbara, CA). These breast tissue expanders are constructed from silicone elastomer and consist of an expansion envelope with a textured surface, and a MAGNA-SITE integrated injection site. The expanders are available in a wide range of styles and sizes to meet diverse surgical needs. Specific styles include: Style 133 FV with MAGNA-SITE injection site, Style 133 LV with MAGNA-SITE injection site, Style 133 MV with MAGNA-SITE injection site.

The MAGNA-SITE injection site and MAGNA-FINDER external locating device contain rare-earth, permanent magnets for an accurate injection system. When the MAGNA-FINDER is passed over the surface of the tissue being expanded, its rare-earth, permanent magnet indicates the location of the MAGNA-SITE injection site.

The Product Information document for the McGhan Medical Breast Tissue Expanders (Style 133 Family of Breast Tissue Expanders with MAGNA-SITE Injection Sites, McGhan Medical/INAMED Aesthetics, Santa Barbara, CA) states: "DO NOT use MAGNA-SITE expanders in patients who already have implanted devices that would be affected by a magnetic field (e.g., pacemakers, drug infusion devices, artificial sensing devices). DO NOT perform diagnostic testing with Magnetic Resonance Imaging (MRI) in patients with MAGNA-SITE expanders in place."

Furthermore, in the Warnings section of the Product Information document, the following is indicated: "Diagnostic testing with Magnetic Resonance Imaging (MRI) is contraindicated in patients with MAGNA-SITE expanders in place. The MRI equipment could cause movement of the MAGNA-SITE breast tissue expander, and result in not only patient discomfort, but also expander displacement, requiring revision surgery. In addition, the MAGNA-SITE magnet could interfere with MRI detection capabilities."

Therefore, MR procedures are deemed unsafe for patients with the McGhan Medical Breast Tissue Expanders, Style 133 Family of Breast Tissue Expanders with MAGNA-SITE Injection Sites) McGhan Medical/INAMED Aesthetics, Santa Barbara, CA).

Zegzula et al. presented a case of bilateral tissue expander infusion port dislodgment associated with an MRI examination. The report involved a 56-year-old woman that underwent bilateral mastectomy and immediate reconstruction with McGhan BIOSPAN tissue expanders. As noted, these implants contain the "MAGNA-SITE" components. Several weeks post-operatively the patient underwent MR imaging of her spine. Subsequently, the infusion ports could not be located with the finder magnet (used to re-fill the tissue expander). A chest radiograph was obtained that demonstrated bilateral dislodgment of the infusion ports. Surgical removal and replacement of the tissue expanders were required. This incident emphasizes the fact that all patients undergoing tissue expansion with implants that contain integral magnetic sports should be thoroughly warned about the potential hazards of MR imaging.

In another incident involving a tissue expander, Duffy and May reported a case of a woman who developed a burning sensation at the site of the tissue expander during an MR procedure. The sensation resolved rapidly once the scan was discontinued. The implications of the symptoms for in this case are unclear. Nevertheless, a patient with a tissue expander that requires an MR procedure should be alerted to the possibility of localized symptoms in the region of the implant during scanning.

REFERENCES

Duffy FJ Jr, May JW Jr. Tissue expanders and magnetic resonance imaging: the "hot" breast implant. Ann Plast Surg 1995;5:647-9.

Fagan LL, Shellock FG, Brenner RJ, Rothman B. Ex vivo evaluation of ferromagnetism, heating, and artifacts of breast tissue expanders exposed to a 1.5-T MR system. J Magn Reson Imaging 1995;5:614-616.

Liang MD, Narayanan K, Kanal E. Magnetic ports in tissue expanders: a caution for MRI. Magn Reson Imaging 1989;7:541-542.

Product Information, Style 133 Family of Breast Tissue Expanders with Magna-Site Injection Sites, McGhan Medical/INAMED Aesthetics, Santa Barbara, CA.

Shellock FG. Pocket Guide to MR Procedures and Metallic Objects: Update 2001, Seventh Edition, Lippincott Williams & Wilkins Healthcare, Philadelphia, 2001.

Shellock FG, Kanal E. Magnetic Resonance: Bioeffects, Safety, and Patient Management. Second Edition, Lippincott-Raven Press, New York, 1996.

Zegzula HD, Lee WP. Infusion port dislodgment of bilateral breast tissue expanders after MRI. Ann Plast Surg 2001;46:46-8.

CARDIAC PACEMAKERS AND IMPLANTABLE CARDIOVERTER DEFIBRILLATORS

Cardiac pacemakers and implantable cardioverter defibrillators (ICDs) are crucial implanted devices for patients with heart conditions and serve to maintain quality of life and substantially reduce morbidity. Expanded indications for cardiac pacemakers and ICDs (e.g., heart failure, obstructive sleep apnea, and other conditions) emphasize that an increasing number of patients will be treated with these devices. Unfortunately, these cardiovascular implants are considered strict contraindications for patients referred for MR procedures (see **Extremity MR System** for additional information). Additionally, individuals with cardiac pacemakers and ICDs should be prevented from entering the MR environment because of potential risks related to exposure to the fringe field of the scanner.

Cardiac pacemakers and ICDs present potential problems to patients undergoing MR procedures from various mechanisms, including:

(1) movement of the implantable pulse generator or lead(s);

(2) temporary or permanent modification of the function of the device;

(3) inappropriate sensing or triggering of the device;

(4) excessive heating of the leads; and

(5) induced currents in the leads.

The effects of the MR environment and MR procedures on the functional and operational aspects of cardiac pacemakers and ICDs vary, depending on several factors including the type of device, how the device is programmed, the static magnetic field strength of the MR system, and the

138

imaging conditions used for the procedure (i.e., the anatomic region imaged, type of surface coil used, the pulse sequence, amount of radiofrequency energy used, etc.).

There have been several cardiac pacemaker studies in which laboratory dogs as well as human subjects have been observed to become tachyarrhythmic and/or hypotensive during MR imaging. The cause of these responses may have been the induction of voltages or currents within the pacemaker-lead-myocardial loop that was sufficient to induce action potentials or contraction of the myocardium and an electrical, as well as physiologic, systole. In fact, investigators have reported observing cardiac pacing at the selected repetition time (TR) of the MR imaging procedure. Rapid pacing rates that yield cardiac outputs incompatible with sustaining life may have been the cause of death in some of the cardiac pacemaker patients that underwent MR procedures, however, this has not been verified.

Excessive heating of the cardiac pacemaker's leads during MR imaging is considered to be one of the more problematic MR safety aspects associated with this type of implant. For example, an investigation conducted by Achenbach et al. reported that it was possible for pacing leads (heating at the electrodes, no cardiac pacemaker attached) to heat up to up to 63.1°C within 90 seconds of scanning under certain conditions using a 1.5-Tesla MR system. Therefore, thermal injury to the endocardium or myocardium must be considered a possible adverse outcome if RF power is transmitted in the vicinity of the pacemaker and/or its leads.

In consideration of the various factors discussed above, a patient or individual with a cardiac pacemaker should not be allowed to enter the MR environment or to undergo an MR procedure. However, it is possible that this recommendation will change as more knowledge and experimental data are acquired pertaining to this issue. There is growing evidence that it may be possible to perform MR procedures safely in certain patients (e.g., such as non-pacemaker dependent patients) under highly controlled MR conditions (see below).

Implantable cardioverter defibrillators (ICDs) are medical devices designed to automatically detect and treat episodes of ventricular fibrillation, ventricular tachycardias, and bradycardia. When an arrhythmia is detected, the device can deliver defibrillation, cardioversion, antitachycardia pacing, or bradycardia pacing therapy.

An ICD typically uses a programmer that has an external magnet to test the battery charger and to activate and deactivate the system. Deactivation of an ICD is usually accomplished by holding a magnet over the device for approximately 30 seconds. As such, deactivations of ICDs have occurred accidentally as a result of patients encountering magnetic fields in home and workplace environments. For example, deactivations of ICDs have occurred in patients from exposures to the magnetic fields found in stereo speakers, bingo wands, and 12-volt battery starters. Therefore, patients and individuals with ICDs should be prevented from inadvertently entering the MR environment.

In general, exposure to an MR system or to an MR procedure would have similar effects on an ICD as that previously described for a cardiac pacemaker, since many of the functional components are comparable. Therefore, patients and individuals with these devices should not be allowed to enter MR environment. In addition, since ICDs also have electrodes placed in the myocardium, patients should not undergo MR procedures because of the inherent risks related to the presence of these conductive materials. Similar to cardiac pacemakers, it is anticipated that, in lieu of developing a truly MR-compatible ICD, safety criteria may be developed for standard ICDs that will entail special programming and monitoring procedures along with potentially being able to perform MR examinations by following strict conditions.

Cardiac Pacemakers and ICDs: Current Experience With the MR Environment and MR Procedures. To date, there have been approximately thirteen fatalities reported for patients with cardiac pacemakers or ICDs that were associated with exposure to the MR environment or related to an MR procedure. In virtually every case, the patient that died apparently entered the MR environment without the staff knowing a pacemaker or ICD was present. Notably, these deaths were poorly characterized, no electrocardiographic data were available for review, it was unknown whether any of these patients was pacemaker dependent, and no mention was made regarding the cause of death.

By comparison, no clinically meaningful irreversible harm has been reported when patients with implantable pacemakers or ICDs were carefully monitored during the MR procedure and/or the device underwent reprogramming prior to the scan. Notably, patients with cardiac pacemakers have been successfully scanned using MR systems operating at static field strengths ranging from 0.35- to 1.5-Tesla without any clinically adverse events. The investigators of these studies recommended certain strategies for performing safe MR procedures. These strategies included programming the device to "off" or asynchronous mode; programming to a bipolar lead configuration, if possible; only scanning non-pacemaker dependent patients, limiting the level of RF power exposure; and only doing MR examinations if the pulse generator is positioned outside of the bore of the MR system (i.e., using a dedicated-extremity MR system). Obviously, careful monitoring was implemented in every case along with proper precautions to ensure patient safety in each case. The presence of resuscitation equipment in close proximity to the MR system room is an absolute requirement. An advanced cardiac life support (ACLS) certified physician must also be in attendance and ready for any untoward consequences.

To date, peer-reviewed literature reports have indicated that well over 200 patients with cardiac pacemakers have been scanned safely, with several large series of patients undergoing MR procedures at 0.5- and 1.5-Tesla. As such, according to Gimbel (2002), this recent evidence suggests that restrictions prohibiting MR procedures in patients with implantable pacemakers and perhaps ICDs might be significantly modified in the near future.

In consideration of the infinite possibilities of cardiac pacing systems, cardiac and lead geometry, as well as the myriad of variables that exist for MR procedures (e.g., variable RF and gradient magnetic fields, etc.) it may not be possible to ensure absolute safety for every pacemaker and MRI interaction. However, given appropriate patient selection as well as continuous monitoring and preparedness for resuscitation efforts, performance of MR procedures on patients with cardiac pacemakers may be achieved with reasonable safety, even at static magnetic field strengths of 1.5-Tesla, as recently reported by Martin et al. (2003).

MR Safety at 1.5 and 3.0-Tesla and Cardiac Pacemakers and ICDs. As previously discussed, one important safety effect of the MR environment on cardiac pacemakers and ICDs is related to magnetic field interactions. Component parts of pacemakers and ICDs, such as batteries, reed-switches, or transformer core materials may contain ferromagnetic materials. Therefore, substantial magnetic field interactions could occur during exposure to the MR environment causing these implants to move or be uncomfortable for patients or individuals. In consideration of these possible scenarios, and as an important part of evaluating MR safety for pacemakers and ICDs, testing for magnetic field interactions has been conducted using MR systems operating at static magnetic field strengths ranging from 0.2-Tesla (i.e., the dedicated-extremity MR system) to 1.5-Tesla. These investigations reported that most modern-day pacemakers pose no serious safety risk with respect to magnetic field interactions at

1.5-Tesla or less, while most ICDs may be problematic due to substantial magnetic field interactions at 1.5-Tesla.

For example, Luechinger et al. investigated magnetic field interactions for thirty-one cardiac pacemakers and thirteen ICDs in association with exposure to a 1.5-Tesla MR system (Gyroscan ACS NT, Philips Medical Systems, Best, The Netherlands). The investigators reported that "newer" cardiac pacemakers had relatively low magnetic force values compared to older devices. With regard to ICDs, with the exception of one newer model (GEM II, 7273 ICD, Medtronic, Minneapolis, MN), all ICDs showed relatively high magnetic field interactions. Luechinger et al. concluded that modern-day pacemakers present no safety risk with respect to magnetic field interactions at 1.5-Tesla, while ICDs may pose problems due to strong magnet-related mechanical forces.

The clinical use of 3.0-Tesla MR systems for brain, musculoskeletal, body and cardiovascular applications is increasing. Because previous investigations performed to determine MR safety for pacemakers and ICDs used MR systems with static magnetic fields of 1.5-Tesla or less, it is crucial to perform ex vivo testing at 3.0-Tesla to characterize magnetic field-related safety for these implants, with full acknowledgment that additional MR safety issues exist for these devices, as described-above.

An important aspect of determining magnetic field interactions for metallic implants involves the measurement of translational attraction. Translational attraction is assessed for metallic implants using the standardized deflection angle test recommended by the American Society for Testing and Materials (ASTM). According to ASTM guidelines, the deflection angle for an implant or device is generally measured at the point of the "highest spatial gradient" for the specific MR system used for testing. Notably, the deflection angle test is commonly performed as an integral part of MR safety testing for metallic implants and devices.

Various types of magnets exist for commercially available 1.5- and 3.0-Tesla MR systems, including magnet configurations that are used for conventional "long-bore" scanners and newer "short-bore" systems. Because of physical differences in the position and magnitude of the highest spatial gradient for different magnets, measurements of deflection angles for implants using long-bore vs. short-bore MR systems can produce substantially different results for magnetic field-related translational attraction, as reported by Shellock et al. In general, the implications are primarily for magnetic field-related translational attraction with regard to long-bore vs. short-bore 3.0-Tesla MR systems (with short-bore scanners producing greater translational attraction for a given implant). Therefore, a study was conducted on fourteen different cardiac pacemakers and four ICDs to evaluate translational attraction for these devices in association with long-bore and short-bore 1.5- and 3.0-Tesla MR systems. Deflection angles were measured based on guidelines from the American Society for Testing and Materials.

In general, deflection angles for the cardiovascular implants that underwent evaluation were significantly ($p < 0.01$) higher on 1.5- and 3.0-Tesla short-bore scanners compared to long-bore MR systems. For the 1.5-Tesla MR systems, three cardiac pacemakers (Cosmos, Model 283-01 Pacemaker, Intermedics, Inc., Freeport, TX; Nova Model 281-01 Pacemaker, Intermedics, Inc., Freeport, TX; Res-Q ACE, Model 101-01 Pacemaker; Intermedics, Inc., Freeport, TX) exhibited deflection angles greater than 45 degrees (i.e., exceeding the recommended ASTM criteria) on both long-bore and short-bore 1.5-Tesla MR systems. The findings indicated that these devices are potentially unsafe for patients from a magnetic field interaction consideration.

On the 3.0-Tesla MR systems, seven implants exhibited deflection angles greater than 45 degrees on the long-bore 3.0-Tesla scanner, while 13 exhibited deflection angles greater than 45 degrees on the short-bore 3.0-T MR system (refer to **The List** for information on the cardiac pacemakers and ICDs that underwent testing). Of note is that the findings for magnetic field-related translational attraction were substantially different comparing the long-bore and short-bore MR systems.

Importantly, other factors exist that may impact MR safety for these cardiac pacemakers and ICDs. Therefore, regardless of the fact that magnetic field interactions may not present a risk for some of the cardiovascular implants that have been tested, these additional potentially hazardous mechanisms should be considered carefully for these devices.

REFERENCES

Achenbach S, Moshage W, Diem B, Bieberle T, Schibgilla V, Bachmann K. Effects of magnetic resonance imaging on cardiac pacemakers and electrodes. Am Heart J 1997;134:467-473.

Alagona P, Toole JC, Maniscalco BS, et al. Nuclear magnetic resonance imaging in a patient with a DDD pacemaker. Pacing Clin Electrophysiol 1989;12:619 (letter).

American Society for Testing and Materials (ASTM) Designation: F 2052. Standard test method for measurement of magnetically induced displacement force on passive implants in the magnetic resonance environment. In: Annual Book of ASTM Standards, Section 13, Medical Devices and Services, Volume 13.01 Medical Devices; Emergency Medical Services. West Conshohocken, PA, 2002, pp. 1576-1580.

Bhachu DS, Kanal E. Implantable pulse generators (pacemakers) and electrodes: safety in the magnetic resonance imaging scanner environment. J Magn Reson Imaging 2000; 12:201-204.

Bonnet CA, Elson JJ, Fogoros RN. Accidental deactivation of the automatic implantable cardioverter defibrillator. Am Heart J 1990;3:696-697.

Duru F, Luechinger R, Scheidegger MB, Luscher TF, Boesiger P, Candinas R. Pacing in magnetic resonance imaging environment: clinical and technical considerations on compatibility. Eur Heart J 2001;22:113-124.

Duru F, Luechinger R, Candinas R. MR imaging in patients with cardiac pacemakers. Radiology 2001;219:856-858.

Erlebacher JA, Cahill PT, Pannizzo F, Knowles RJR. Effect of magnetic resonance imaging on DDD pacemakers. Am J Cardio 1986;57:437-440.

Fetter J, Aram G, Holmes DR, Gray JE, Hayes DL. The effects of nuclear magnetic resonance imagers on external and implantable pulse generators. Pacing Clin Electrophysiol 1984;7:720-727.

Fontaine JM, Mohamed FB, Gottlieb C, Callans DJ, Marchlinski FE. Rapid ventricular pacing in a pacemaker patient undergoing magnetic resonance imaging. Pacing Clin Electrophysiol 1998;21:1336-1339.

Garcia-Bolao I, Albaladejo V, Benito A, Alegria E, Zubieta J. Magnetic resonance imaging in a patient with a dual chamber pacemaker. Acta Cardiol 1998;53:33-35.

Gimbel JR. Letter to the Editor. Pacing Clin Electrophysiol 2003;26:1.

Gimbel JR, Johnson D, Levine PA, Wilkoff BL. Safe performance of magnetic resonance imaging on five patients with permanent cardiac pacemakers. Pacing Clin Electrophysiol 1996;19:913-919.

Gimbel JR. Implantable pacemaker and defibrillator safety in the MR environment: new thoughts for the new millennium. RSNA Special Cross-Specialty Categorical Course in Diagnostic Radiology: Practical MR Safety Considerations for Physicians, Physicists, and Technologists 2001;69-76.

Hayes DL, Holmes DR, Gray JE. Effect of 1.5 Tesla nuclear magnetic resonance imaging scanner on implanted permanent pacemakers. J Am Coll Cardiol 1987;10:782-786.

Holmes DJ, Hayes DL, Gray JE, Merideth J. The effects of magnetic resonance imaging on implantable pulse generators. Pacing Clin Electrophysiol 1986;9:360-370.

Inbar S, Larson J, Burt T, Mafee M, Ezri M. Case report: nuclear magnetic resonance imaging in a patient with a pacemaker. Am J Med Sci 1993;3:174-175.

Juralti NM, Sparker J, Gimbel JR, et al. Strategies for the safe performance of magnetic resonance imaging in selected pacemaker patients (abstract) Circulation 2001;104 (Suppl.):3020.

Lauck G, von Smekal A, Wolke S, Seelos KC, Jung W, Manz M, et al. Effects of nuclear magnetic resonance imaging on cardiac pacemakers. Pacing Clin Electrophysiol 1995;18:1549-55.

Luechinger RC. Safety Aspects of Cardiac Pacemakers in Magnetic Resonance Imaging. Swiss Institute of Technology, Zurich, Dissertation, 2002

Luechinger RC, Duru F, Zeijlemaker VA, Scheidegger MB, Boesiger P, Candinas R. Pacemaker reed-switch behavior in 0.5, 1.5, and 3.0 Tesla magnetic resonance units: Are reed switches always closed in strong magnetic fields? Pacing Clin Electrophysiol 2002;25:1419-1423.

Luechinger RC, Duru F, Scheidegger MB, Boesiger P, Candinas R. Force and torque effects of a 1.5 Tesla MRI scanner on cardiac pacemakers and ICDs. Pacing Clin Electrophysiol 2001;24:199-205.

Martin ET, Coman JA, Owen W, Shellock FG. Cardiac pacemakers and MRI: safe evaluation of 47 patients using a 1.5-Tesla MR system without altering pacemaker or imaging parameters. Proceedings of the International Society for Magnetic Resonance in Medicine (abstract) 2003;11:2445.

Pavlicek W, Geisinger M, Castle L, Borkowski G, Meaney T, Bream B, et al. The effects of nuclear magnetic resonance on patients with cardiac pacemakers. Radiology 1983;147:149-153.

Shellock FG, Magnetic Resonance Procedures: Health Effects, Safety, and Patient Management CRC Press, LLC, Boca Raton, FL, 2001.

Shellock FG, Kanal E. Magnetic resonance: Bioeffects, Safety and Patient Management, 2nd edition. Lippincott-Raven, New York, 1996.

Shellock FG, O'Neil M, Ivans V, Kelly D, O'Connor M, Toay L, Crues JV. Cardiac pacemakers and implantable cardiac defibrillators are unaffected by operation of an extremity MR system. AJR Am J Roentgenol 1999;72:165-170.

Shellock FG, Tkach JA, Ruggieri PM, Masaryk TJ. Cardiac pacemakers, ICDs, and loop recorder: Evaluation of translational attraction using conventional ("long-bore") and "short-bore" 1.5- and 3.0-Tesla MR systems. Journal of Cardiovascular Magnetic Resonance 2003;5:387-397.

Sommer T, Vahlhaus C, Lauck G, Smekal A, Reinke M, Hofer U, et al. MR imaging and cardiac pacemakers: in vitro evaluation and in vivo studies in 51 patients at 0.5 T. Radiology 2000;215:869-879.

Vahlhaus C, Sommer T, Lewalter T, Schimpf, R, Schumacher B, Jung W, et al. interference with cardiac pacemakers by magnetic resonance imaging: Are there irreversible changes at 0.5 Tesla? Pacing Clin Electrophysiol 2001;24(Pt. I):489-95.

Zimmermann BH, Faul DD. Artifacts and hazards in NMR imaging due to metal implants and cardiac pacemakers. Diagn Imaging Clin Med 1984;53:53-56.

CARDIOVASCULAR CATHETERS AND ACCESSORIES

Cardiovascular catheters and accessories are indicated for use in the assessment and management of critically-ill or high-risk patients including those with acute heart failure, cardiogenic shock, severe hypovolemia, complex circulatory abnormalities, acute respiratory distress syndrome, pulmonary hypertension, certain types of arrhythmias and other various medical emergencies. In these cases, cardiovascular catheters are used to measure intravascular pressures, intracardiac pressures, cardiac output, and oxyhemoglobin saturation. Secondary indications include venous blood sampling and therapeutic infusion of solutions or medications. In addition, some cardiovascular catheters are designed for temporary cardiac pacing and intra-atrial or intraventricular electrocardiographic monitoring.

Because patients with cardiovascular catheters and associated accessories may require evaluation using MR procedures or these devices may be considered for use during MR-guided procedures, it is imperative that a thorough *ex vivo* assessment of MR-safety be conducted for these devices to ascertain the potential risks of their use in the MR environment. For example, MR imaging, angiography, and spectroscopy procedures may play an important role in the diagnostic evaluation of these patients. Furthermore, the performance of certain MR-guided interventional procedures may require the utilization of cardiovascular catheters and accessories to monitor patients during biopsies, interventions, or treatments.

There is at least one report of a cardiovascular catheter (Swan-Ganz Triple Lumen Thermodilution Catheter) that "melted" in a patient undergoing MR imaging. This catheter contained a wire made from a conductive material that was considered to be responsible for this problem. Thus, there are realistic concerns pertaining to the use of similar devices during MR examinations. Therefore, an investigation was performed using *ex vivo* testing techniques to evaluate various cardiovascular catheters and accessories with regard to magnetic field interactions, heating, and artifacts associated with MR imaging.

A total of fifteen different cardiovascular catheters and accessories (Abbott Laboratories, Morgan Hill, CA) were selected for evaluation because they represent a wide-variety of the styles and types of devices that are commonly-used in the critical care setting (i.e., the basic structures of these devices are comparable to those made by other manufacturers). Of these devices, the 3-Lumen CVP Catheter, CVP-PVC Catheter (used for central venous pressure monitoring, administration of fluids, and venous blood sampling; polyurethane and polyvinyl chloride, respectively), Thermoset-Iced, and Thermoset-Room (used as accessories for determination of cardiac output using the thermodilution method; plastic), and Safe-set with In-Line Reservoir (used for in-line blood sampling; plastic) were determined to have no metallic components (Personal communications, Ann McGibbon, Abbott Laboratories, 1997). Therefore, these devices were deemed safe for patients undergoing MR procedures and were not included in the overall ex vivo tests for MR safety. The remaining ten devices were evaluated for the presence of potential problems in the MR environment.

Excessive heating of implants or devices made from conductive materials has been reported to be a hazard for patients who undergo MR procedures. This is particularly a problem for devices that are in the form of a loop or coil because current can be induced in this shape during operation of the MR system, to the extent that a first, second or third-degree burn can be produced.

The additional physical factors responsible for this hazard have not been identified or well-characterized (i.e., the imaging parameters, specific gradient field effects, size of the loop associated with excessive heating, etc.). For this reason, the study examining cardiovascular catheters and accessories did not attempt to investigate the effect of various "coiled" catheter shapes on the development of substantial heating during an MR procedure, especially since there are many factors in addition to the shape of the catheter with a conductive component that can also influence the amount of heating that occurs during an MR procedure.

The thermodilution Swan-Ganz catheter and other similar cardiovascular catheters are constructed of nonferromagnetic materials that include conductive wires. A report indicated that a portion of a Swan-Ganz thermod-ilution catheter that was outside the patient melted during an MR imaging procedure. It was postulated that the high-frequency electromagnetic fields generated by the MR system caused eddy current-induced heating of either the wires within the thermodilution catheter or the radiopaque material used in the construction of the catheter.

This incident suggests that patients with this catheter or similar device that has conductive wires or other component parts, could be potentially injured during an MR procedure. Furthermore, heating of the wire or lead of a temporary pacemaker (e.g., the RV Pacing Lead) is of at least a theo-retical concern for any similar wire in the bore of an MR system. Cardiac pacemaker leads are typically intravascular for most of their length and heat transfer and dissipation from the leads into the blood may prevent dangerous levels of lead heating to be reached or maintained for the intravascular segments of pacemaker leads.

However, for certain segments of these leads it is at least theoretically pos-sible that sufficient power deposition or heating may be induced within these leads to result in local tissue injury or burn during an MR procedure. An *ex vivo* study conducted by Achenbach et al. substantiates this con-tention, whereby temperature increases of up to 63.1°C were recorded at the tips of pacemaker electrodes during MR imaging performed in phan-toms.

Because of possible deleterious and unpredictable effects, patients referred for MR procedures with cardiovascular catheters and accessories that have internally or externally-positioned conductive wires or similar components should not undergo MR procedures because of the possible associated risks, unless MR-safety testing information demonstrates oth-erwise. Further support of this recommendation is based on the fact that inappropriate use of monitoring devices during MR procedures is often the cause of patient injuries. For example, burns have resulted in the MR environment in association with the use of devices that utilize conductive wires.

REFERENCES

Achenbach S, Moshage W, Diem B, Bieberle T, Schibgilla V, Bachmann K. Effects of magnetic resonance imaging on cardiac pacemakers and electrodes. Am Heart J 1997;134:467-473

Dempsey MF, Condon B, Hadley DM. Investigation of the factors responsible for burns during MRI. J Magn Reson Imaging 2001;13:627-631.

ECRI, Health devices alert. A new MRI complication? May 27, 1988.

Shellock FG. Pocket Guide to MR Procedures and Metallic Objects: Update 2001, Seventh Edition, Lippincott Williams & Wilkins Healthcare, Philadelphia, 2001.

Shellock FG. Magnetic Resonance Procedures: Health Effects and Safety. CRC Press, LLC, Boca Raton, FL, 2001.

Shellock FG, Kanal E. Magnetic Resonance: Bioeffects, Safety, and Patient Management. Second Edition, Lippincott-Raven Press, New York, 1996.

Shellock FG, Shellock VJ. Cardiovascular catheters and accessories: Ex vivo testing of ferromagnetism, heating, and artifacts associated with MRI. J Magn Reson Imaging 1998;8:1338-1342.

CAROTID ARTERY
VASCULAR CLAMPS

Each of the carotid artery vascular clamps tested in association with exposure to a 1.5-Tesla MR system displayed positive magnetic field interactions. However, only the Poppen-Blaylock carotid artery vascular clamp is considered contraindicated for patients undergoing MR procedures due to the existence of substantial ferromagnetism. The other carotid artery clamps were considered safe for patients exposed to MR systems because they were deemed "weakly" ferromagnetic. With the exception of the Poppen-Blaylock clamp, patients with metallic carotid artery vascular clamps have been imaged by MR systems with static magnetic fields up to 1.5-Tesla without experiencing any discomfort or neurological sequelae.

REFERENCES

Shellock FG. Pocket Guide to MR Procedures and Metallic Objects: Update 2001, Seventh Edition, Lippincott Williams & Wilkins Healthcare, Philadelphia, 2001.

Shellock FG. Magnetic Resonance Procedures: Health Effects and Safety. CRC Press, LLC, Boca Raton, FL, 2001.

Shellock FG, Kanal E. Magnetic Resonance: Bioeffects, Safety, and Patient Management. Second Edition, Lippincott-Raven Press, New York, 1996.

Teitelbaum GP, Lin MCW, Watanabe AT, et al. Ferromagnetism and MR imaging: safety of carotid vascular clamps. AJNR Am J Neuroradiol 1990;11:267-272.

COCHLEAR IMPLANTS

Cochlear implants are electronically-activated devices. Consequently, an MR procedure is typically contraindicated for a patient with this type of implant because of the possibility of injuring the patient and/or altering or damaging the function of the device. In general, all individuals with cochlear implants should be prevented from entering the MR environment, as well.

Investigations have been conducted to determine if there are any situations during which a patient with a cochlear implant could safely undergo an MR procedure. Two studies were performed to evaluate the safety of using MR procedures in patients with specially modified devices: the Nucleus Mini-22 Cochlear Implant and the Multichannel Auditory Brainstem Implant.

Tests were conducted to assess the operation of these cochlear implants in the MR environment as well as to determine magnetic field interactions, artifacts, induced current, and heating during MR imaging. The reports indicated that a margin of safety exists with regard to the MR environment for these devices, as long as the specific recommendations indicated the product labeling are adhered to (refer to the guidelines for the safe performance of MR imaging shown in the product labels for the Nucleus Mini-22 Cochlear Implant and the Multichannel Auditory Brainstem Implant).

In vitro experiments were conducted to determine MR safety for the cochlear implant, the Combi 40/40+ Multichannel System (MedEl, Innsbruck, Austria). This cochlear implant underwent testing in association with 0.2- and 1.5-Tesla MR systems. According to the results, partial demagnetization of the cochlear implant occurred within the 1.5-Tesla MR system, while the 0.2-Tesla MR system produced no significant alteration in the magnetic component of this implant. Partial demagnetization could be avoided by orienting the patient's head parallel to the magnetic field of the 1.5-Tesla MR system, however, this may not be practical for a patient undergoing a clinical MR procedure.

In general, electromagnetic interference related to the use of the 1.5-Tesla MR system remained within acceptable limits for the Combi 40/40+ Multichannel System. Of great concern were the relative amounts of magnetic field translational attraction and torque acting on the cochlear implant that have important implications for the safe performance of MR procedures using the 1.5-Tesla MR system. By comparison, MR safety issues were minimal with the use of the 0.2-Tesla MR system. The investigators recommended that MR imaging may be performed in a patient with the Combi 40/40+ Multichannel System only if there is a strong medical indication.

Additional *in vitro* work was conducted on the Combi 40/40+ cochlear implant to determine MR safety within a wide range of clinical applications using a 1.5-Tesla MR system. Torque, translational force, demagnetization, artifacts, induced voltages, and heating were assessed under extreme MR conditions. Of the MR safety issues evaluated for this implant, only the torque on the internal magnet of this cochlear implant was considered to be problematic. Thus, some form of external stabilization is required (e.g., tape, support bandage, etc.) for the Combi 40/40+ cochlear implant for a patient undergoing an MR procedure at 1.5-Tesla.

The overall test findings for the Combi 40/40+ cochlear implant indicated that an MR procedure should only be performed in a patient with this device if there is a strong medical necessity. An assessment of the relative risks involved versus the risk of not providing the diagnostic information from an MR procedure for the specific patient is required.

COCHLEAR NUCLEUS 24 AUDITORYBRAINSTEM IMPLANT (ABI) SYSTEM (Cochlear Corporation, Englewood, CO). The Nucleus 24 Auditory Brainstem Implant (ABI) system is designed to provide useful hearing to individuals with Neurofibromatosis Type 2, who become deaf following surgery to remove bilateral auditory nerve tumors. The ABI system includes both surgically implanted and externally worn components. Externally worn components include the SPrint (body-worn) speech processor and a microphone/headset. Together, the external components transform acoustical signals in the environment into an electrical code. This coded information is then transmitted from the system's headset to a surgically implanted receiver/stimulator, where it is decoded and delivered to a multi-channel electrode array. The system's 21-electrode array is surgically placed on the surface of the cochlear nucleus, which is the first auditory center within the brainstem. Electrical stimulation of the cochlear nucleus results in the signal's transmission to the brain, where it is interpreted as sound.

Magnetic Resonance Imaging (MRI)

Magnetic Resonance Imaging (MRI) is contraindicated except under the circumstances described below. Do not allow patients with a Nucleus 24 ABI to be in the room where an MRI scanner is located except under the following special circumstances.

The Nucleus 24 ABI has a removable magnet and specific design characteristics to enable it to withstand MRI up to 1.5-Tesla, but not higher. If the ABI's magnet is in place, it must be removed surgically before the patient undergoes an MR procedure.

The patient must take off the speech processor and headset before entering a room where an MRI scanner is located.

If the implant's magnet is still in place, tissue damage may occur if the recipient is exposed to MRI. Once the magnet is surgically removed, the metal in the auditory brainstem implant will affect the quality of the MRI. Image shadowing may extend as far as 6-cm from the implant, thereby, resulting in loss of diagnostic information in the vicinity of the implant.

All Nucleus ABI devices have removable magnets. However, if the physician is unsure that the patient has a Nucleus 24 ABI, an x-ray may be used to check the radiopaque lettering on the device. There are three platinum letters printed on the receiver/stimulator portion of each Nucleus auditory brainstem implant. If the middle letter is a "G", the implant is a Nucleus 24 ABI. Once the magnet has been removed, MRI can be performed.

If you require additional information about removal of the magnet, please contact Cochlear Corporation [www.Cochlear.com; (800) 523-5798]

(*Excerpted with permission from the Package Insert, Nucleus 24 Cochlear Implant System, Cochlear Corporation, Englewood, CO).

REFERENCES

Chou H-K, McDougall JA, Can KW. Absence of radiofrequency heating from auditory implants during magnetic resonance imaging. Bioelectromagnetics 1995;16:307-316.

Cochlear, Package Insert, Nucleus 24 Cochlear Implant, Cochlear Corporation, Englewood, CO.

Heller JW, Brackmann DE, Tucci DL, Nyenhuis JA, Chou H-K. Evaluation of MRI compatibility of the modified nucleus multi-channel auditory brainstem and cochlear implants. Am J Otol 1996;17:724-729.

Ouayoun M, Dupuch K, Aitbenamou C, Chouard CH. Nuclear magnetic resonance and cochlear implant. Ann Otolaryngol Chir Cervicofac 1997;114:65-70.

Shellock FG. Pocket Guide to MR Procedures and Metallic Objects: Update 2001, Seventh Edition, Lippincott Williams & Wilkins Healthcare, Philadelphia, 2001.

Shellock FG. Magnetic Resonance Procedures: Health Effects and Safety. CRC Press, LLC, Boca Raton, FL, 2001.

Shellock FG, Kanal E. Magnetic Resonance: Bioeffects, Safety, and Patient Management. Second Edition, Lippincott-Raven Press, New York, 1996.

Teissl C, Kremser C, Hochmair ES, Hochmair-Desoyer IJ. Cochlear implants: *in vitro* investigation of electromagnetic interference at MR imaging-compatibility and safety aspects. Radiology 1998;208:700-708.

Teissl C, Kremser C, Hochmair ES, Hochmair-Desoyer IJ. Magnetic resonance imaging and cochlear implants: compatibility and safety aspects. J Magn Reson Imaging 1999;9:26-38.

COILS, STENTS, AND FILTERS

Various types of intravascular and other types of coils, stents, and filters have been evaluated for safety with MR systems. Several of these demonstrated magnetic field interactions in association with scanners. Fortunately, the devices that exhibit positive magnetic field interactions typically become incorporated securely into the vessel wall primarily due to tissue ingrowth within approximately six to eight weeks after implantation (Note: To date, MR safety for drug-eluting stents has not been determined). Therefore, it is unlikely that any of these implants would become moved or dislodged as a result of exposure to static magnetic fields of MR systems operating at 1.5-Tesla or less.

Other similar devices made from nonferromagnetic materials, such as the LGM IVC filter (Vena Tech) used for caval interruption or the Wallstent biliary endoprosthesis [Schneider (USA), Inc.] used for treatment of biliary obstruction, are considered to be safe for patients undergoing MR procedures. Notably, it is unnecessary to wait any period of time after surgery to perform an MR procedure in a patient with a metallic implant that is made from a nonferromagnetic material (see **Guidelines for the Management of the Post-Operative Patient Referred for a Magnetic Resonance Procedure**). In fact, there are various published reports in the peer-reviewed literature that describe placement of vascular stents using MR-guidance, even with high-field-strength (1.5-Tesla) MR systems.

The Guglielmi detachable coil (GDC) used for endovascular embolization, was evaluated for MR-safety. Importantly, because of the coiled-shape of the GDC, potential heating during MR imaging was suspected. Therefore, a study was performed using *ex vivo* testing techniques to determine the MR-safety of the Guglielmi detachable coil with respect to magnetic field interactions, heating, and artifacts. The results indicated that there were no magnetic field interactions, the temperature increase was minimal during extreme MR imaging conditions, and the artifacts involved a mild signal void relative to the size and shape of the GDC. Subsequently, more than 100 patients with GDCs underwent MR imaging without incident. Other embolization coils made from Nitinol, platinum, or platinum and iridium have been evaluated and found to be safe for patients undergoing MR procedures.

Patients with the specific coils, stents, and filters indicated in **The List** have had procedures using MR systems operating at static magnetic field strengths of 1.5-Tesla or less without reported injuries or other problems. Nevertheless, an MR procedure should not be performed if there is any possibility that the coil, stent, or filter is not positioned properly or firmly in place. Additionally, it should be duly noted that not all stents are safe for patients undergoing MR procedures, particularly since, to date, not all stents have undergone MR safety testing.

A study by Taal et al. supports the fact that not all stents are safe for patients undergoing MR procedures. This investigation was performed to evaluate potential problems for four different types of stents: the Ultraflex (titanium alloy), the covered Wallstent (Nitinol), the Gianturco stent (Cook), and the modified Gianturco stent (Song) -the last two being made of stainless steel. Taal et al. reported "an appreciable attraction force and torque" found for both types of Gianturco stents. In particular, "the Gianturco (Cook) stent pulled toward the head with a force of 7 g…however, it is uncertain whether this is a potential risk for dislodgment." In consideration of these results the investigators advised, "…specific information on the type of stent is necessary before a magnetic resonance imaging examination is planned."

MR Safety at 3.0-Tesla and Coils and Stents. Several different coils and stents have been evaluated at 3.0-Tesla. Of these implants, two displayed magnetic field interactions that exceeded the ASTM guideline for MR safety (i.e., the deflection angles were greater than 45°). However, similar to other coils and stents, tissue ingrowth may be sufficient to prevent these implants from posing a substantial risk to a patient or individual in the 3.0-Tesla MR environment. Furthermore, certain stents have hooks or barbs to prevent migration after placement that may also help to retain the implant in place. Thus, these issues warrant further study or analysis.

REFERENCES

Ahmed S, Shellock FG. Magnetic resonance imaging safety: implications for cardiovascular patients. Journal of Cardiovascular Magnetic Resonance 2001;3:171-181.

American Society for Testing and Materials (ASTM) Designation: F 2052. Standard test method for measurement of magnetically induced displacement force on passive implants in the magnetic resonance environment. In: Annual Book of ASTM Standards, Section 13, Medical Devices and Services, Volume 13.01 Medical Devices; Emergency Medical Services. West Conshohocken, PA, 2002, pp. 1576-1580.

Bueker A, et al. Real-time MR fluoroscopy for MR-guided iliac artery stent placement. J Magn Reson Imaging 2000;12:616-622.

Hennemeyer CT, Wicklow K, Feinberg DA, Derdeyn CP. In vitro evaluation of platinum Guglielmi detachable coils at 3-T with a porcine model: safety issues and artifacts. Radiology 2001;219:732-737.

Hug J, Nagel E, Bornstedt A, Schackenburg B, Oswald H, Fleck E. Coronary arterial stents: safety and artifacts during MR imaging. Radiology 2000;216:781-787.

Girard MJ, Hahn P, Saini S, Dawson SL, Goldberg MA, Mueller PR. Wallstent metallic biliary endoprosthesis: MR imaging characteristics. Radiology 1992;184:874-876.

Kiproff PM, Deeb DL, Contractor FM, Khoury MB. Magnetic resonance characteristics of the LGM vena cava filter: technical note. Cardiovasc Intervent Radiol 1991;14:254-255.

Leibman CE, Messersmith RN, Levin DN, et al. MR imaging of inferior vena caval filter: safety and artifacts. AJR Am J Roentgenol 1988;150:1174-1176.

Manke C, Nitz WR, Djavidani B, et al. MR imaging-guided stent placement in iliac arterial stenoses: A feasibility study. Radiology 2001;219:527-534.

Marshall MW, Teitelbaum GP, Kim HS, et al. Ferromagnetism and magnetic resonance artifacts of platinum embolization microcoils. Cardiovasc Intervent Radiol 1991;14:163-166.

Rutledge JM, Vick GW, Mullins CE, Grifka RG. Safety of magnetic resonance immediately following Palmaz stent implant: a report of three cases. Catheter Cardiovasc Interv 2001;53:519-523.

Shellock FG. Biomedical implants and devices: assessment of magnetic field interactions with a 3.0-Tesla MR system. J Magn Reson Imaging 2002;16:721-732.

Shellock FG. Pocket Guide to MR Procedures and Metallic Objects: Update 2001, Seventh Edition, Lippincott Williams & Wilkins Healthcare, Philadelphia, 2001.

Shellock FG. Magnetic Resonance Procedures: Health Effects and Safety. CRC Press, LLC, Boca Raton, FL, 2001.

Shellock FG, Detrick MS, Brant-Zawadski M. MR-compatibility of the Guglielmi detachable coils. Radiology 1997;203:568-570.

Shellock FG, Kanal E. Magnetic Resonance: Bioeffects, Safety, and Patient Management. Second Edition, Lippincott-Raven Press, New York, 1996.

Shellock FG, Morisoli S, Kanal E. MR procedures and biomedical implants, materials, and devices: 1993 update. Radiology 1993;189:587-599.

Shellock FG, Shellock VJ. Stents: Evaluation of MRI safety. Am J Roentgenol 1999;173:543-546.

Spuentrup E, et al. Magnetic resonance-guided coronary artery stent placement in a swine model. Circulation 2002;105:874-879.

Taal BG, Muller SH, Boot H, Koop W. Potential risks and artifacts of magnetic resonance imaging of self-expandable esophageal stents. Gastrointestinal Endoscopy 1997;46;424-429.

Teitelbaum GP, Bradley WG, Klein BD. MR imaging artifacts, ferromagnetism, and magnetic torque of intravascular filters, stents, and coils. Radiology 1988;166:657-664.

Teitelbaum GP, Lin MCW, Watanabe AT, et al. Ferromagnetism and MR imaging: safety of cartoid vascular clamps. AJNR Am J Neuroradiol 1990;11:267-272.

Teitelbaum GP, Ortega HV, Vinitski S, et al. Low artifact intravascular devices: MR imaging evaluation. Radiology 1988;168:713-719.

Teitelbaum GP, Raney M, Carvlin MJ, et al. Evaluation of ferromagnetism and magnetic resonance imaging artifacts of the Strecker tantalum vascular stent. Cardiovasc Intervent Radiol 1989;12:125-127.

Watanabe AT, Teitelbaum GP, Gomes AS, et al. MR imaging of the bird's nest filter. Radiology 1990;177:578-579.

COZMO PUMP, INSULIN PUMP

According to the User Manual (11/22/02) for the Cozmo™ Pump (Deltec, Inc., St. Paul, MN), which is a device used to administer insulin, the following is stated regarding Magnetic Resonance Imaging (MRI)- "Caution: Avoid strong electromagnetic fields, like those present with Magnetic Resonance Imaging (MRI) and direct x-ray, as they can affect how the pump works. If you cannot avoid them, you must take the pump off."

REFERENCE

http://www.delteccozmo.com/new/html

CRANIAL FLAP FIXATION CLAMPS AND SIMILAR DEVICES

After performing a craniotomy, bone flaps are typically fixed with wire, suture material, or small plates and screws. Problems related to cranial bone flap fixation after craniotomy are more common with the trend for performing smaller craniotomies that are frequently utilized for minimally invasive surgical procedures. The use of small plates and screws for fixation of cranial bone flaps has improved the overall attachment process and end result. However, this technique requires a considerable amount of time and expense compared to using wire and suture techniques.

In consideration of the various problems with bone flap fixation, a special metallic implant system, named the Craniofix (Aesculap, Inc., Central Valley, PA) was developed for fixation of cranial bone flaps after craniotomy. MR safety tests conducted to assess magnetic field interaction, heating, and artifacts indicated that the clamps used for the cranial bone flap fixation system present no risk to the patient in the MR environment of MR systems operating at 1.5-Tesla or less. Furthermore, the quality of the diagnostic MR images was acceptable, particularly for conventional spin echo or fast spin echo pulse sequences. Other devices used for similar applications have also been evaluated MR safety.

MR Safety at 3.0-Tesla and Cranial Fixation Implants and Devices.
Several different cranial or burr hole fixation implants and devices made from titanium or titanium alloy have been tested at 3.0-Tesla. These were found to be safe for patients undergoing MR procedures using MR systems operating at this field strength.

REFERENCES

Shellock FG. Pocket Guide to MR Procedures and Metallic Objects: Update 2001, Seventh Edition, Lippincott Williams & Wilkins Healthcare, Philadelphia, 2001.

Shellock FG. Magnetic Resonance Procedures: Health Effects and Safety. CRC Press, LLC, Boca Raton, FL, 2001.

Shellock FG. Biomedical implants and devices: assessment of magnetic field interactions with a 3.0-Tesla MR system. J Magn Reson Imaging 2002;16:721-732.

Shellock FG, Kanal E. Magnetic Resonance: Bioeffects, Safety, and Patient Management. Second Edition, Lippincott-Raven Press, New York, 1996.

Shellock FG, Shellock VJ. Evaluation of cranial flap fixation clamps for compatibility with MR imaging. Radiology 1998;207:822-825.

DENTAL IMPLANTS, DEVICES, AND MATERIALS

Many of the dental implants, devices, materials, and objects evaluated for ferromagnetic qualities exhibited measurable deflection forces, but only the ones with magnetically-activated components present a potential problem for patients during MR procedures (see **Magnetically-Activated Implants and Devices**). The other dental implants, devices, and materials are held in place with sufficient counter-forces to prevent them from causing problems for patients by being moved or dislodged by exposure to MR systems operating with static magnetic fields of 1.5-Tesla or less.

REFERENCES

Gegauff A, Laurell KA, Thavendrarajah A, et al. A potential MRI hazard: forces on dental magnet keepers. J Oral Rehabil 1990;17:403-410.

Lissac MI, Metrop D, Brugigrad, et al. Dental materials and magnetic resonance imaging. Invest Radiol 1991;26:40-45.

New PFJ, Rosen BR, Brady TJ, et al. Potential hazards and artifacts of ferromagnetic and nonferromagnetic surgical and dental materials and devices in nuclear magnetic resonance imaging. Radiology 1983;147:139-148.

Shellock FG. Pocket Guide to MR Procedures and Metallic Objects: Update 2001, Seventh Edition, Lippincott Williams & Wilkins Healthcare, Philadelphia, 2001.

Shellock FG. Magnetic Resonance Procedures: Health Effects and Safety. CRC Press, LLC, Boca Raton, FL, 2001.

Shellock FG. Ex vivo assessment of deflection forces and artifacts associated with high-field strength MRI of "mini-magnet" dental prostheses. Magn Reson Imaging 1989;7 (Suppl 1):38.

Shellock FG, Crues JV. High-field strength MR imaging and metallic biomedical implants: an *ex vivo* evaluation of deflection forces. Am J Roentgenol 1988;151:389-392.

Shellock FG, Kanal E. Magnetic Resonance: Bioeffects, Safety, and Patient Management. Second Edition, Lippincott-Raven Press, New York, 1996.

DIAPHRAGMS

Contraceptive diaphragms may have metallic rings that serve to maintain them in position during use. Thus, certain contraceptive diaphragms display positive magnetic field interactions in association with exposure to MR systems. However, MR examinations have been performed in patients with these devices without complaints or adverse sensations related to movement of the diaphragms. Furthermore, there is no danger of heating a contraceptive diaphragm during an MR procedure under conditions currently recommended by the United States Food and Drug Administration. Therefore, the presence of a diaphragm is not a contraindication for a patient undergoing an MR examination using an MR system operating at 1.5-Tesla or less.

REFERENCES

Shellock FG. Pocket Guide to MR Procedures and Metallic Objects: Update 2001, Seventh Edition, Lippincott Williams & Wilkins Healthcare, Philadelphia, 2001.

Shellock FG. Magnetic Resonance Procedures: Health Effects and Safety. CRC Press, LLC, Boca Raton, FL, 2001.

Shellock FG, Kanal E. Magnetic Resonance: Bioeffects, Safety, and Patient Management. Second Edition, Lippincott-Raven Press, New York, 1996.

ECG ELECTRODES

The use of MR-safe electrocardiogram (ECG) electrodes is strongly recommended to ensure patient safety and proper recording of the ECG in the MR environment. Accordingly, ECG electrodes have been specially developed for use during MR procedures to protect the patient from potentially hazardous conditions and to minimize MRI-related artifacts. **The List** provides a compilation of ECG electrodes that have been evaluated for MR-safety using MR systems operating with static magnetic fields up to 1.5-Tesla.

REFERENCES

Shellock FG. MRI and ECG electrodes. Signals, No. 29, Issue 1, pp. 10-14, 1999.

Shellock FG. Pocket Guide to MR Procedures and Metallic Objects: Update 2001, Seventh Edition, Lippincott Williams & Wilkins Healthcare, Philadelphia, 2001.

Shellock FG. Magnetic Resonance Procedures: Health Effects and Safety. CRC Press, LLC, Boca Raton, FL, 2001.

Shellock FG, Kanal E. Magnetic Resonance: Bioeffects, Safety, and Patient Management. Second Edition, Lippincott-Raven Press, New York, 1996.

ENTERRA THERAPY, GASTRIC ELECTRICAL STIMULATION (GES)

The Enterra Therapy, Gastric Electrical Stimulation (GES) System Medtronic (Minneapolis, MN), is a neurostimulation system indicated for treatment of patients with chronic, intractable (drug refractory) nausea and vomiting secondary to gastroparesis of diabetic or idiopathic etiology. Gastroparesis is a stomach disorder in which food moves through the stomach more slowly than normal. In some patients, this condition results in severe, chronic nausea and vomiting that cannot be adequately controlled by available drugs. These patients have difficulty eating and may require some form of tube feeding to ensure adequate nutrition. Enterra Therapy uses mild electrical pulses to stimulate the stomach. This electrical stimulation helps to control the symptoms associated with gastroparesis, including nausea and vomiting.

Enterra Therapy has been used successfully in people for several years. Worldwide clinical trials have been conducted in patients with severe symptoms of nausea and vomiting associated with gastroparesis. The results (Medtronic FDA Application H990014) showed that the therapy successfully reduced symptoms of nausea and vomiting in patients for whom drug treatments did not work. Patients also experienced improvements in other upper GI symptoms, solid food intake, and a reduction in hypoglycemic attacks, as well as significant improvements in health related quality of life.

The Enterra Therapy system is comprised of the following components: a neurostimulator, an implantable intramuscular lead, and an external programming system (i.e., *Medtronic ITREL 3 Model 7425G Neurostimulator* - The ITREL 3 is a battery powered implantable device that is commercially available in the U.S. *Medtronic Model 4301 Lead -*

The 4301 is a unipolar lead intended for use with an implantable ITREL neurostimulator. The 4301 lead connects directly to the ITREL 3 device. It is available in 35-cm and 50-cm lengths. Two leads are used in each patient. *Programmer System* - The programmer for the ITREL 3 system consists of the Model 7432 physician programmer and the Model 7457 MemoryMod Software Cartridge.)

Magnetic Resonance Imaging (MRI). Use of MRI may result in dislodgment or heating of the neurostimulator and/or the leads used for gastric electrical stimulation. The voltage induced through the lead and neurostimulator may cause uncomfortable "jolting" or "shocking" levels of stimulation. Patients should not be exposed to the electromagnetic fields that MRI produces.

[Information excerpted with permission from Medtronic, Minneapolis, MN; http://www.medtronic.com/neuro/enterra/]

ESSURE DEVICE

The Essure Device (Conceptus, San Carlos, CA) is a metallic implant developed for permanent female contraception. This implant is a dynamically expanding micro-coil that is placed in the proximal section of the fallopian tube using a non-incisional technique. Subsequently, the device elicits an intended benign tissue response, resulting in tissue in-growth into the device, anchoring it firmly into the fallopian tube. This benign tissue response is local, fibrotic, and occlusive in nature. Accordingly, the presence of this implant is intended to alter the function and architecture of the fallopian tube, resulting in permanent contraception. The Essure Device is composed of 316L stainless steel, platinum, iridium, nickel-titanium alloy, silver solder, and polyethylene terephthalate (PET) fibers. This implant has the following dimensions: inner coil length, 2.9 to 3.1-cm; outer coil diameter after deployment, 1.5 to 2.0-mm.

The MR safety assessment of this device involved testing for magnetic field interactions (1.5-Tesla), heating, induced electrical currents, and artifacts using previously described techniques. There were no magnetic field interactions, the highest temperature changes were $\leq +0.6°C$, and the induced electrical currents were minimal. Furthermore, artifacts should not create a substantial problem for diagnostic MR imaging unless the area of interest is in the exact same position as where this implant is located. Thus, the findings of this investigation indicated that it should be safe for patients with the Essure Device to undergo MR procedures using MR systems operating with static magnetic fields of 1.5-Tesla or less.

MR Safety at 3.0-Tesla and the Essure Device. The Essure Device was evaluated for MR safety at 3.0-Tesla and found to be safe for patients undergoing MR procedures at this static magnetic field strength.

REFERENCES

American Society for Testing and Materials (ASTM) Designation: F 2052. Standard test method for measurement of magnetically induced displacement force on passive implants in the magnetic resonance environment. In: Annual Book of ASTM Standards, Section 13, Medical Devices and Services, Volume 13.01 Medical Devices; Emergency Medical Services. West Conshohocken, PA, 2002, pp. 1576-1580.

Shellock FG. Biomedical implants and devices: assessment of magnetic field interactions with a 3.0-Tesla MR system. J Magn Reson Imaging 2002;16:721-732.

Shellock FG. New metallic implant used for permanent female contraception: evaluation of MR safety. Am J Roentgenol 2002;178:1513-1516.

FIBER-OPTIC PACING LEAD USED FOR PHOTONIC TEMPORARY PACEMAKER

Cardiac pacemakers are the most common electronically-activated, contraindicated implants found in patients referred for magnetic resonance (MR) imaging procedures. As such, this contraindication prevents MR examinations from being used in many patients that could benefit from this important diagnostic modality.

As previously stated, cardiac pacemakers present potential problems to patients undergoing MR procedures from several mechanisms. These problems may result in serious injuries or lethal consequences for patients. With specific regard to the cardiac pacing lead, there are concerns related to magnetic field-induced movement and substantial heating that may occur. In addition, the electrically-conductive lead may pick up electromagnetic interference from the MR system, which can impair the pacemaker's performance or produce a life-threatening situation for the patient. In consideration of the above, it would be desirable to have a cardiac pacing lead that could function safely in the MR environment.

Recently, new technology has been developed that involves the stimulation of the heart by means of a fiber-optic lead that replaces the standard metallic lead of a cardiac pacemaker. The fiber-optic cardiac pacing lead has specially-designed components that incorporate a low power, semiconductor laser to regulate the patient's heartbeat (Biophan Technologies, Inc., Rochester, NY). This innovation essentially eliminates possible dangers associated with having a conductive pacing wire in a patient undergoing an MR examination.

171

The fiber-optic cardiac pacing lead (Biophan Technologies, Inc., Rochester, NY) is intended for use via connection to the Temporary Photonic Pulse Generator (Model X-801, Biophan Technologies, Inc., Rochester, NY). The cardiac pacing lead is made from 200-um, fiber-optic cable. The distal end of the lead has two electrodes designed to stimulate the heart. Within the ring and tip portions of the lead are a power converter, resistor, and capacitor. Inside the ring electrode is a power converter that changes light energy into electrical energy for heart stimulation. The pulse generator for this device (Temporary Photonic Pulse Generator, Model X-801) produces a 1-millisecond pulse (variable from 0.1 to 30-msec), which drives a 150-milliwatt gallium-arsenide laser. The light pulse is connected to the distal end of the fiber-optic lead where it illuminates a band of six gallium-arsenide photo diodes. The diodes are electrically connected in series to produce a voltage pulse of 4 volts, which drive "tip" and "ring" electrodes of the lead to stimulate the heart.

The results of MR-safety tests conducted for the fiber-optic cardiac pacing lead indicated there were only minor magnetic field interactions associated with exposure to a 1.5-Tesla MR system. In addition, there was essentially no MRI-related heating for the fiber-optic cardiac pacing lead in association with MR imaging conducted at a whole-body-averaged SAR of 1.5 W/kg. Based on this information, it appears that this device will not present an additional hazard or risk to a patient undergoing an MR procedure under the conditions used for this evaluation.

By comparison, in an *in vitro* evaluation of 44 commercially-available pacemaker leads, Sommer et al. reported the maximum temperature change measured at the lead tip was 23.5°C above baseline in association with MR imaging performed at 0.5-Tesla and a whole body averaged SAR of 1.3 W/kg for 10-min. Additionally, Achenbach et al. reported a peak temperature change of 63.1°C measured for a temporary pacing electrode that occurred within 90-sec. of MR imaging (the SAR was not reported). Furthermore, MR imaging at 1.5-T and an SAR of 3.0 W/kg has been shown to cause severe necrosis in the mucous membranes of dogs with transesophageal cardiac pacing leads *in situ*.

Previous attempts to develop an MR-safe cardiac pacing lead have been reported by Jerzewski et al. and Hofman et al. Jerzewski et al. modified a commercially available pacing catheter by leaving out the stainless steel catheter shaft braiding and using 90% platinum/10% iridium electrodes at the tip. The electrical wiring was made of nearly pure copper of reduced conductance. Thus, this pacing device still involved the use of conductive metallic materials. MR imaging was performed on rabbits with this temporary pacing lead at 1.5-Tesla. These laboratory animals were primarily monitored for extra-systolic cardiac contractions. Jerzewski et al. did not assess magnetic field interactions or heating for the cardiac pacing lead but rather evaluated imaging artifacts (which is not an MR safety issue). Before it can be claimed that this cardiac pacing system is safe for MR imaging, the issue of tissue heating around this pacing catheter should be addressed.

In an investigation performed in laboratory dogs, Hofman et al. studied the feasibility of transesophageal cardiac pacing during MR imaging at 1.5-Tesla and concluded that tissue around the catheter tip may become heated. As previously stated, severe necrosis in the mucous membranes of dogs with transesophageal pacing leads *in situ* has been found during MR imaging in combination with exposure to high levels of RF energy. Therefore, while it may be possible to perform transesophageal atrial pacing during MR imaging, it requires relatively low levels of RF energy, which is likely to be impractical for most anticipated clinical uses of this technology.

In view of the previously published reports on cardiac pacemakers and, more specifically, pacing leads, the information for the fiber-optic cardiac pacing lead is particularly compelling from an MR safety viewpoint. This unique technology may be applied to other devices that require leads but are known to present potential hazards to patients undergoing MR procedures (e.g., neurostimulation systems).

One final consideration for the fiber-optic cardiac pacing lead is the issue of electromagnetic interference (e.g., inappropriate or rapid pacing due to pulsed gradient magnetic fields and/or pulsed RF from the operating MR system, with the pacing lead acting as an antenna). From a theoretical consideration, because of the "fiber-optic nature" of this specially-designed pacing lead, there should be no problems related to this device acting like an antenna during MR procedures since it should be immune from such problems. Other devices (pulse oximeters, cutaneous blood flow monitors, electrocardiagraphic systems, thermometry systems, etc.) have likewise incorporated fiber-optic interfaces to the patients to successfully prevent EMI-related problems from occurring in the MR environment. However, additional investigation directed at addressing the EMI aspects of the fiber-optic cardiac pacing lead in the MR setting is warranted.

REFERENCES

Achenbach S, et al. Effects of magnetic resonance imaging on cardiac pacemakers and electrodes. Am Heart J 1997;134:467-473.

American Society for Testing and Materials (ASTM) Designation: F 2052. Standard test method for measurement of magnetically induced displacement force on passive implants in the magnetic resonance environment. In: Annual Book of ASTM Standards, Volume 13.01 Medical Devices; Emergency Medical Services. West Conshohocken, PA, 2002, pp. 1576-1580.

Duru F, Luechinger R, Candinas R. MR imaging in patients with cardiac pacemakers. Radiology 2001;219:856-858.

Erlebacher JA, Cahill PT, Pannizzo F, et al. Effect of magnetic resonance imaging on DDD pacemakers. Am J Cardiol 1986; 57: 437-440.

ECRI, Health Devices Alert, A new MRI complication? Health Devices Alert, May 27, 1988, pp. 1.

Fetter J, Aram G, Holmes DR, et al. The effects of nuclear magnetic resonance imagers on external and implantable pulse generators. Pacing Clin Electrophysiol 1984; 7:720-727.

Gimbel JR. Implantable pacemaker and defibrillator safety in the MR environment: new thoughts for the new millennium. In, 2001 Syllabus, Special Cross-Specialty Categorical Course in Diagnostic Radiology: Practical MR Safety Considerations for Physicians, Physicists, and Technologists, Radiological Society of North America, pp. 69-76, 2001.

Greatbatch W, Miller V, Shellock FG. Magnetic resonance safety testing of a newly-developed, fiber-optic cardiac pacing lead. J Magn Reson Imaging 2002;16:97-103.

Hayes DL, Holmes DR, Gray JE. Effect of a 1.5 Tesla magnetic resonance imaging scanner on implanted permanent pacemakers. J Am Coll Cardiol 1987;10:782-786.

Hofman MB, de Cock CC, van der Linden JC, et al. Transesophageal cardiac pacing during magnetic resonance imaging: feasibility and safety considerations. Magn Reson Med 1996;35:413-422.

Holmes DR, Hayes DL, Gray JE, et al. The effects of magnetic resonance imaging on implantable pulse generators. Pacing Clin Electrophysiol 1986;9:360-370.

Jerewski A, et al. Development of an MRI-compatible catheter for pacing the heart: initial in vitro and in vivo results. J Magn Reson Imaging 1996; 6:948-949.

Konings MK, et al. Heating around intravascular guidewires by resonating RF waves. J Magn Reson Imaging 2000;12:79-85.

Ladd ME, Quick HH. Reduction of resonant RF heating in intravascular catheters using coaxial chokes. Magn Reson Med 2000;615-619.

Pavlicek W, Geisinger M, Castle L, et al. The effects of nuclear magnetic resonance on patients with cardiac pacemakers. Radiology 1983;147:149-153.

Rezai AR, Finelli D, Nyenhuis JA, Hrdlick G, Tkach J, Ruggieri P, Stypulkowski PH, Sharan A, Shellock FG. Neurostimulator for deep brain stimulation: Ex vivo evaluation of MRI-related heating at 1.5-Tesla. J Magn Reson Imaging 2002;15:241-250.

Shellock et al. Cardiac pacemakers and implantable cardiac defibrillators are unaffected by operation of an extremity MR system. AJR American Journal of Roentgenology. 1999;172:165-172.

Sommer et al. MR Imaging and cardiac pacemakers: in vitro evaluation and in vivo studies in 51 patients at 0.5-T. Radiology 2000;215:869-879.

Vahlhaus C, Sommer T, Lewalter T, et al. Interference with cardiac pacemakers by magnetic resonance imaging: are there irreversible changes at 0.5 Tesla? Pacing Clin Electrophysiol 2001;24(4 Pt 1):489-495.

Zaremba L. FDA guidance for MR system safety and patient exposures: current status and future considerations. In: Magnetic resonance Procedures: Health Effects and Safety. CRC Press, Boca Ration, FL, pp. 183-196, 2001.

FOLEY CATHETERS WITH AND WITHOUT TEMPERATURE SENSORS

Most Foley catheters used to drain the bladder have no or few metallic components. Accordingly, these devices are safe for patients undergoing MR procedures. However, certain Foley catheters have temperature sensors to permit recording of the temperature of the urine in the bladder, which is a sensitive means of determining "deep" body or core temperature. This type of Foley catheter typically has a thermistor or thermocouple located on or near the tip of the device and a wire that runs the length of the catheter to a connector that plugs into a temperature monitor. Of note is that a Foley catheter with a temperature sensor should never be connected to the temperature monitor during the MR procedure because this equipment has not been shown to be safe or compatible in the MR environment.

Several Foley catheters with temperature sensors have been evaluated for safety in the MR environment by determining magnetic field interactions, artifacts, and heating. In general, the findings of this assessment indicated that it would be safe to perform MR procedures in patients with certain Foley catheters with temperature sensors as long as highly-specific recommendations are followed.

Similar to any device with a wire component, the position of the wire of the Foley catheter with a temperature sensor has an important effect on the amount of heating that develops during an MR procedure. Accordingly, a Foley catheter with a temperature sensor must be positioned in a straight configuration (without a loop) down the center of the MR system and away from RF coils, to prevent possible excessive heating associated with an MR procedure. Furthermore, only conventional pulse sequences should be used (e.g., no echo planar techniques, magnetization transfer contrast, etc.) while imaging with an MR system with a static magnetic field of 1.5-Tesla or less. Additional recommendations for the Foley catheters with temperature sensors that have undergone MR safety testing include, the following:

(1) If the Foley catheter with a temperature sensor has a removable catheter connector cable, it should be disconnected prior to the MR procedure.

(2) Remove all electrically conductive material from the bore of the MR system that is not required for the procedure (i.e., unused surface coils, cables, etc.).

(3) Keep electrically conductive material that must remain in the bore of the MR system from directly contacting the patient by placing thermal and/or electrical insulation (including air) between the conductive material and the patient.

(4) Position the Foley catheter with a temperature sensor in a straight configuration to prevent cross points and conductive loops.

(5) MR imaging should be performed using MR systems with static magnetic fields of 1.5-Tesla or less and conventional. Pulse sequences, techniques (e.g., echo planar techniques) or conditions that produce exposures to high levels of RF energy (i.e., exceeding a whole body averaged specific absorption rate of 1.0 W/kg) or exposure to gradient fields that exceed 20-T/second, or any other unconventional MR technique should be avoided.

(6) Monitor the patient continuously using a verbal means (e.g., intercom). If the patient reports feeling warm or hot in association with the presence of the Foley catheter with a temperature sensor, discontinue the MR procedure immediately.

In addition to the above, the **Guidelines to Prevent Excessive Heating and Burns Associated with Magnetic Resonance Procedures** should be considered and implemented, as needed.

REFERENCES

Dempsey MF, Condon B, Hadley DM. Investigation of the factors responsible for burns during MRI. J Magn Reson Imaging 2001;13:627-631.

Shellock FG. Pocket Guide to MR Procedures and Metallic Objects: Update 2001, Seventh Edition, Lippincott Williams & Wilkins Healthcare, Philadelphia, 2001.

Shellock FG. Magnetic Resonance Procedures: Health Effects and Safety. CRC Press, LLC, Boca Raton, FL, 2001.

Shellock FG, Kanal E. Magnetic Resonance: Bioeffects, Safety, and Patient Management. Second Edition, Lippincott-Raven Press, New York, 1996.

FREEHAND® SYSTEM IMPLANTABLE FUNCTIONAL NEUROSTIMULATOR (FNS)

The FREEHAND Implantable Functional Neurostimulator System (FNS)(NeuroControl, Cleveland, OH) is a radiofrequency (RF) powered motor control neuroprosthesis that consists of both implanted and external components. It utilizes low levels of electrical current to stimulate the peripheral nerves that innervate muscles in the forearm and hand providing functional hand grasp patterns. The NeuroControl FREEHAND System consists of the following subsystems:

- The **Implanted Components** include the Implantable Receiver-Stimulator, Epimysial and Intramuscular Electrodes, and Connectors (sleeves and springs).

- The **External Components** include the External Controller, Transmit Coil, Shoulder Position Sensor, Battery Charger, and Remote On/Off Switch.

- The **Programming System** consists of the Pre-configured Personal Computer loaded with the Programming Interface Software, the Interface Module, and a Serial Cable.

- The **Surgical Implementation Components** include the **Electrode Positioning Kit** (the Surgical Stimulator, Epimysial Probe, Anode Plate, and Clip Lead) and

- The **Intramuscular Electrode Insertion Tool Kit** (the Intramuscular Probe and the Cannula). The Intramuscular Electrode is preloaded in the Lead Carrier with Carrier Cover.

179

The NeuroControl FREEHAND System is intended to improve a patient's ability to grasp, hold, and release objects. It is indicated for use in patients who:

- are tetraplegic due to C5 or C6 level spinal cord injury (ASIA Classification)

- have adequate functional range of motion of the upper extremity

- have intact lower motor neuron innervation of the forearm and hand musculature and are skeletally mature

WARNINGS

The NeuroControl FREEHAND System may only be prescribed, implanted, or adjusted by clinicians who have been trained and certified in its implementation and use.

- **Monopolar Electrosurgical Instruments** should not be used on the implanted upper extremity. These tools could damage the Implantable Receiver-Stimulator. Bipolar electrosurgical instruments can be used safely in coagulating mode.

- **Electrostatic Discharge (ESD)** may damage the FREEHAND Implantable Receiver-Stimulator during intraoperative handling. Handle only as instructed in the Clinician's Manual.

- **Neuromuscular Blocking Agents** (long-acting) should not be administered during implantation surgery. These agents render nerves unresponsive to electrical stimulation and may compromise the proper installation of the device.

PRECAUTIONS

- **Surface Stimulation:** Electrical surface stimulation (muscle stimulator, EMG, or TENS) should be used with caution on or near the implanted upper extremity as it may damage the system. Contact NeuroControl prior to applying surface stimulation.

- **X-rays, mammography, ultrasound:** X-ray imaging (e.g., CT or mammography), and ultrasound have not been reported to affect the function of the Implantable Receiver-Stimulator or Epimysial Electrodes. However, the implantable components may obscure the view of other anatomic structures.

- **MRI:** Testing of the FREEHAND System in a 1.5-Tesla scanner with a maximum spatial gradient of 450 gauss/cm or less, exposed to a whole body averaged Specific Absorption Rate (SAR) of 1.1 W/kg, for a 30 minute duration resulted in localized temperature rises no more than 2.7°C in a gel phantom (without blood flow) and translational force less than that of a 3-gram mass and torque of 0.063 N-cm (significantly less than produced by the weight of the device). A patient with a FREEHAND System may undergo an MR procedure using a shielded or unshielded MR system with a static magnetic field of 1.5-Tesla <u>only</u> and a maximum spatial gradient of 450 gauss/cm or less. The implantable components may obscure the view of other nearby anatomic structures. Artifact size is dependent on a variety of factors including the type of pulse sequence used for imaging (e.g. larger for gradient echo pulse sequences and smaller for spin echo and fast spin echo pulse sequences), the direction of the frequency encoding, and the size of the field of view used for imaging. The use of non-standard scanning modes to minimize image artifact or improve visibility should be applied with caution and with the Specific Absorption Rate (SAR) not to exceed an average of 1.1 W/kg and gradient magnetic fields no greater than 20 Tesla/sec. The use of Transmit Coils other than the scanner's Body Coil or a Head Coil is prohibited. Testing of the function of each electrode should be conducted prior to MRI scanning to ensure no leads are broken. Do not expose patients to MRI if any lead is broken or if integrity cannot be established as excessive heating may result in a broken lead. The external components of the FREEHAND System must be removed prior to MRI scanning. Patients must be continuously observed during the MR procedure and instructed to report any unusual sensations (e.g., warming, burning, or neuromuscular stimulation). Scanning must be discontinued immediately if any unusual sensation occurs. Contact NeuroControl Corporation for additional information.

- **Antibiotic prophylaxis:** Standard antibiotic prophylaxis for patients with an implant should be utilized to protect the patient when invasive procedures (e.g., oral surgery) are performed.

- **Ultrasound:** Therapeutic ultrasound should not be performed over the area of the Implantable Receiver-Stimulator or Epimysial Electrodes as it may damage the system.

- **Diathermy:** Therapeutic diathermy should not be used in patients with the NeuroControl FREEHAND System as it may damage the system.

- **Therapeutic Radiation:** The electronic components in the FREE-HAND System may be damaged by therapeutic ionizing radiation. The damage that occurs may not be immediately detectable. Any changes in sensation or muscle contraction should be reported to the physician. If the changes are painful or uncomfortable the patient should stop using the FREEHAND System pending review by the clinician.

- **Invasive Procedures:** To avoid unintentional damage to implanted components, invasive procedures such as drawing blood or administering an intravenous infusion should be avoided on the implanted arm or in the area of the Implantable Receiver-Stimulator or near sites of the Epimysial Electrodes.

- **Serum CPK levels:** Exercise and muscle activity are known to cause changes in certain blood enzymes measured by standard laboratory and clinical tests, such as serum CPK. Exercise, whether volitional or induced by electrical stimulation, may produce elevated serum CPK levels. If a FREEHAND System patient has elevated CPK, fractionation is indicated to differentiate between CK-MM from skeletal muscle and CK-MB from cardiac muscle that could be the result of cardiac injury.

- **Drug Interactions:** Muscle inhibitors and muscle relaxants may affect the strength of muscle contraction achieved using the FREE-HAND System. It is recommended that these medications be stabilized prior to implementing the FREEHAND System so that muscle response to electrical stimulation can be accurately evaluated.

- **Electrostatic Discharge (ESD)** exposure can cause loss of current amplitude programmability which also produces stimulus current amplitude higher than default. This does not result in any increased safety concerns from stimulation, has not caused compromise in hand function, and device operation will remain stable in this mode of operation. If a patient indicates that an increase in grasp strength is suddenly observed, this malfunction mode should be considered and evaluated. Changes in hand grasp can be managed by reprogramming grasp patterns, usually lowering stimulus pulse widths.

- **Pacemaker Warning Areas:** Patients should avoid areas posted with a warning to persons who have an implanted pacemaker. Contact NeuroControl Corporation for additional information.

- **Studies have not been conducted** on the use of the NeuroControl FREEHAND System in patients with the following conditions:

- children who are skeletally immature (usually males < 16 years, females < 15 years)

- prior history of a major chronic systemic infection or other illness that would increase the risk of surgery

- poorly controlled autonomic dysreflexia

- seizures and balance disorders

- pregnancy

Risks and benefits in patients with any of these conditions should be carefully evaluated before using the FREEHAND System.

- **Safety critical tasks:** Patients should be advised to avoid performing tasks which may be critical to their safety, e.g., controlling an automobile (throttle or brake), handling an object that could injure the patient (scald or burn), etc.

- The patient should be advised to avoid the use of a compression cuff for measuring blood pressures on the arm in which the FREEHAND System is implanted.

- **Post-operatively**, the patient should be advised to regularly check the condition of his or her skin across the hand, across the volar and dorsal aspects of the forearm, and across the chest where the FREEHAND System Receiver-Stimulator, leads and Electrodes are located for signs of redness, swelling, or breakdown. If skin breakdown becomes apparent, patients should contact their clinician immediately. The clinician should treat the infection, taking into consideration the extra risk presented by the presence of the implanted materials.

- **Keep it dry:** The user should avoid getting the external components, cables, and attachments of the FREEHAND System wet.

- The patient and caregiver should be advised to **inspect the cables and connectors** regularly for fraying or damage and replace components when necessary.

[Excerpted with permission from the Package Insert, FREEHAND System, NeuroControl Corporation, Cleveland, OH]

GUIDEWIRES

Advances in interventional MR procedures have resulted in the need for development of MR-safe and MR-compatible guidewires used for endovascular therapy, drainage procedures, and other similar applications. Conventional guidewires are made from stainless steel or Nitinol, materials known to be conductive. Radiofrequency fields used for MR procedures may induce substantial electrical currents in conductive guidewires, leading to excessive temperature increases and potential injuries.

Liu et al. studied the theoretical and experimental aspects of RF heating resonance phenomenon of an endovascular guidewire. A Nitinol-based guidewire (Terumo, Tokyo, Japan) was inserted into a vessel phantom and imaged using 1.5- and 0.2-Tesla MR systems with continuous temperature monitoring at the guidewire tip. The guidwire was deployed in the phantom in a "straight" manner. The heating effects due to different experimental conditions were examined. A model was developed for the resonant current and the associated electric field produced by the guidewire acting as an antenna. Temperature increases of up to 17°C were measured while imaging the guidewire at an off-center position in the 1.5-Tesla MR system. Power absorption produced by the resonating wire decreased as the repetition time was increased. No temperature rise was measured during MRI performed using the 0.2-Tesla MR system. Thus, considering the potential utility of low-field, open MR systems for MR-guided endovascular interventions, it is important to be aware of the safety of such applications for metallic guidewires and the potential hazards associated with using a guidewire with a high-field-strength MR system.

An investigation conducted by Konings et al. examined the unwanted radiofrequency (RF) heating of an endovascular guidewire frequently used in interventional MR procedures. A Terumo guidewire was partly immersed in an oblong saline bath to simulate an endovascular intervention. The temperature rise of the guidewire tip during MR imaging was measured using a Luxtron fluoptic thermometry system. Starting from

184

a baseline level of 26°C, the tip of the guidewire reached temperatures up to 74°C after 30 seconds of scanning using a 1.5-Tesla MR system. Touching the guidewire caused sudden heating at the point of contact, which in one instance produced a skin burn.

The excessive heating of a linear conductor, like the guidewire that underwent evaluation, could only be explained by resonating RF waves. According to Konings et al., the capricious dependencies of this resonance phenomenon on environmental factors have severe consequences for predictability and safety guidelines of interventional MR procedures involving guidewires.

REFERENCES

Konings MK, Bartels LW, Smits HJ, Bakker CJ. Heating around intravascular guidewires by resonating RF waves. J Magn Reson Imaging 2000;12:79-85.

Lin C-Y, Farahani K, Lu DSK, Shellock FG. Safety of MRI-guided endovascular guidewire applications. Proceedings of the International Society of Magnetic Resonance Imaging, 1999;7:1015.

Shellock FG. Magnetic Resonance Procedures: Health Effects and Safety. CRC Press, LLC, Boca Raton, FL, 2001.

Shellock FG. Radiofrequency-induced heating during MR procedures: A review. Journal of Magnetic Resonance Imaging. 2000;12: 30-36.

HALO VESTS AND CERVICAL FIXATON DEVICES

Halo vests and cervical fixation devices may be constructed from ferro-magnetic, nonferromagnetic, or a combination of metallic components and other materials. Although some commercially available halo vests or cervical fixation devices are composed entirely of nonferromagnetic materials, there is a theoretical hazard of inducing electrical current in the ring portion of any halo or cervical fixation device made from conductive materials.

Additionally, there is a potential for the patient's tissue to be involved in part of this current loop, such that the possibility of inducing a burn or electrical injury exists. The induced current within a ring or conductive loop is of additional concern because of eddy current induction and the potential for image degradation. Adjusting the phase encoding direction of the pulse sequence so that it is parallel to the axis of the halo vest may reduce artifacts associated with the production of eddy currents during MR imaging.

At present, there are no reports of injuries associated with MR procedures performed in patients with halo vests or cervical fixation devices. However, one incident of "electrical arcing" without injury was reported in the 1988 Society for Magnetic Resonance Imaging Safety Survey (Phase I Study; Kanal E, personal communication, 1988). Because of safety and image quality issues, MR procedures should only be performed on patients with specially designed halo vests or cervical fixation devices made from nonferromagnetic and nonconductive materials that have been demonstrated to create little or no interaction with the electromagnetic fields generated by MR systems.

186

There have been anecdotal reports of patients with halo vests or cervical fixation devices experiencing sensations of heat during MR imaging procedures. This has also been presumed to be a problem for certain stereotactic headframes used in the MR environment. However, MR safety testing experiments demonstrated that there was no heating for at least one such device indicated in The List (Unpublished Observations, F.G. Shellock, 1996).

A study was conducted to evaluate the possible heating of halo vests and cervical fixation devices during MR imaging performed at 1.5-Tesla MR system and various pulse sequences typically used to image the cervical spine. The data indicated that no substantial heating was detected. Of interest is that there appeared to be subtle motions of the halo ring associated with the use of a magnetization transfer contrast (MTC) pulse sequence, as shown by recordings obtained using a motion sensitive, laser-Doppler flow monitor.

Apparently, the specific imaging parameters used for the MTC pulse sequence produced sufficient vibration of the halo ring to create the sensation of heating. These rapid vibrations may have been felt by the subject and interpreted as a "heating" sensation. This is likely to occur when the frequency and/or amount of vibration is at a certain level that stimulates nerve receptors located in the subcutaneous region that detect sensations of pain and temperature changes.

The aforementioned is merely a hypothesis based on the available experimental data and requires further investigation to substantiate this theory. However, additional support for this premise comes from a report by Hartwell and Shellock. In this case, a halo ring and vest (removed from a patient who complained of severe "burning" in a front skull pin during MR imaging) was evaluated for heating and other potential problems associated with MR imaging. This work was performed in conjunction with the neurosurgeon who applied the device to the patient. The halo ring and vest were connected similar to the manner used on the patient. A fluid-filled Plexiglas phantom was placed within the vest. The device was then placed within a 1.5-Tesla MR system and imaging was performed using the same parameters that were associated with the "burning" sensation experienced by the patient. The neurosurgeon remained within the MR system to visually observe cervical fixation device and to maintain physical contact (i.e., touching the skull pins and other components) with it during the MR procedure.

No perceivable temperature change was noted for any of the metallic components during MR imaging. However, the metallic components of the cervical fixation device (e.g., halo ring, vertical supports, vest bolts, etc.) vibrated substantially during MR imaging. Furthermore, when the skull pins were held firmly during MR imaging, there was a so-called "drilling" sensation, which could be interpreted as a "burning" effect. Nevertheless, the skull pins remained cool to the touch throughout the MR procedure.

In consideration of the above information, it is inadvisable to permit patients with certain cervical fixation devices to undergo MR procedures using an MTC pulse sequence until this problem can be further characterized to avoid unwanted patient responses, regardless of the lack of safety concern related to excessive heating. Other comparable pulse sequences should likewise be avoided when performing MR imaging of patients with halo vests and cervical fixation devices until the precise cause of this problem is determined and fully characterized. Additionally, all instructions for use and patient application information provided by halo vest and cervical fixation device manufacturers should be carefully followed.

REFERENCES

Ballock RT, Hajed PC, Byrne TP, et al. The quality of magnetic resonance imaging, as affected by the composition of the halo orthosis. J Bone Joint Surg 1989;71-A:431-434.

Clayman DA, Murakami ME, Vines FS. Compatibility of cervical spine braces with MR imaging. A study of nine nonferrous devices. AJNR Am J Neuroradiol 1990;11:385-390.

Hartwell CG, Shellock FG. MRI of cervical fixation devices: Sensation of heating caused by vibration of metallic components. J Magn Reson Imaging 1997;7:771.

Hua J, Fox RA. Magnetic resonance imaging of patients wearing a surgical traction halo. J Magn Reson Imaging 1996;:1:264-267.

Malko JA, Hoffman JC, Jarrett PJ. Eddy-current-induced artifacts caused by an "MR-compatible" halo device. Radiology 1989;173:563-564.

Shellock FG. MR imaging and cervical fixation devices: assessment of ferromagnetism, heating, and artifacts. Magn Reson Imaging 1996;14:1093-1098.

Shellock FG. Pocket Guide to MR Procedures and Metallic Objects: Update 2001, Seventh Edition, Lippincott Williams & Wilkins Healthcare, Philadelphia, 2001.

Shellock FG. Magnetic Resonance Procedures: Health Effects and Safety. CRC Press, LLC, Boca Raton, FL, 2001.

Shellock FG, Kanal E. Magnetic Resonance: Bioeffects, Safety, and Patient Management. Second Edition, Lippincott-Raven Press, New York, 1996.

Shellock FG, Slimp G. Halo vest for cervical spine fixation during MR imaging. AJR Am J Roentgenol 1990;154:631-632.

HEARING AIDS AND OTHER HEARING SYSTEMS

External hearing aids are included in the category of electronically-activated implants or devices that may be found in patients referred for MR procedures. Exposure to the magnetic fields used for MR examinations can easily damage these devices. Therefore, a patient or other individual with an external hearing aid must not enter the MR environment due to the possible risk of damage to the device. Fortunately, to prevent damage, an external hearing aid can be readily identified and removed from the patient or individual prior to permitting entrance to the MR environment.

Other hearing devices exist that have external components as well as pieces that are surgically implanted in the middle ear. Hearing devices with external and internal components may be especially problematic for patients and individuals in the MR environment. Typically, these devices are used to treat patients with moderate to severe sensorineural hearing loss. The SOUNDTEC Direct Drive Hearing System (SOUNDTEC, Inc., Oklahoma City, OK) has an external component that changes sound into an electronic signal that is sent to an implanted magnet attached to the middle ear bones. This causes the middle ear bones to vibrate, sending sound to the brain. Because the strong magnetic field of an MR system may affect this device, a patient with the SOUNDTEC Direct Drive Hearing System is not allowed to undergo an MR procedure until the device has been surgically removed. Similarly, patients with the Vibrant Soundbridge (Symphonix Devices, Inc., San Jose, CA), which is also a hearing device with an implanted magnetic component, may not have MR procedures of any type. Furthermore, patients and individuals with these particular hearing devices are not allowed into the MR environment because of possibly damaging the components.

REFERENCES

Shellock FG. Pocket Guide to MR Procedures and Metallic Objects: Update 2001, Seventh Edition, Lippincott Williams & Wilkins Healthcare, Philadelphia, 2001.

Shellock FG. Magnetic Resonance Procedures: Health Effects and Safety. CRC Press, LLC, Boca Raton, FL, 2001.

Shellock FG, Kanal E. Magnetic Resonance: Bioeffects, Safety, and Patient Management. Second Edition, Lippincott-Raven Press, New York, 1996.

HEART VALVE PROSTHESES AND ANNULOPLASTY RINGS

Many heart valve prostheses and annuloplasty rings have been evaluated for MR safety, especially with regard to the presence of magnetic field interactions associated with exposure to MR systems operating at field strengths of as high as 3.0-Tesla. Of these, the majority displayed measurable yet relatively minor magnetic field interactions. That is, because the actual attractive forces exerted on the heart valve prostheses and annuloplasty rings were minimal compared to the force exerted by the beating heart (i.e., approximately 7.2 N), an MR procedure is not considered to be hazardous for a patient that has any of the heart valve prostheses or annuloplasty rings that have been tested, to date. This recommendation includes the Starr-Edwards Model Pre-6000 heart valve prosthesis previously suggested to be a potential risk for a patient undergoing an MR procedure. With respect to clinical MR procedures, there has never been a report of a patient incident or injury related to the presence of a heart valve prosthesis or annuloplasty ring.

Condon and Hadley reported the theoretical possibility of a previously unconsidered electromagnetic interaction with heart valves that contain metallic disks or leaflets. Basically, any metal (not just ferromagnetic metals) moving through a magnetic field will develop a magnetic field that opposes the original magnetic field. This phenomenon is referred to as the "Lenz effect". In theory, "resistive pressure" may develop with the potential to inhibit both the opening and closing aspects of the mechanical heart valve prosthesis. The Lenz effect is proportional to the strength of the static magnetic field. Accordingly, there may be problems for patients with heart valves that have metal leaflets undergoing MR procedures on MR systems greater than 1.5-Tesla, although this has never been demonstrated nor reported.

MR Safety at 3.0-Tesla and Heart Valve Prostheses and Annuloplasty Rings. Findings obtained at 3.0-Tesla for various heart valve prostheses and annuloplasty rings that underwent testing indicated that one annuloplasty ring (Carpentier-Edwards Physio Annuloplasty Ring, Mitral Model 4450, Edwards Lifesciences, Irvine, CA) showed relatively minor magnetic field interactions. Therefore, similar to heart valves prostheses and annuloplasty rings tested 1.5-Tesla, because the actual attractive forces exerted on these implants are deemed minimal compared to the force exerted by the beating heart (i.e., approximately 7.2 N), MR procedures at 3.0-Tesla or less are not considered to be hazardous for patients or individuals that have these devices.

Additional heart valves and annuloplasty rings from the Medtronic Heart Valve Division have undergone MR safety testing at 3.0-Tesla (work conducted by E. Kanal for Medtronic Heart Valve Division, Medtronic, Inc., Minneapolis, MN). These implants were tested for magnetic field interactions and artifacts using a shielded, 3.0-Tesla MR system. According to information provided by the manufacturer (Medtronic Heart Valves, Technical Service, Medtronic, Inc., Minneapolis, MN; Personnel Communication, Kathryn M. Bayer, Senior Technical Consultant,), these specific products are safe for patients undergoing MR procedures using scanners operating up to 3.0-Tesla. No adverse effects have been experienced with MR imaging (even when imaging close to the devices) using an MR system operating up to 3.0-Tesla.

REFERENCES

Ahmed S, Shellock FG. Magnetic resonance imaging safety: implications for cardiovascular patients. Journal of Cardiovascular Magnetic Resonance 2001;3:171-181.

Condon B, Hadley DM. Potential MR hazard to patients with metallic heart valves: the Lenz effect. Journal of Magnetic Resonance Imaging 2000;12:171-176.

Edwards, M-B, Taylor KM, Shellock FG. Prosthetic heart valves: evaluation of magnetic field interactions, heating, and artifacts at 1.5-Tesla. Journal of Magnetic Resonance Imaging. 2000;12:363-369.

Frank H, Buxbaum P, Huber L, et al. *In vitro* behavior of mechanical heart valves in 1.5-T superconducting magnet. Eur J Radiol 1992;2:555-558.

Hassler M, Le Bas JF, Wolf JE, et al. Effects of magnetic fields used in MRI on 15 prosthetic heart valves. J Radiol 1986;67:661-666.

Medtronic Heart Valves, Medtronic, Inc., Minneapolis, MN, Permission to publish 3-Tesla MR testing information for Medtronic Heart Valves provided by Kathryn M. Bayer, Senior Technical Consultant, Medtronic Heart Valves, Technical Service.

Pruefer D, et al. In vitro investigation of prosthetic heart valves in magnetic resonance imaging: evaluation of potential hazards. J Heart Valve Disease 2001;10:410-414.

Randall PA, Kohman LJ, Scalzetti EM, et al. Magnetic resonance imaging of prosthetic cardiac valves *in vitro* and *in vivo*. Am J Cardiol 1988;62:973-976.

Shellock FG. Magnetic Resonance Procedures: Health Effects and Safety. CRC Press, LLC, Boca Raton, FL, 2001.

Shellock FG. Pocket Guide to MR Procedures and Metallic Objects: Update 2001, Seventh Edition, Lippincott Williams & Wilkins Healthcare, Philadelphia, 2001.

Shellock FG. Biomedical implants and devices: assessment of magnetic field interactions with a 3.0-Tesla MR system. J Magn Reson Imaging 2002;16:721-732.

Shellock FG. Prosthetic heart valves and annuloplasty rings: assessment of magnetic field interactions, heating, and artifacts at 1.5-Tesla. Journal of Cardiovascular Magnetic Resonance 2001;3:159-169.

Shellock FG, Morisoli SM. Ex vivo evaluation of ferromagnetism, heating, and artifacts for heart valve prostheses exposed to a 1.5-Tesla MR system. J Magn Reson Imaging 1994;4:756-758.

Shellock FG, Shellock VJ. MRI Safety of cardiovascular implants: evaluation of ferromagnetism, heating, and artifacts. Radiology 2000;214:P19H.

Soulen RL. Magnetic resonance imaging of prosthetic heart valves [Letter]. Radiology 1986;158:279.

Soulen RL, Budinger TF, Higgins CB. Magnetic resonance imaging of prosthetic heart valves. Radiology 1985;154:705-707.

HEMOSTATIC CLIPS, OTHER CLIPS, FASTENERS, AND STAPLES

To date, the various hemostatic vascular clips, other types of clips, fasteners, and staples evaluated for magnetic field interactions were not attracted by static magnetic fields of MR systems operating at 1.5-Tesla or less. These implants were made from nonferromagnetic materials such as tantalum, commercially pure titanium, and nonferromagnetic forms of stainless steel. Additionally, some forms of ligating, hemostatic, or other types of clips are made from biodegradable materials. Therefore, patients that have the hemostatic vascular clips, other clips, fasteners, and staples indicated in **The List** are not at risk for injury during MR procedures. Notably, there has never been a report of an injury to a patient in association with a hemostatic vascular clip, other type of clip, fastener, or staple in the MR environment. Patients with nonferromagnetic versions of these implants may undergo MR procedures immediately after they are placed surgically.

Vascular graphs frequently have clips or fasteners applied that may present problems for MR imaging because of the associated artifacts. Weishaupt et al. evaluated the artifact size on three-dimensional MR angiograms as well as the MR safety for 18 different commercially available hemostatic and ligating clips. All of the clips were safe at 1.5-Tesla insofar as there was no heating or magnetic field interactions measured for these implants.

MR Safety at 3.0-Tesla and Hemostatic Clips, Other Clips, Fasteners, and Staples. At 3.0-Tesla, a variety of hemostatic clips, other clips, fasteners, and staples have been evaluated for MR safety. The Surgiclip spring made from carbon steel (United States Surgical, North Haven, CT) showed a deflection angle of 90° and a qualitative torque of +4. However, considering the "intended in vivo use" of this device, the closing force may provide substantial counterforce to prevent it from being moved or dislodged. However, this remains to be determined by further experimental findings. In lieu of this information, this implant is currently categorized as "unsafe" at 3.0-Tesla.

REFERENCES

Brown MA, Carden JA, Coleman RE, et al. Magnetic field effects on surgical ligation clips. Magn Reson Imaging 1987;5:443-453.

Barrafato D, Henkelman RM. Magnetic resonance imaging and surgical clips. Can J Surg 1984;27:509-512.

Shellock FG. Pocket Guide to MR Procedures and Metallic Objects: Update 2001, Seventh Edition, Lippincott Williams & Wilkins Healthcare, Philadelphia, 2001.

Shellock FG. Magnetic Resonance Procedures: Health Effects and Safety. CRC Press, LLC, Boca Raton, FL, 2001.

Shellock FG. MR imaging of metallic implants and materials: a compilation of the literature. Am J Roentgenol 1988;151:811-814.

Shellock FG. Biomedical implants and devices: assessment of magnetic field interactions with a 3.0-Tesla MR system. J Magn Reson Imaging 2002;16:721-732.

Shellock FG, Crues JV. High-field strength MR imaging and metallic biomedical implants: an ex vivo evaluation of deflection forces. AJR Am J Roentgenol 1988;151:389-392.

Shellock FG, Kanal E. Magnetic Resonance: Bioeffects, Safety, and Patient Management. Second Edition, Lippincott-Raven Press, New York, 1996.

Shellock FG, Morisoli S, Kanal E. MR procedures and biomedical implants, materials, and devices: 1993 update. Radiology 1993;189:587-599.

Shellock FG, Swengros-Curtis J. MR imaging and biomedical implants, materials, and devices: an updated review. Radiology 1991;180:541-550.

Weishaupt D, Quick HH, Nanz D, Schmidt M, Cassina PC, Debatin JF. Ligating clips for three-dimensional MR angiography at 1.5-T: *in vitro* evaluation. Radiology 2000;214:902-907.

INTRAUTERINE CONTRACEPTIVE DEVICES

Intrauterine contraceptive devices (IUD) may be made from nonmetallic materials (e.g., plastic) or a combination of nonmetallic and metallic materials. Copper is typically the metal used in an IUD. The "Copper T" and "Copper 7" both have a fine copper coil wound around a portion of the IUD. Testing conducted to determine the MR-safety aspects of IUDs with metallic components indicated that these objects are safe for patients in the MR environment using MR systems operating at 1.5-Tesla or less. This includes the Multiload Cu375, the Nova T (containing copper and silver), and the Gyne T IUDs. An artifact may be seen for the metallic component of the IUD, however, the extent of this artifact is relatively small because of the low magnetic susceptibility of copper.

The Mirena intrauterine system (IUS), is a hormone releasing device that contains levonorgestrel to prevent pregnancy (Berlex Laboratories, Wayne, NJ). This T-shaped device is made entirely from nonmetallic materials that include polyethylene, barium sulfate (i.e., which makes it radiopaque), and silicone. Thus, the Mirena is safe for patients undergoing MR procedures using MR systems operating at all static magnetic field strengths.

REFERENCES

Hess T, Stepanow B, Knopp MV. Safety of intrauterine contraceptive devices during MR imaging. Eur Radiol 1996;6:66-68.

Mark AS, Hricak H. Intrauterine contraceptive devices: MR imaging. Radiology 1987;162:311-314.

Mirena (levonorgestrel-releasing intrauterine system (IUS), Product Insert information, Berlex Laboratories, Wayne, NJ

Shellock FG. Pocket Guide to MR Procedures and Metallic Objects: Update 2001, Seventh Edition, Lippincott Williams & Wilkins Healthcare, Philadelphia, 2001.

Shellock FG. Magnetic Resonance Procedures: Health Effects and Safety. CRC Press, LLC, Boca Raton, FL, 2001.

Shellock FG, Kanal E. Magnetic Resonance: Bioeffects, Safety, and Patient Management. Second Edition, Lippincott-Raven Press, New York, 1996.

INTERSTIM THERAPY - SACRAL NERVE STIMULATION (SNS) FOR URINARY CONTROL

InterStim Therapy - Sacral Nerve Stimulation (SNS) for Urinary Control (Medtronic, Minneapolis, MN) is a therapy for managing urinary urge incontinence, nonobstructive urinary retention, and significant symptoms of urgency-frequency in patients who have failed or could not tolerate more conservative treatments. In properly selected patients, InterStim Therapy can be dramatically successful in reducing or eliminating symptoms. The implantable InterStim System uses mild electrical stimulation of the sacral nerve that influences the behavior of the bladder, sphincter, and pelvic floor muscles.

Indications. The Medtronic InterStim® Therapy for Urinary Control is indicated for the treatment of urinary urge incontinence, urinary retention, and significant symptoms of urgency-frequency in patients who have failed or could not tolerate more conservative treatments.

Contraindications. Patients are contraindicated for implant of the InterStim Therapy System if they have not demonstrated an appropriate response to test stimulation or are unable to operate the neurostimulator. Also, diathermy (e.g., shortwave diathermy, microwave diathermy or therapeutic ultrasound diathermy) is contraindicated because diathermy's energy can be transferred through the implanted system (or any of the separate implanted components), which can cause tissue damage and can result in severe injury or death. Diathermy can damage parts of the neurostimulation system.

199

Precautions/Adverse Events. Safety and effectiveness have not been established for: bilateral stimulation, patients with neurological disease origins such as multiple sclerosis, pregnancy and delivery, or for pediatric use under the age of 16.

The system may be affected by or adversely affect cardiac demand pacemakers or therapies, cardioverter defibrillators, electrocautery, external defibrillators, ultrasonic equipment, radiation therapy, magnetic resonance imaging (MRI), theft detectors and screening devices.

Adverse events related to the therapy, device, or procedure can include: pain at the implant sites, lead migration, infection or skin irritation, technical or device problems, transient electric shock, adverse change in bowel or voiding function, numbness, nerve injury, seroma at the neurostimulator site, change in menstrual cycle, and undesirable stimulation or sensations.

[Reprinted with permission from Medtronic, Minneapolis, MN]

REFERENCE

http://www.medtronic.com/neuro/interstim/interstim_warning.html

ISOMED IMPLANTABLE CONSTANT-FLOW INFUSION PUMP

Exposure of IsoMed® Implantable Constant-Flow Infusion Pumps (Medtronic, Minneapolis, MN) to Magnetic Resonance Imaging (MRI) fields of 1.5-T (Tesla) has demonstrated no impact on pump performance and a limited effect on the quality of the diagnostic information. Testing on the IsoMed pump has established the following with regard to MR safety and diagnostic issues:

Implant Heating During MRI Scans

Specific Absorption Rate (SAR): Presence of the pump can potentially cause a two-fold increase of the local temperature rise in tissues near the pump. During a 20-minute pulse sequence in a 1.5-T (Tesla) GE Signa Scanner with a whole-body average SAR of 1.0 W/kg, a temperature rise of 1 degree Celsius in a static phantom was observed near the pump implanted in the "abdomen" of the phantom. The temperature rise in a static phantom represents a worst case for physiological temperature rise and the 20-minute scan time is representative of a typical imaging session. Implanting the pump in other locations may result in higher temperature rises in tissues near the pump.

In the unlikely event that the patient experiences uncomfortable warmth near the pump, the MRI scan should be stopped and the scan parameters adjusted to reduce the SAR to comfortable levels.

Peripheral Nerve Stimulation

Time-Varying Gradient Magnetic Fields: Presence of the pump may potentially cause a two-fold increase of the induced electric field in tissues near the pump. With the pump implanted in the abdomen, using pulse sequences that have dB/dt up to 20 T/s, the measured induced electric field near the pump is below the threshold necessary to cause stimulation.

In the unlikely event that the patient reports stimulation during the scan, the proper procedure is the same as for patients without implants-stop the MRI scan and adjust the scan parameters to reduce the potential for nerve stimulation.

Static Magnetic Field

For magnetic fields up to 1.5-T, the magnetic force and torque on the IsoMed pump will be less than the force and torque due to gravity. In the unlikely event that the patient reports a slight tugging sensation at the pump implant site, an elastic garment or wrap may be used to prevent the pump from moving and reduce the sensation the patient may experience.

Image Distortion

The IsoMed pump will cause image dropout on MRI images in the region surrounding the pump. The extent of image artifact depends on the pulse sequence chosen with gradient echo sequences generally causing the most image dropout. Spin echo sequences will cause image dropout in a region approximately 50% larger than the pump itself, about 12-cm across, but with little image distortion or artifact beyond that region.

Minimizing Image Distortion

MR image artifact may be minimized by careful choice of pulse sequence parameters and location of the angle and location of the imaging plane. However, the reduction in image distortion obtained by adjustment of pulse sequence parameters will usually be at a cost in signal-to-noise ratio. These general principles should be followed:

- Use imaging sequences with stronger gradients for both slice and read encoding directions. Employ higher bandwidth for both RF pulse and data sampling.

- Choose an orientation for read-out axis that minimizes the appearance of in-plane distortion.

- Use spin echo (SE) or gradient echo (GE) MR imaging sequences with a relatively high data sampling bandwidth.

- Use shorter echo time (TE) for gradient echo technique, whenever possible.

- Be aware that the actual imaging slice shape can be curved in space due to the presence of the field disturbance of the pump (as stated above).

- Identify the location of the implant in the patient and when possible, orient all imaging slices away from the implanted pump.

[Reprinted with permission from Medtronic, Minneapolis, MN]

MAGNETICALLY-ACTIVATED IMPLANTS AND DEVICES

Various types of implants and devices incorporate magnets as a means of activating the implant. The magnet may be used to retain the implant in place (e.g., certain prosthetic devices), to guide a ferromagnetic object into a specific position, to permit important functional aspects of the implant (e.g., hearing devices with internal components), change the operation of the implant (e.g., certain percutaneous adjustable pressure valves), or to program the device (e.g., pacemakers, ICDs, etc.). Because there is a high likelihood of perturbing the function, demagnetizing, or displacing these implants, MR procedures typically should not be performed in patients with these implants or devices. However, in some cases, patients with magnetically-activated implants and devices may undergo MR procedures as long as certain specific precautions are followed.

Implants and devices that use magnets (e.g., certain types of dental implants, magnetic sphincters, magnetic stoma plugs, magnetic ocular implants, otologic implants, and other similar prosthetic devices) may be damaged by the magnetic fields of MR systems which, in turn, may necessitate surgery to replace or reposition them. For example, Schneider et al. reported that an MR examination is capable of demagnetizing the permanent magnet associated with an otologic implant (i.e., the Audiant magnet). Obviously, this has important implications for patients undergoing MR procedures.

Whenever possible, and if this can be done without risk to the patient (i.e., from the retained magnetic "keeper" or similar component), a magnetically-activated implant or device (e.g., an externally applied prosthesis or magnetic stoma plug) should be removed from the patient prior to the MR procedure. This will permit the examination to be performed safely.

Knowledge of the specific aspects of the magnetically-activated implant or device is essential to recognize potential problems and to guarantee that an MR procedure may be performed on a patient without problems or an injury.

Extrusion of an eye socket magnetic implant in a patient imaged with a 0.5-Tesla MR system has been described. This type of magnetic prosthesis is used in a patient after enucleation. A removable eye prosthesis adheres with a magnet of opposite polarity to a permanent implant sutured to the rectus muscles and conjunctiva by magnetic attraction through the conjunctiva. This "magnetic linkage" enables the eye prosthesis to move in a coordinated fashion with that of normal eye movement. In the reported incident, the static magnetic field of the MR system produced sufficient attraction of the ferromagnetic portion of the magnetic prosthesis to cause it to extrude through the tissue, thus, injuring the patient.

Certain dental prosthetic appliances utilize magnetic forces to retain the implant in place. The magnet may be contained within the prosthesis and attached to a ferromagnetic post implanted in the mandible or vise versa. An MR procedure may be performed safely in a patient with this type of dental magnet appliance as long as it has been determined that it is properly attached to supporting tissue.

Patients with hydrocephalus or other disorders are often treated with a percutaneous adjustable pressure valve that may have a magnetically-activated component that allows a change to be made to the resistance required to open the valve. This is accomplished using an externally applied magnet.

Pressure adjustable valves permit noninvasive readjustment of the opening pressure of an implanted shunt to cerebrospinal fluid hydrodynamics. Changing the resistance of this type of valve by MR-induced dysfunction of this implant without recognizing it could cause problems, including acute hydrocephalus, for the patient. Currently, it is recommended that patients with percutaneous adjustable pressure valves (e.g., the Sophy or Codman-Medos programmable valves) have the specific valve checked immediately before and after the MR procedure to determine if exposure to the MR system caused a change in the valve setting. This should be done by an appropriate healthcare professional. If a change occurred as a result of the MR procedure, the neurosurgeon or other individual responsible for the medical management of the patient and familiar with the operation of this type of device should reset the percutaneous adjustable pressure valve to its original setting. Fortunately, there are no known risks or hazards associated with the Sophy or Codman-Medos programmable valves with respect to movement, torque, or heating in the MR environment.

REFERENCES

Fransen P, Dooms G, Thauvoy C. Safety of the adjustable pressure ventricular valve in magnetic resonance imaging: problems and solutions. Neuroradiology 1992;34:508-509.

Gaston A, Marsault C, Lacaze A, et al. External magnetic guidance of endovascular catheters with a superconducting magnet: preliminary trials. J Neuroradiol 1988;15:137-147.

Grady MS, Howard MA, Molloy JA, et al. Nonlinear magnetic stereotaxis: three dimensional in vivo remote magnetic manipulation of a small object in canine brain. Med Phys 1990;17:405-415.

Liang MD, Narayanan K, Kanal E. Magnetic ports in tissue expanders: a caution for MRI. Magn Reson Imaging 1989;7:541-542.

Ortler M, Kostron H, Felber S. Transcutaneous pressure-adjustable valves and magnetic resonance imaging: an ex vivo examination of the Codman-Medos programmable valve and the Sophy adjustable pressure valve. Neurosurgery 1997;40:1050-1057.

Ranney DF, Huffaker HH. Magnetic microspheres for the targeted controlled release of drugs and diagnostic agents. Ann NY Acad Sci 1987;507:104-119.

Schneider ML, Walker GB, Dormer KJ. Effects of magnetic resonance imaging on implantable permanent magnets. Am J Otol 1995;16:687-689.

Shellock FG. Pocket Guide to MR Procedures and Metallic Objects: Update 2001, Seventh Edition, Lippincott Williams & Wilkins Healthcare, Philadelphia, 2001.

Shellock FG. Magnetic Resonance Procedures: Health Effects and Safety. CRC Press, LLC, Boca Raton, FL, 2001.

Shellock FG. Ex vivo assessment of deflection forces and artifacts associated with high-field strength MRI of "mini-magnet" dental prostheses. Magn Reson Imaging 1989;7 (Suppl 1):38.

Young DB, Pawlak AM. An electromagnetically controllable heart valve suitable for chronic implantation. ASAIO Trans 1990;36:M421-M425.

Yuh WTC, Hanigan MT, et al. Extrusion of a eye socket magnetic implant after MR imaging examination: potential hazard to a patient with eye prosthesis. J Magn Reson Imaging 1991;1:711-713

MEDICAL DEVICES AND ACCESSORIES FOR THE MR ENVIRONMENT

The increasing capabilities of magnetic resonance (MR) studies to impact medical diagnosis and prognosis has dramatically increased the number of procedures performed worldwide. As such, many more patients, especially those in high-risk or special population groups, are undergoing MR examinations for an ever-widening spectrum of medical indications.

As the number of MR procedures increases, so does the potential for incidents and accidents in the MR environment. Of note is that the installation of clinical MR systems operating at 3.0-Tesla or higher is growing rapidly. As such, hazards related to ferromagnetic projectiles may also increase in relation to the use of these powerful MR systems.

In addition to the diagnostic imaging uses of MR technology, as Jolesz et al. have stated, continuous progress has been made to expand the utilization of MR imaging beyond diagnosis into intervention. This has resulted in the development and performance of innovative procedures including percutaneous biopsy (e.g., breast, bone, brain, abdominal), endoscopic surgery of the abdomen, spine, and sinuses, open brain surgery, and MR-guided monitoring of thermal therapies (i.e., laser-induced, RF-induced, and cryomediated procedures).

Various vendors and manufacturers, prompted by recommendations and requests from MR healthcare professionals, have recognized the need for developing specialized medical devices, equipment, accessories, and instruments necessary for use in the MR environment and for interventional MR procedures. Accordingly, there are now numerous patient support devices and accessories that have been developed and that have

207

undergone thorough evaluation to assess appropriate use in the MR environment or during interventional MR procedures (see **Appendix I**)

Basic Patient Management Accessories and Equipment. All new and existing MRI facilities should be prepared to handle patients and everyday situations (e.g., maintenance) in the MR environment by obtaining a selection of nonmagnetic or other suitable accessories or equipment. For example, useful items for an out-patient facility include nonmagnetic equipment such as a wheelchair (one or more), stretcher or gurney, step stool, IV pole, laundry cart, stethoscope, blood pressure manometer, storage or utility care, fire extinguisher, and custodial cart.

MRI facilities that handle both out-patients and in-patients should additionally consider obtaining a nonmagnetic patient slider board, physiologic monitoring equipment (e.g., fiber-optic pulse oximeter), nonmagnetic oxygen tank (including nonmagnetic regulator, cart or stand), portable suction, and other appropriate devices and accessories. These afore-mentioned accessories and equipment are available from a variety of vendors (see **Appendix I**).

REFERENCES

Jolesz FA, et al. Compatible instrumentation for intraoperative MRI: expanding resources. J Magn Reson Imaging 1998;8:8-11.

Keeler EK, et al. Accessory equipment considerations with respect to MRI compatibility. J Magn Reson Imaging 1998;8:12-18.

Shellock FG. Magnetic Resonance Procedures: Health Effects and Safety. CRC Press, Boca Raton, FL, 2001.

MEDTRONIC MINIMED 2007 IMPLANTABLE INSULIN PUMP SYSTEM

The Medtronic MiniMed 2007 Implantable Insulin Pump System may offer treatment advantages for diabetes patients who have difficulty maintaining consistent glycemic control. Patients that have not responded well to intensive insulin therapy, including multiple daily insulin injections or continuous subcutaneous insulin infusion using an external pump, may be primary candidates for the Medtronic MiniMed 2007 System. The Medtronic MiniMed 2007 System delivers insulin into the peritoneal cavity in short, frequent bursts or "pulses", similar to how pancreatic beta cells secrete insulin.

Medtronic MiniMed 2007 Implantable Insulin Pump System and MRI. The Medtronic MiniMed 2007 Implantable Insulin Pump is designed to withstand common electrostatic and electromagnetic interference but must be removed prior to undergoing an MR procedure. Any magnetic field exceeding 600 Gauss will interfere with the proper functioning of the pump for as long as the pump remains in that field. Fields much higher than that, such as those emitted by an MRI machine can and may cause irreparable damage to the pump.

The infusion sets (MMT-11X, MMT-31X, MMT-32X, MMT-37X, MMT-39X) on the other hand, contain no metal components and are safe to be used and can remain attached to the user during an MR procedure. The only exceptions would be the Polyfin infusion sets. The Polyfin infusion sets (MMT-106 AND MMT-107, MMT-16X, MMT-30X, MMT-36X) consists of a surgical steel needle that remains in the subcutaneous tissue. These infusion sets should be removed prior to any MR procedure .

(Reprinted with permission from Medtronic, Product Information for the Medtronic MiniMed 2007 Implantable Insulin Pump System.)

REFERENCE

http://www.minimed.com/doctors/md_products_2007.shtml

MISCELLANEOUS IMPLANTS AND DEVICES

Many different miscellaneous implants, materials, devices, and objects have been tested for ferromagnetic qualities and other aspects of MR-safety or MR-compatibility. For example, various types of firearms have been tested for safety in the MR environment. These firearms exhibited strong ferromagnetism. In fact, two of the six firearms tested discharged in a reproducible manner while in the MR system room. Obviously, firearms should remain outside of the MR environment and MR system room to prevent problems or possible injuries.

MR-guided biopsy, therapeutic, and minimally invasive surgical procedures are important clinical applications that are performed on conventional, open-architecture, or the "double-donut" MR systems specially designed for this work. These procedures present challenges with regard to the instruments and devices that are needed to support these interventions.

Metallic surgical instruments and other devices potentially pose hazards (e.g., "missile" effects) or other problems (i.e., image distortion that can obscure the area of interest and either affect adequate visualization of the abnormality or prevent performance of the procedure) that must be addressed to apply MR-guided techniques effectively. Various manufacturers have used "weakly" ferromagnetic, nonferromagnetic or nonmetallic materials to make special instruments for interventional MR procedures.

At least one manufacturer has used ceramic material as a means of constructing prototype devices that include scalpels, cranial drill bits, scissors, and tweezers that have been determined to be MR-compatible. Ceramic instruments have been shown to have particularly good qualities for the MR environment insofar as there was no magnetic field attraction, negligible heating, and no substantial image distortion determined by the *ex vivo* testing techniques for this material.

Other medical products and devices have been developed with metallic components that are either entirely nonferromagnetic (e.g., stereotactic head-frame, Compass International, Inc., Rochester, MN) or made from metals (e.g., titanium, non-magnetic types of stainless steel, etc.) that are acceptable for use in the MR environment.

MR Safety at 3.0-Tesla and Miscellaneous Implants and Devices. Several implants and devices (refer to **The List** to determine specific implants and devices) have been tested for MR safety in association with 3.0-Tesla MR systems. Of these, several exhibited measurable magnetic field interactions but none were at a level considered to present a hazard to a patient undergoing an MR procedure at this static magnetic field strength.

REFERENCES

American Society for Testing and Materials (ASTM) Designation: F 2052. Standard test method for measurement of magnetically induced displacement force on passive implants in the magnetic resonance environment. In: Annual Book of ASTM Standards, Section 13, Medical Devices and Services, Volume 13.01 Medical Devices; Emergency Medical Services. West Conshohocken, PA, 2002, pp. 1576-1580.

Fransen P, Dooms G, Thauvoy C. Safety of the adjustable pressure ventricular valve in magnetic resonance imaging: problems and solutions. Neuroradiology 1992;34:508-509.

Go KG, Kamman RL, Mooyaart EL. Interaction of metallic neurosurgical implants with magnetic resonance imaging at 1.5-Tesla as a cause of image distortion and of hazardous movement of the implant. Clin Neurosurg 1989;91:109-115.

Kanal E, Shaibani A. Firearm safety in the MR imaging environment. Radiology 1994;193:875-876.

Lufkin R, Jordan S, Lylyck P, et al. MR imaging with topographic EEG electrodes in place. AJNR Am J Neuroradiol 1988;9:953-954.

Marra S, Leonetti JP, Konior RJ, Raslan W. Effect of magnetic resonance imaging on implantable eyelid weights. Ann Otol Rhinol Laryngol 1995;104:448-452.

Ortler M, Kostron H, Felber S. Transcutaneous pressure-adjustable valves and magnetic resonance imaging: an ex vivo examination of the Codman-Medos programmable valve and the Sophy adjustable pressure valve. Neurosurgery 1997;40:1050-1057.

Planert J, Modler H, Vosshenrich R. Measurements of magnetism-forces and torque moments affecting medical instruments, implants, and foreign objects during magnetic resonance imaging at all degrees of freedom. Medical Physics 1996;23:851-856.

Shellock FG. Biomedical implants and devices: assessment of magnetic field interactions with a 3.0-Tesla MR system. J Magn Reson Imaging 2002;16:721-732.

Shellock FG. MR-compatibility of an endoscope designed for use in interventional MR procedures. AJR Am J Roentgenol 1998;71:1297-1300.

Shellock FG. Pocket Guide to MR Procedures and Metallic Objects: Update 2001, Seventh Edition, Lippincott Williams & Wilkins Healthcare, Philadelphia, 2001.

Shellock FG. Magnetic Resonance Procedures: Health Effects and Safety. CRC Press, LLC, Boca Raton, FL, 2001.

Shellock FG. MR safety update 2002: Implants and devices. J Magn Reson Imaging 2002;16:485-496.

Shellock FG. Metallic neurosurgical implants: assessment of magnetic field interactions, heating, and artifacts at 1.5-Tesla. Radiology 2001;218:611.

Shellock FG. Surgical instruments for interventional MRI procedures: assessment of MR safety. J Magn Reson Imaging 2001;13:152-157.

Shellock FG, Kanal E. Magnetic Resonance: Bioeffects, Safety, and Patient Management. Second Edition, Lippincott-Raven Press, New York, 1996.

Shellock FG, Shellock VJ. Ceramic surgical instruments: evaluation of MR-compatibility at 1.5-Tesla. J Magn Reson Imaging 1996;6:954-956.

Shellock FG, Shellock VJ. Evaluation of MR compatibility of 38 bioimplants and devices. Radiology 1995;197:174.

To SYC, Lufkin RB, Chiu L. MR-compatible winged infusion set. Comput Med Imag Graph 1989;13:469-472.

Zhang J, Wilson CL, Levesque MF, Behnke EJ, Lufkin RB. Temperature changes in nickel-chromium intracranial depth electrodes during MR scanning. AJNR Am J Neuroradiol 1993;14:497-500.

MRI CONTRAST AGENT INJECTION SYSTEMS

MRI contrast agent injection systems deliver a precisely timed bolus that results in a highly reproducible, contrast-enhanced MRI examination. The controlled, power injection of MRI contrast agents is gaining in popularity for a variety of clinical applications including examinations of abdominal organs, vascular anatomy, and dynamic MRI studies of the breast. These specialized power injectors must be able to operate in the MR environment without affecting magnet homogeneity, degrading signal-to-noise, or causing artifacts. To date, two devices are available for power delivery of MRI contrast agents: the Optistar MR Contrast Delivery System (Mallinckrodt, St. Louis, MO; www.Mallinckrodt.com) and the Spectris MR Injection System (Medrad, Inc., Indianola, PA; www.Medrad.com).

REFERENCES

Earls JP, Rofsky NM, DeCorato DR, Krinsky GA, Weinreb JC. Breath-hold single-dose gadolinium-enhanced three-dimensional MR aortography: usefulness of a timing examination and MR power injector. Radiology 1996;201:705-710.

Earls JP, Rofsky NM, DeCorato DR, Krinsky GA, Weinreb JC. Hepatic arterial-phase dynamic gadolinium-enhanced MR imaging: optimization with a test examination and a power injector. Radiology 1997;202:268-273.

Korst MB, et al. Accuracy of normal-dose contrast-enhanced MR angiography in assessing renal artery stenosis and accessory renal arteries. AJR Am J Roentgenol 2000;174:629-634.

Mitsuzaki K, Yamashita Y, Ogata I, Tang Y, Namimoto T, Takahashi M. Optimal protocol for injection of contrast material at MR angiography: study of healthy volunteers. Radiology 1999;213:913-918

Runge VM, Williams NM. Dynamic contrast-enhanced magnetic resonance imaging in a model of splenic metastasis. Investigative Radiology 1998;33:45-50.

Volk M, et al. Time-resolved contrast-enhanced MR angiography of renal artery stenosis: diagnostic accuracy and interobserver variability. AJR Am J Roentgenol 2000;174:1583-1588.

NEUROSTIMULATION SYSTEMS

The incidence of patients receiving implanted neurostimulation systems for treatment of various neurological disorders and other conditions is increasing. Because of the inherent design and intended function of neurostimulation systems, the electromagnetic fields used for MR procedures may produce various problems with these devices. For example, altered function of a neurostimulation system that results from exposure to the electromagnetic fields of an MR system may cause discomfort, pain, or injury to the patient. Under certain MR operational conditions, damage to the nerve fibers secondary to excessive heating of the implanted electrodes of the neurostimulation system may occur, as described by Rezai et al. Therefore, the present policy regarding a patient with a neurostimulation system is that the individual should not undergo an MR procedure unless comprehensive testing has been conducted to define parameters and guidelines for the safe use of the MR procedure for a given neurostimulation system. Importantly, the exact safety criteria for the particular neurostimulation system with regard to the implantable pulse generator (IPG), leads, electrodes, and operational conditions for the device and the MR system conditions must be defined.

NeuroCybernetic Prosthesis (NCP® System*, Cyberonics, Houston, TX)

Vagus nerve stimulation (VNS) with the Cyberonics NeuroCybernetic Prosthesis (NCP®) System is the first new approach to the treatment of epilepsy in over 100 years. After 15 years of research and clinical studies, VNS was approved by the United States Food and Drug Administration (FDA) on July 16, 1997 as an add-on therapy in reducing the frequency of seizures in adults and adolescents over 12 years of age with partial onset seizures refractory to antiepileptic medications.

216

VNS consists of electrical signals that are applied to the vagus nerve in the neck for transmission to the brain. The vagus nerve averages 22 inches in length in adults and is located in the upper body. The vagus nerve is one of the primary communication lines from the major organs of the body to the brain.

The vagus nerve has proven to be a good way to communicate with the brain for VNS because:

(1) there are few if any pain fibers in the vagus nerve;

(2) over 80% of the electrical signals are applied to the vagus;

(3) nerves in the neck are sent upwards to the brain; and

(4) the stimulation lead may be attached to the vagus nerve in a surgical procedure which does not involve the brain.

The Model 100 NCP Pulse Generator is an implantable, multiprogrammable Pulse Generator that delivers electrical signals to the vagus nerve. Constant current, capacitively coupled, charge-balanced signals are transmitted from the Generator to the vagus nerve via the NCP Lead (Model 300 Series). The Model 100 NCP Pulse Generator is housed in a hermetically sealed titanium case. Feedthrough capacitors are used to filter electromagnetic interference from the Pulse Generator circuitry.

The major components and functions of the generator for this devices are as follows: a microprocessor, a voltage regulator, a 76.8 kHz crystal oscillator, one antenna to transmit information and another antenna to receive information, communication circuitry, DC-DC voltage generation and control circuitry, constant current control circuitry, a dual pole magnetic reed switch for manual activation of the Generator and for inhibition of the output pulses, and a lithium thionyl chloride cell to provide power for stimulation and circuit operation.

The lithium thionyl chloride battery chemistry has the low impedance and high energy density characteristics required for the rapid pulsing needed in peripheral nerve stimulation, and similar batteries have been previously used in cardiac pacemakers, implantable spinal cord stimulators, and implantable drug pumps. The implantable lead delivers electrical signals from the Generator to the vagus nerve. The lead has two helical electrodes with a helical anchor tether for placement around the nerve and two 5-millimeter (mm) connectors for attachment to the generator.

The helix of the lead is available in two sizes of inner diameter (2.0-mm and 3.0-mm) to allow for appropriate fit on different sized nerves. The helical design is soft, pliable, and expands or contracts with changes in nerve diameter, which may occur immediately post implant. These design features allow the 2-mm inside diameter helical electrode to fit most vagus nerves. The Model 300 NCP Bipolar Lead is insulated with silicone rubber and is bifurcated at each end. The lead wire is quadrifilar MP35N, and the electrode is a platinum ribbon.

The Model 100 NCP Pulse Generator has a number of programmable settings that allow the physician to optimize the treatment for a patient. Those settings include pulse width, output current, signal frequency, signal ON time, signal OFF time, magnet-activated ON time, magnet-activated pulse width and magnet-activated output current. Cyberonics provides a magnet that may be used to either manually initiate stimulation or to turn OFF the device.

The Model 100 NCP Pulse Generator has telemetry capability that supplies information about its operating characteristics, such as parameter settings, lead impedance and history of magnet use. The Generator has a number of characteristics intended to enhance operational reliability and safety, such as electromagnetic interference (EMI) filter capacitors, a series battery resistor to limit temperature rise in the event of short circuit, defibrillation protection diodes, direct current-blocking capacitors on both Leads that prevent direct current (DC) from being applied to the patient, a software watchdog timer to prevent continuous stimulation, and protection against voltage dips on the battery that could disrupt microprocessor memory.

The NeuroCybernetic Prosthesis, NCP, Pulse Generator, Model 100 (Cyberonics, Houston, TX) received approval of an MR-safe labeling claim from the United States Food and Drug Administration, allowing an MR procedure to be conducted in a patient with this device, as long as strict guidelines are followed.

These guidelines are (Note: This information is reprinted with permission from the Product Information, NeuroCybernetic Prosthesis, NCP, Pulse Generator, Model 100, Cyberonics):

Magnetic resonance imaging (MRI) should not be performed with a magnetic resonance body coil in the transmit mode. The heat induced in the Bipolar Lead by an MRI body scan can cause injury.

(1) If an MRI should be done, use only a transmit and receive type of head coil. Magnetic and radiofrequency (RF) fields produced by MRI may change the Pulse Generator settings (change to reset parameters)

or activate the device. Stimulation has been shown to cause the adverse events reported in the "Adverse Events" section of this manual. MRI compatibility was demonstrated using a 1.5-T General Electric Signa Imager only. Testing on this imager as performed on a phantom indicated that the following Pulse Generator and MRI settings can be used safely without adverse events.

(2) Pulse Generator output programmed to 0 mA for the MRI procedure, and afterward, retested by performing the Lead Test diagnostics and reprogrammed to the original settings.

- Head coil type: transmit and receive only

- Static magnetic field strength: ≤ 2.0-Tesla

- Specific-rate absorption (SAR): 1.3 W/kgfor a 154.5-lb (70-kg) patient

- Time-varying intensity: 10-Tesla/sec

(3) Use caution when other MRI systems are used, since adverse events may occur because of different magnetic field distributions.

(4) No scan in which the radiofrequency (RF) is transmitted by the body coil should be done on a patient who has the NCP System. Thus, protocols must not be used which utilize local coils that are RF-receive only, with RF-transmit performed by the body coil. Note that some RF head coils are receive only, and that most other local coils, such as knee and spinal coils, are also RF-receive only. These coils must not be used in patients with the NCP System."

[*Information for the NCP is provided with permission from Gayle Nesom of Cyberonics and obtained from Cyberonics web site, www.cyberonics.com.]

Neurostimulation System Used for Deep Brain Stimulation (DBS)

There is a heightened interest in the use of chronic deep brain stimulation (DBS) of the thalamus, globus pallidus, and the subthalamic nucleus for the treatment of medically refractory movement disorders and other types of neurological conditions. Thus, the number of patients receiving neurostimulation systems is growing rapidly.

It is desirable to be able to use MR imaging in patients with neurostimulators as well as to use MR-guidance or stereotactic techniques to optimally place DBS electrodes. Additionally, the nature of neurological conditions often necessitates further patient examinations using MR procedures after the electrodes are implanted. However, only a limited number of studies have been conducted to address MR safety issues for neurostimulation systems used for DBS. The possible MR safety issues that exist for neurostimulation systems include magnetic field interactions, heating, induced electrical currents and functional disruption of the operational aspects of these devices.

Therefore, investigations have been performed by Rezai et al., Finelli et al., and Ruggieri et al. to assess MRI-related heating for the Activa Tremor Control System (Medtronic, Minneapolis, MN) approved for deep brain stimulation (DBS). Different configurations of bilateral neurostimulators (i.e., implantable pulse generators), extensions, and leads were evaluated along with different uses of a 1.5-Tesla MR system using various transmit/receive body and head radiofrequency coil combinations to assess worst-case and clinically relevant scenarios. These investigations were conducted with the intent of developing guidelines that would permit patients with implanted neurostimulation systems and DBS electrodes to undergo MR imaging examinations safely.

The Activa Tremor Control System (Medtronic, Minneapolis, MN) is a fully implantable, multiprogrammable neurostimulation system designed to deliver electrical stimulation to the thalamus or other brain structures. The implantable system is comprised of the neurostimulator (or implantable pulse generator, IPG), DBS lead, and an extension that connects the lead to the IPG. This system delivers high frequency electrical stimulation to a multiple contact electrode placed in a deep brain site. For example, DBS may be used on the ventral intermediate nucleus of the thalamus for control of essential of Parkinsonian tremor. On a research investigation basis, the same system components are used within other deep brain structures to treat other conditions.

Based on the research, experimental, and clinical findings obtained to date, the following are MR-safety guidelines for the Activa System (includes Model 7426 Soletra and Model 7424 Itrel II neurostimulators; Model 7482 and Model 7495 extensions; Model 3387 and Model 3389 DBS leads; Medtronic, Minneapolis, MN), as specified in the Product Information for this device by Medtronic, MN (2001):

Magnetic Resonance Imaging

Based on tests to date, some MRI procedures can be performed safely with an implanted Activa System. MRI systems used to safely perform MRI include MRI systems operating at 1.5 Tesla (specific MRI machines include Siemens Magnetom 1.5-T VISION, Picker International 1.5-T Edge, and GE Signa 1.5-T Echospeed). The safety of other MRI machines used with implanted Activa Systems is not known.

• Use only a transmit and receive type RF head coil to minimize the exposure of the lead/neurostimulator system to the MRI RF fields. Do not use a whole body RF coil.

• Select imaging parameters to perform MRI at a specific absorption rate (SAR) that does not exceed 0.4 W/kg in the head.

• Carefully weigh any decision to perform magnetic resonance imaging (MRI) scans on patients who require the neurostimulator to control tremor. Image quality during MRI scans may be reduced, because the tremor may return when the brain stimulator is turned off. Use of MRI could possibly result in movement, heating or damage to the implanted Activa System. The MRI image around the implanted lead may be distorted and shadowed. Induced voltages in the neurostimulator and/or lead may occur, possibly causing uncomfortable ("jolting" or "shocking") levels of stimulation. Clinicians should carefully weigh the decision to use MRI in patients with an implanted Activa System.

MRI and Activa Therapy - Activa Clinical Experience

Due to the variability of clinical MRI systems, the safety of patients or the functioning of devices exposed to MRI systems cannot be unequivocally ensured. However, 39 patients in the Medtronic sponsored clinical study with an implanted Activa System have safely undergone 1.5-Tesla MRI procedures. These patients were implanted with the Itrel II Model 7424 Neurostimulator, the Model 7495 Extension, and the Model 3387/3389 DBS Lead. For comparison purposes, the Activa System comprising the Soletra Model 7426 Neurostimulator, the 7495/7482 Extension, and the 3387/89 DBS lead is expected to have similar outcomes.

- Use only a transmit and receive type RF head coil to minimize the exposure of the lead/neurostimulator system to the MRI RF fields. Do not use a whole body RF coil.

- Select imaging parameters to perform MRI at a specific absorption rate (SAR) that does not exceed 0.4 W/kg in the head.

To validate the in-vitro information regarding safety and efficacy, patient medical records were reviewed at four North American centers that participated in the Medtronic-sponsored clinical investigations of Activa therapies. These centers had 39 patients who underwent a total of 55 MRI procedures with implanted components of the Activa System. At least one neurostimulator was in place for 27 of these procedures.

The MRI procedures were used to verify lead implant within the target (40 procedures); to assess other medical conditions (7 procedures); to localize the brain target contralateral to an implanted lead or to replace a lead (8 procedures). There were no reported adverse events associated with MRI procedures used in conjunction with the implanted Activa lead, regardless of implantation of extensions and neurostimulators.

Risks of MRI and Activa Safety Results

The potential risks of performing MRI on patients with implanted neurostimulation systems and the specific safety information related to the Activa System include:

- Magnetic field interactions and mechanical forces

- Heating effects around the neurostimulator or lead electrodes from Electromagnetic Interference (EMI)

- Inadvertent reed switch activation from magnetic fields

- Neurostimulator damage

- Image distortion and artifacts

Magnetic Field Interactions and Mechanical Forces

Potential Risk: The neurostimulator may experience mechanical forces within or near the static magnetic field of the MRI system due to small amounts of magnetic material contained in the neurostimulator. This may cause the neurostimulator to move within the implant pocket and/or may place mechanical stress on tissues and/or the lead. Compromised tissues (such as recently sutured tissue) may be susceptible to further injury from these forces. Patients may feel a tugging sensation at the site of the neurostimulator implant.

Summary of Test Results: Implanted leads and extensions should not experience magnetic field related mechanical forces since they are made from nonmagnetic material. Based on MRI safety testing conducted using an MRI system with a static magnetic field of 1.5-Tesla, patient-equivalent phantoms and various device configurations, the magnetic forces acting on the Activa neurostimulator are less than the force of gravity.

Heating Effects

Potential Risk: As with many biomedical implants (e.g. joint replacements, spinal fixation rods, etc.), heating associated with MRI may occur in the Activa System. Tissue damage is a risk because of the potential for temperature rise in the system.

Summary of Test Results: To date, no clinically significant heating has been reported in patients with Activa Systems in the Medtronic sponsored clinical investigations. Clinically significant heating of up to 15° C was observed in a patient-equivalent phantom during testing in a 1.5-Tesla MRI. To minimize heating the recommendations in "**MRI Operation/Settings**" must be followed.

Inadvertent Reed Switch Activation/Electromagnetic Interference (EMI)

Potential Risk: The magnetic fields of the MRI may activate the magnetic reed switch within the neurostimulator. This may cause the neurostimulator to switch between On and Off (in addition, the Model 7424 Itrel II Neurostimulator may also switch between normal and Mag Amp modes). The MRI gradient and/or RF fields could cause extraneous electrical current to be induced through the lead/extension that the patient may feel.

Summary of Test Results: MRI testing using a 1.5-Tesla MRI System and a patient-equivalent phantom show that the neurostimulator reed switch may be activated by the MRI resulting in on/off switching of the neurostimulator (or also for Itrel II, changing between Normal and Mag Amp modes). In the Medtronic-sponsored clinical trial, there were no reports of shocking or jolting resulting from induced voltages on the lead system.

Neurostimulator Damage

Potential Risk: Induced voltages on the lead/extension system could damage the electronic circuitry and result in a nonfunctioning neurostimulator, requiring replacement. Induced voltages could also cause the neurostimulator to lose its programmed parameter values, which would require subsequent reprogramming. In addition, the neurostimulator could lose its serial number, which cannot be reprogrammed by the physician programmer but does not affect therapeutic use of the neurostimulator.

Summary of Test Results: Medtronic-sponsored in-vitro testing using a 1.5 Tesla MRI system (GE Signa 1.5-T) did not result in damage or reprogramming to the Activa neurostimulator or associated leads and extensions.

Image Distortion and Artifacts

Potential Risk: The neurostimulator and lead/extension may distort the MRI image or cause artifacts that may block viewing of tissue located near them.

Summary of Test Results: MRI testing using a 1.5-Tesla MRI System and a patient-equivalent phantom show that there may be substantial image distortion or artifacts near the DBS lead system and neurostimulator. Proper selection of various imaging parameters, while not exceeding the recommendations in "MRI Operation/Settings" will help reduce image distortion or artifacts.

MRI Guidelines

Pre-MRI Preparation

Because of the need to change the operating parameters for the Activa System, an appropriate health care professional with access to a Medtronic neurological physician programmer should assist and prepare the patient with this device for the MRI procedure.

• If the neurostimulator has already been implanted, record the patient's current therapeutic settings, set the neurostimulator amplitude to 0 volts (normal and magnet amplitude for the Model 7424 neurostimulator) and turn the neurostimulator output to "Off".

- Disconnect all external leads (screening cables) from any percutaneous extensions. Any parts of the percutaneous extensions that exit the body should be wrapped in a thermally and electrically insulating material of approximately 0.5-inch thickness or greater. These coils/leads should be kept out of contact with the patient's skin to avoid the risk of thermal burns from RF energy.

- Instruct the patient to alert the MRI system operator of any problems (heating, shocks, etc.) so the operator can terminate the MRI procedure if needed.

Implant Recommendations

- Implant the minimum length lead and extension possible to minimize induced RF voltage in the lead system.

- Avoid, if possible, implanting the neurostimulator in the abdomen. This requires the use of longer length leads/extensions that can increase the amplitude intensity of the induced RF voltage on the lead system.

MRI Operation/Settings

- Use only MRI systems operating at a static magnetic field strength of 1.5-Tesla.

- Use only a transmit and receive type RF head coil to minimize the exposure of the lead/neurostimulator system to the MRI RF fields. Do not use a whole body RF coil.

- Select imaging parameters to perform MRI at a specific absorption rate (SAR) that does not exceed 0.4 W/kg in the head.

- Carefully perform continuous verbal and visual monitoring of the patient throughout the MRI procedure.

- Discontinue the MRI if the patient experiences any pain or discomfort, or if you observe heating or other problems with the implanted components.

Post-MRI Recommendations Operation/Settings

- Verify the neurostimulator is functional.

- Reprogram the stimulation parameters to pre-MRI values.

Caution: An MRI procedure should not be performed in a patient with an Activa System that has a broken lead wire because tissue damage may result from localized heating at the break. If a broken lead wire is suspected, lead impedance should be checked on all electrodes in unipolar mode. If any electrode impedance is > 2000 ohms and battery current is <10 μA, then an x-ray should be obtained prior to an MRI to verify the presence of a broken wire.

[*The information for the Activa System was reprinted with permission, Medtronic, Minneapolis, MN, www.Medtronic.com.]

REFERENCES

Dormont D, Cornu P, Pidoux B, Bonnet AM, Biondi A, Oppenheim C, Hasboun D, Damier P, et al. Chronic thalamic stimulation with three-dimensional MR stereotactic guidance. AJNR Am J Neuroradiol 1997;18:1093-1097.

Finelli DA, Rezai AR, Ruggieri P, Tkach J, Nyenhuis J, Hridlicka G, Sharan A, Stypulkowski PH, Shellock FG. MR-related heating of deep brain stimulation electrodes: an *in vitro* study of clinical imaging sequences. AJNR Am J Neuroradiol 2002;23:1795-1802.

Finelli D, Rezai AR, Rugieri P, Tkach J, Nyenhuis JA, Shellock FG. Neurostimulation systems used for deep brain stimulation: *in vitro* assessment of MRI-related heating at 1.5-Tesla. Radiology 2002;222:586.

Gleason CA, Kaula NF, Hricak H, et al. The effect of magnetic resonance imagers on implanted neurostimulators. Pacing Clin Electrophysiol 1992;15:81-94.

Liem LA, van Dongen VC. Magnetic resonance imaging and spinal cord stimulation systems. Pain 1997;70:95-97.

Product Information, DBS, Implant Manual, Activa System. Medtronic, Minneapolis, MN, 2001 (www.Medtronic.com).

Rezai AR, et al. Thalamic stimulation and functional magnetic resonance imaging: localization of cortical and subcortical activation with implanted electrodes. J Neurosurg 1999;90;583-590.

Rezai AR, Finelli D, Ruggieri P, Tkach J, Nyenhuis JA, Shellock FG. Neurostimulators: Potential for excessive heating of deep brain stimulation electrodes during MR imaging. Journal of Magnetic Resonance Imaging 2001;14:488-489.

Rezai AR, Finelli D, Nyenhuis JA, Hrdlick G, Tkach J, Ruggieri P, Stypulkowski PH, Sharan A, Shellock FG. Neurostimulator for deep brain stimulation: Ex vivo evaluation of MRI-related heating at 1.5-Tesla. Journal of Magnetic Resonance Imaging 2002;15:241-250.

Rise MT. Instrumentation for neuromodulation. Archives Of Medical Research 2000;31:237-247.

Ruggieri P, Finelli DA, Rezai AR, Tkach J, Sharan A, Nyenhuis JA, Hrdlicka G, Stypulkowski PH, Shellock FG. Neurostimulation systems used for deep brain stimulation: in vitro evaluation of MRI related heating at 1.5-Tesla. Proceedings of the International Society for Magnetic Resonance in Medicine, Book of Abstracts 2002;10:843.

Shellock FG. Pocket Guide to MR Procedures and Metallic Objects: Update 2001, Seventh Edition, Lippincott Williams & Wilkins Healthcare, Philadelphia, 2001.

Shellock FG. Magnetic Resonance Procedures: Health Effects and Safety. CRC Press, LLC, Boca Raton, FL, 2001.

Shellock FG. MR imaging and electronically-activated devices. Radiology. 219:294-295, 2001.

Shellock FG. MR safety update 2002: Implants and devices. Journal of Magnetic Resonance Imaging 2002;16:485-496.

Shellock FG, Kanal E. Magnetic Resonance: Bioeffects, Safety, and Patient Management. Second Edition, Lippincott-Raven Press, New York, 1996.

Smith CD, Kildishev AV, Nyenhuis JA, Foster KS, Bourland JD, Interactions of MRI magnetic fields with elongated medical implants. J Appl Physics 2000; 87:6188-6190.

Smith CD, Nyenhuis JA, Kildishev AV. Chapter 16. Health effects of induced electrical currents: Implications for implants. In: Magnetic resonance: health effects and safety, FG Shellock, Editor, CRC Press, Boca Raton, FL, 2001; pp. 393-413.

Tronnier VM, Stauber A, Hahnel S, Sarem-Aslani A. Magnetic resonance imaging with implanted neurostimulators: an *in vitro* and *in vivo* study. Neurosurgery 1999;44:118-125.

Zonenshayn M, Mogilner AY, Rezai AR. Neurostimulation and functional brain stimulation. Neurological Research 2000;22;318-325.

OCULAR IMPLANTS AND DEVICES

Of the different ocular implants and devices that have been tested for MR safety, the Fatio eyelid spring, the retinal tack made from martensitic (i.e., ferromagnetic) stainless steel (Western European), the Troutman magnetic ocular implant, and the Unitek round wire eyelid spring demonstrated positive magnetic field interactions in association with exposure to a 1.5-Tesla MR system.

A patient with a Fatio eyelid spring or round wire eyelid spring may experience discomfort but would probably not be injured as a result of exposure to the magnetic fields of an MR system. Patients have undergone MR procedures with eyelid wires after having a protective plastic covering placed around the globe along with a firmly applied eye patch.

The retinal tack made from martensitic stainless steel and Troutman magnetic ocular implant may injure a patient undergoing an MR procedure although no such case has ever been reported (see **Magnetically-Activated Implants and Devices** for additional information pertaining to magnetic ocular implants or prostheses).

REFERENCES

Albert DW, Olson KR, Parel JM, et al. Magnetic resonance imaging and retinal tacks. Arch Ophthalmol 1990;108:320-321.

de Keizer RJ, Te Strake L. Intraocular lens implants (pseudophakoi) and steelwire sutures: a contraindication for MRI? Doc Ophthalmol 1984;61:281-284.

Joondeph BC, Peyman GA, Mafee MF, et al. Magnetic resonance imaging and retinal tacks [Letter]. Arch Ophthalmol 1987;105:1479-1480.

228

Marra S, Leonetti JP, Konior RJ, Raslan W. Effect of magnetic resonance imaging on implantable eyelid weights. Ann Otol Rhinol Laryngol 1995;104:448-452.

Roberts CW, Haik BG, Cahill P. Magnetic resonance imaging of metal loop intraocular lenses. Arch Ophthalmol 1990;108:320-321.

Seiff SR, Vestel KP, Truwit CL. Eyelid palpebral springs in patients undergoing magnetic resonance imaging: an area of possible concern [Letter]. Arch Ophthalmol 1991;109:319.

Shellock FG. Pocket Guide to MR Procedures and Metallic Objects: Update 2001, Seventh Edition, Lippincott Williams & Wilkins Healthcare, Philadelphia, 2001.

Shellock FG. Magnetic Resonance Procedures: Health Effects and Safety. CRC Press, LLC, Boca Raton, FL, 2001.

Shellock FG, Kanal E. Magnetic Resonance: Bioeffects, Safety, and Patient Management. Second Edition, Lippincott-Raven Press, New York, 1996.

Shellock FG, Myers SM, Schatz CJ. Ex vivo evaluation of ferromagnetism determined for metallic scleral "buckles" exposed to a 1.5-T MR scanner. Radiology 1992;185:288-289.

ORTHOPEDIC IMPLANTS, MATERIALS, AND DEVICES

Most of the orthopedic implants, materials, and devices evaluated for MR safety are made from nonferromagnetic materials and, therefore, are safe for patients undergoing MR procedures. Only the Perfix interference screw used for reconstruction of the anterior cruciate ligament has been found to be highly ferromagnetic. However, because this interference screw is firmly imbedded in bone for its specific application, it is held in place with sufficient retentive forces to prevent movement or dislodgment. Patients with the Perfix interference screw have safely undergone MR procedures using MR systems operating at 1.5-Tesla.

The presence of the Perfix interference screw causes extensive image distortion during MR imaging of the knee. Therefore, one of the other nonferromagnetic interference screws that are commercially available should be used for reconstruction of the anterior cruciate ligament if MR imaging is to be utilized for subsequent evaluation of the knee. Patients with the orthopedic implants, materials, and devices indicated in **The List** have undergone MR procedures using MR systems operating with static magnetic fields up to 1.5-Tesla without incident.

MR Safety at 3.0-Tesla and Orthopedic Implants, Materials, and Devices. A variety of orthopedic implants have been evaluated for magnetic field interactions at 3.0-Tesla (see **The List**). All of these are considered to be safe based on findings for deflection angles, qualitative torque measurements, and the intended *in vivo* uses of these devices.

230

REFERENCES

American Society for Testing and Materials (ASTM) Designation: F 2052. Standard test method for measurement of magnetically induced displacement force on passive implants in the magnetic resonance environment. In: Annual Book of ASTM Standards, Section 13, Medical Devices and Services, Volume 13.01 Medical Devices; Emergency Medical Services. West Conshohocken, PA, 2002, pp. 1576-1580.

Lyons CJ, Betz RR, Mesgarzadeh M, et al. The effect of magnetic resonance imaging on metal spine implants. Spine 1989;14:670-672.

Mechlin M, Thickman D, Kressel HY, et al. Magnetic resonance imaging of postoperative patients with metallic implants. AJR Am J Roentgenol 1984;143:1281-1284.

Mesgarzadeh M, Revesz G, Bonakdarpour A, et al. The effect on medical metal implants by magnetic fields of magnetic resonance imaging. Skeletal Radiol 1985;14:205-206.

Shellock FG. Magnetic Resonance Procedures: Health Effects and Safety. CRC Press, LLC, Boca Raton, FL, 2001.

Shellock FG. Pocket Guide to MR Procedures and Metallic Objects: Update 2001, Seventh Edition, Lippincott Williams & Wilkins Healthcare, Philadelphia, 2001.

Shellock FG. Biomedical implants and devices: assessment of magnetic field interactions with a 3.0-Tesla MR system. J Magn Reson Imaging 2002;16:721-732.

Shellock FG, Crues JV. High-field-strength MR imaging and metallic bioimplants: an *in vitro* evaluation of deflection forces and temperature changes induced in large prostheses [Abstract]. Radiology 1987;165:150.

Shellock FG, Kanal E. Magnetic Resonance: Bioeffects, Safety, and Patient Management. Second Edition, Lippincott-Raven Press, New York, 1996.

Shellock FG, Mink JH, Curtin S, et al. MRI and orthopedic implants used for anterior cruciate ligament reconstruction: assessment of ferromagnetism and artifacts. J Magn Reson Imaging 1992;2:225-228.

Shellock FG, Morisoli S, Kanal E. MR procedures and biomedical implants, materials, and devices: 1993 update. Radiology 1993;189:587-599.

OTOLOGIC IMPLANTS

A patient who has any of the cochlear implants or other hearing devices with implanted or internal components shown in **The List** should not be exposed to the magnetic fields of the MR system, unless MR safety investigations have demonstrated otherwise. Additionally, these devices are activated by electronic and/or magnetic mechanisms that could be problematic for patients or individuals who enter the MR environment or undergo MR procedures. For additional information about these and other similar implants, refer to **Chochlear Implants** and **Hearing Aids and Other Hearing Devices.**

Of the remaining otologic implants that have been evaluated for the presence of ferromagnetism, the McGee stapedectomy piston prosthesis, made from platinum and chromium-nickel alloy stainless steel, is ferromagnetic. This particular otologic implant has been recalled by the manufacturer. Patients who received these devices have been issued warnings to avoid MR procedures. The specific item and lot numbers of the McGee implants that were recalled and considered to be contraindication for MR procedures are as follows (Personal Communication, Winston Geer, Smith & Nephew Richards Inc., Barlett, TN, 1995):

Item No.	Lot Number:
14-0330	1W91100, 4UO9690
14-0331	4U09700
14-0332	1W91110, 4U58540, 4U86300
14-0333	4U09710, 1W34390, 2WR4073
14-0334	4U09720, 1W34390, 2WR4073
14-0335	1W34400, 4U09730
14-0336	3U18350, 3U50470, 4UR2889
14-0337	3U18370, 4UR2889
14-0338	3U18390, 4U02900, 4UR1453
14-0339	3U18400, 3U50500
14-0340	3U18410, 3U50500
14-0341	3U41200, 4UR2889

REFERENCES

Applebaum EL, Valvassori GE. Effects of magnetic resonance imaging fields on stapedectomy prostheses. Arch Otolaryngol 1985;11:820-821.

Applebaum EL, Valvassori GE. Further studies on the effects of magnetic resonance fields on middle ear implants. Ann Otol Rhinol Laryngol 1990;99:801-804.

Leon JA, Gabriele OF. Middle ear prothesis: significance in magnetic resonance imaging. Magn Reson Imaging 1987;5:405-406.

Nogueira M, Shellock FG. Otologic bioimplants: Ex vivo assessment of ferromagnetism and artifacts at 1.5-Tesla. AJR Am J Roentgenol 1995;163:1472-1473.

Shellock FG. Guide to MR Procedures and Metallic Objects: Update 2001, Seventh Edition, Lippincott Williams & Wilkins Healthcare, Philadelphia, 2001.

Shellock FG. Magnetic Resonance Procedures: Health Effects and Safety. CRC Press, LLC, Boca Raton, FL, 2001.

Shellock FG, Kanal E. Magnetic Resonance: Bioeffects, Safety, and Patient Management. Second Edition, Lippincott-Raven Press, New York, 1996.

Shellock FG, Morisoli S, Kanal E. MR procedures and biomedical implants, materials, and devices: 1993 update. Radiology 1993;189:587-599.

White DW. Interaction between magnetic fields and metallic ossicular prostheses. Am J Otol 1987;8:290-292.

OXYGEN TANKS AND GAS CYLINDERS

According to Chaljub et al., accidents related to ferromagnetic oxygen tanks and other gas cylinders that can become projectiles may be increasing. In fact, missile-related accidents for these objects have resulted in at least one fatality, several injuries, substantial damage to MR systems, and down-time (i.e., loss of revenue) for MRI centers.

Therefore, MRI facilities should devise an appropriate policy for delivery of oxygen or other gases to patients undergoing MR procedures. In lieu of utilizing pipes to directly deliver gases to patients, the use of non-magnetic (usually aluminum) gas cylinders is one means of preventing "missile effect" hazards in the MR environment. Various sizes of non-magnetic oxygen tanks and cylinders for other gases are commercially available from a variety of vendors, including Magmedix (Magmedix, Gardner, MA; www.Magmedix.com).

MRI centers should have a sufficient number of nonmagnetic oxygen tanks in the immediate and general area to prevent responding emergency staff members from introducing ferromagnetic objects into the MR environment. In fact, some hospital-based MR centers have nonmagnetic oxygen tanks used throughout their buildings to prevent projectile accidents.

Nonmagnetic tanks *must be prominently labeled* to avoid confusion with magnetic cylinders. Furthermore, all healthcare workers that work in and around the MR environment must be informed regarding the fact that only nonmagnetic oxygen and other gas cylinders are allowed into the MR system room.

Nonmagnetic oxygen regulators, flow meters, cylinder carts, cylinder stands, cylinder holders for wheelchairs, and suction devices are also commercially available to provide safe respiratory support of patients in the MR environment.

234

REFERENCES

Chaljub G, et al. Projectile cylinder accidents resulting from the presence of ferromagnetic nitrous oxide or oxygen tanks in the MR suite. American Journal of Roentgenology 2001;177:27-30.

ECRI. Patient Death Illustrates the Importance of Adhering to Safety Precautions in Magnetic Resonance Environments. ECRI, Plymouth Meeting, PA, Aug. 6, 2001.

Jolesz FA, et al. Compatible instrumentation for intraoperative MRI: expanding resources. Journal of Magnetic Resonance Imaging, 1998;8:8-11.

Keeler EK, et al. Accessory equipment considerations with respect to MRI compatibility. Journal of Magnetic Resonance Imaging, 1998;8:12-18.

Shellock FG. Magnetic Resonance Procedures: Health Effects and Safety. CRC Press, Boca Raton, FL, 2001.

PATENT DUCTUS ARTERIOSUS (PDA), ATRIAL SEPTAL DEFECT (ASD), AND VENTRICULAR SEPTAL DEFECT (VSD) OCCLUDERS

Metallic cardiac occluders are implants used to treat patients with patient ductus arteriosus (PDA), atrial septal defect (ASD), or ventricular septal defect (VSD) heart conditions. As long as the proper size occluder is used, the amount of retention provided by the folded-back, hinged arms of the device is sufficient to keep it in place, acutely. Eventually, tissue growth covers the cardiac occluder and facilitates retention.

The metallic PDA, ASD, and VSD occluders that have been tested for magnetic qualities were made from either 304V stainless steel or MP35N. Occluders made from 304V stainless steel were found to be "weakly" ferromagnetism, whereas those made from MP35N were nonferromagnetic in association with a 1.5-Tesla MR system.

Patients with cardiac occluders made from MP35N (i.e., a nonferromagnetic alloy) may undergo MR procedures at 1.5-Tesla or less any time after placement of these implants. However, patients with cardiac occluders made from 304V stainless steel are recommended to wait approximately 6 to 8 weeks after placement of these devices before undergoing MR procedures. This "wait period" permits tissue ingrowth to provide additional retentive forces for the occluders made from weakly ferromagnetic materials. If there is any question about the integrity of the retention aspects of a metallic cardiac occluder made from a ferromagnetic material, the patient or individual should not be allowed into the MR environment or to undergo an MR procedure.

REFERENCES

Shellock FG, Morisoli SM. Ex vivo evaluation of ferromagnetism and artifacts for cardiac occluders exposed to a 1.5-Tesla MR system. J Magn Reson Imaging 1994;4:213-215.

Shellock FG. Pocket Guide to MR Procedures and Metallic Objects: Update 2001, Seventh Edition, Lippincott Williams & Wilkins Healthcare, Philadelphia, 2001.

Shellock FG. Magnetic Resonance Procedures: Health Effects and Safety. CRC Press, LLC, Boca Raton, FL, 2001.

Shellock FG, Kanal E. Magnetic Resonance: Bioeffects, Safety, and Patient Management. Second Edition, Lippincott-Raven Press, New York, 1996.

PELLETS AND BULLETS

The majority of pellets and bullets tested for MR safety were found to be composed of nonferromagnetic materials. Ammunition that proved to be ferromagnetic tended to be manufactured in foreign countries and/or used for military applications. Shrapnel typically contains steel and, therefore, presents a potential hazard for patients undergoing MR procedures. Because pellets, bullets, and shrapnel are typically contaminated with ferromagnetic materials, the risk vs. benefit of performing an MR procedure in a patient should be carefully considered. Additional consideration should be given to whether or not the metallic object is located near or in a vital anatomic structure, with the assumption that the object is likely to be ferromagnetic.

In an effort to reduce lead poisoning in "puddling" type ducks, the federal government requires many of the eastern United States to use steel shotgun pellets instead of lead. The presence of steel shotgun pellets presents a potential hazard to patients undergoing MR procedures and causes severe imaging artifacts at the immediate position of these metallic objects.

In one case, a small metallic BB located in a subcutaneous site caused painful symptoms in a patient exposed to the magnetic fields of the MR system, although no serious injury occurred. In consideration of this information, MR healthcare professionals should exercise caution whenever deciding to perform MR procedures in patients with pellets, bullets, shrapnel or any other similar ballistic objects.

An investigation by Smugar et al. was conducted to determine whether neurologic problems developed in patients with intraspinal bullets or bullet fragments in association with MR imaging performed at 1.5-Tesla. Patients were queried during scanning for symptoms of discomfort, pain, or changes in neurological status. Additionally, detailed neurological examinations were performed prior to MRI, post MRI, and at the patients' discharge. Based on these findings, Smugar et al. concluded that patients

238

with complete spinal chord injury may undergo MR imaging if they have intraspinal bullets or fragments without concern for affects on their physical or neurological status. Thus, metallic fragments in the spinal canals of paralyzed patients are believed to represent only a relative contraindication to MR procedures.

REFERENCES

Shellock FG. Pocket Guide to MR Procedures and Metallic Objects: Update 2001, Seventh Edition, Lippincott Williams & Wilkins Healthcare, Philadelphia, 2001.

Shellock FG. Magnetic Resonance Procedures: Health Effects and Safety. CRC Press, LLC, Boca Raton, FL, 2001.

Shellock FG, Kanal E. Magnetic Resonance: Bioeffects, Safety, and Patient Management. Second Edition, Lippincott-Raven Press, New York, 1996.

Smugar SS, Schweitzer ME, Hume E. MRI in patients with intraspinal bullets. J Magn Reson Imaging 1999;9:151-153.

Teitelbaum GP. Metallic ballistic fragments: MR imaging safety and artifacts [Letter]. Radiology 1990;177:883.

Teitelbaum GP, Yee CA, Van Horn DD, et al. Metallic ballistic fragments: MR imaging safety and artifacts. Radiology 1990;175:855-859.

PENILE IMPLANTS

Several different types of penile implants and prostheses have been evaluated for MR safety. Of these, two (i.e., the Duraphase and Omniphase models) demonstrated substantial ferromagnetic qualities when exposed to a 1.5-Tesla MR system. However, it is unlikely that a penile implant would severely injure a patient undergoing an MR procedure because of the relative degree of the magnetic field interactions associated with a 1.5-Tesla MR system. This is especially the case considering the manner in which this type of device is utilized. Nevertheless, it would undoubtedly be uncomfortable and disconcerting for a patient with a ferromagnetic penile implant to undergo an MR examination. For this reason, subjecting a patient with the Duraphase or Omniphase penile implant to the MR environment or an MR procedure is inadvisable.

MR Safety at 3.0-Tesla and Penile Implants. Several different penile implants have been tested for MR safety in association with 3.0-Tesla MR systems. Findings for these specific penile implants indicated that they either exhibited no magnetic field interactions or relatively minor or "weak" magnetic field interactions. Accordingly, these specific penile implants are considered safe for patients undergoing MR procedures using MR systems operating at 3.0-Tesla.

REFERENCES

Shellock FG. Magnetic Resonance Procedures: Health Effects and Safety. CRC Press, LLC, Boca Raton, FL, 2001.

Shellock FG. Pocket Guide to MR Procedures and Metallic Objects: Update 2001, Seventh Edition, Lippincott Williams & Wilkins Healthcare, Philadelphia, 2001.

Shellock FG, Crues JV, Sacks SA. High-field magnetic resonance imaging of penile prostheses: *in vitro* evaluation of deflection forces and imaging artifacts [Abstract]. In: Book of Abstracts, Society of Magnetic Resonance in Medicine. Berkeley, CA: Society of Magnetic Resonance in Medicine, 1987;3:915.

Shellock FG, Kanal E. Magnetic Resonance: Bioeffects, Safety, and Patient Management. Second Edition, Lippincott-Raven Press, New York, 1996.

RETAINED CARDIAC PACING WIRES AND TEMPORARY CARDIAC PACING WIRES

A patient referred for an MR procedure may have cardiac pacing wires retained after cardiac surgery (i.e., referred to as "retained cardiac pacing wires"). Alternatively, the patient may have cardiac pacing wires that are not connected to an implanted pulse generator (e.g., connected to an external pulse generator and used for temporary arrhythmia treatment). Careful consideration must be given to these cases prior to performance of MR procedures.

A study by Hartnell et al. reported that patients with retained temporary epicardial pacing wires, cut short at the skin (i.e., after they were no longer needed during the postoperative period), did not experience changes in baseline electrocardiographic rhythms or symptoms during MR procedures. The investigation by Hartnell et al. is of particular importance because the presence of retained cardiac pacing wires was previously considered to be a relative contraindication for MR procedures due to the theoretical risk of inducing electrical current or heating that, in turn, could injure patients. The study by Hartnell et al. utilized 1.0- and 1.5-Tesla MR systems operating with conventional pulse sequences. Therefore, it would be prudent to use similar MR techniques and parameters as described by Hartnell et al. for patients with retained cardiac pacing wires until additional investigations are conducted to further define MR safety criteria for these devices. Of note is that there has never been a report of an incident or injury related to retained cardiac pacing wires in association with an MR procedure.

242

An investigation has been conducted to assess temporary pacing wires (i.e., commonly used with external pulse generators) to characterize them for MR safety. An *ex vivo* assessment of magnetic field interactions, artifacts, and heating associated with the presence of the temporary pacing wires was performed on the following temporary pacing wires (note that this information is highly specific to these two types of temporary pacing wires, *only*):

(1) Temporary Cardiac Pacing Wire, TPW-62, 0 (3.5 metric), (316L SS), Ethicon, Inc., Somerville, NJ

(2) Temporary Cardiac Pacing Wire With Wave, TPW92, 2-0 (3.0 metric), (316L SS), Ethicon, Inc., Somerville, NJ

The findings from this evaluation resulted in specific safety guidelines that would permit a patient with the temporary pacing wires that were evaluated to undergo an MR procedure.

Based on the results of the *ex vivo* tests, there should be no risk with respect to movement or dislodgment for these temporary cardiac pacing wires using an MR system operating at 1.5-Tesla or less. A small portion of the temporary cardiac pacing wire showed magnetic field interactions (i.e., the straight wire portion). However, this part of the pacing wire is positioned outside of the patient's body and can be maintained in a fixed position using tape or other similar method during exposure to the MR environment.

The findings from the artifact evaluation indicated that the presence of the temporary cardiac pacing wires should not greatly affect the diagnostic use of MRI, as long as the area of interest is not in the exact same position where the temporary cardiac pacing wires are located. The experiment performed to determine MRI-related heating demonstrated that there were relatively minor temperature increases in the temporary cardiac pacing wires that were considered to be physiologically inconsequential. Furthermore, the overall test results indicated that there would be no hazard or risk to a patient with these specific temporary pacing wires in association with an MR procedure as long as the exposure to RF energy does not exceed a whole body averaged specific absorption rate of 1.1 W/kg.

Specific safety guidelines for performing an MR procedure in patients with the temporary cardiac pacing wires that underwent testing are, as follows:

(1) The temporary cardiac pacing wires must be disconnected from the external pulse generator prior to the MR procedure (i.e., the patient's

heart cannot be paced during the MR procedure). The pulse generator must not be brought into or used in the MR system room.

(2) The ends of the temporary cardiac pacing wire (i.e., the straight leads that connect to the pulse generator) should be taped with electrician's tape to insulate them. The ends of the temporary cardiac pacing wires should then be securely attached to the patient using adhesive tape or other similar means.

(3) The temporary cardiac pacing wire should be placed on the patient in a "straight line" configuration, without any loops and fixed in this position using adhesive tape or other means and directed down the center of the MR system. In addition, the **Guidelines to Prevent Excessive Heating and Burns Associated with Magnetic Resonance Procedures** should be considered and implemented, as needed.

(4) **Static Magnetic Field of the MR System and Pulse Sequences:** MR imaging should only be performed using MR systems with static magnetic fields of 1.5-Tesla or less and conventional techniques. Standard spin echo, fast spin echo, and gradient echo pulse sequences may be used. Pulse sequences (e.g., echo planar techniques) or conditions that produce exposures to high levels of RF energy (i.e., exceeding a whole body averaged specific absorption rate of 1.1 W/kg) or exposure to gradient fields that exceed 20 Tesla/second, or any other unconventional MRI technique should be avoided.

(5) **Gradient Magnetic Fields of the MR System:** Pulse sequences (e.g., echo planar imaging techniques or other rapid imaging pulse sequences), gradient coils or other techniques and procedures that exceed a gradient magnetic field of 20 Tesla/sec must not be used for MR procedures. The use of unconventional or non-standard MR imaging techniques must be avoided.

(6) **Radiofrequency (RF) Fields of the MR System:** MR procedures must not exceed exposures to RF fields greater than a whole body averaged specific absorption rate (SAR) of 1.1 W/kg for 15-minutes of imaging. The use of unconventional or non-standard MR imaging techniques must be avoided.

(7) **MRI Artifacts:** Artifacts for temporary cardiac pacing wires have been characterized using a 1.5-Tesla MR system and various pulse sequences. In general, the artifact size is dependent on the type of pulse sequence used for imaging, the direction of the frequency encoding direction, and the size of the field of view.

(8) The patient should be continuously observed and monitored during the MR procedure and instructed to report any unusual sensations to the MR system operator. If any unusual sensation is experienced by the patient, the MR procedure must be discontinued immediately.

REFERENCES

Hartnell GG, et al. Safety of MR imaging in patients who have retained metallic materials after cardiac surgery. Am J Roentgenol 1997;168: 1157-1159.

Shellock FG. Magnetic Resonance Procedures: Health Effects and Safety. CRC Press, LLC, Boca Raton, FL, 2001.

Shellock FG. Pocket Guide to MR Procedures and Metallic Objects: Update 2001, Seventh Edition, Lippincott Williams & Wilkins Healthcare, Philadelphia, 2001.

Shellock FG, Kanal E. Magnetic Resonance: Bioeffects, Safety, and Patient Management. Second Edition, Lippincott-Raven Press, New York, 1996.

Shellock FG, Shellock VJ. Cardiovascular catheters and accessories: Ex vivo testing of ferromagnetism, heating, and artifacts associated with MRI. J Magn Reson Imaging 1998;8:1338-1342.

SURGICAL INSTRUMENTS AND DEVICES

Interventional magnetic resonance (MR) procedures have evolved into clinically viable techniques for a variety of minimally invasive surgical and therapeutic applications. This has resulted in the development and performance of innovative MR procedures that include percutaneous biopsy (e.g., breast, bone, brain, abdominal), endoscopic surgery of the abdomen, spine, and sinuses, open brain surgery, and MR-guided monitoring of thermal therapies (i.e., laser-induced, RF-induced, and cryomediated procedures).

Surgical instruments and devices are an obvious necessity for interventional MR procedures. Besides the typical MR safety concerns, there are possible hazards in the interventional MR environment related to the surgical instruments and devices that must be addressed to ensure the safety of MR healthcare practitioners and patients. Many of the conventional instruments and devices are made from metallic materials that can create substantial problems in association with interventional MR procedures.

The interventional MR safety and compatibility issues that exist for surgical instruments and devices include unwanted movement caused by magnetic field interactions (e.g., the "missile effect", translational attraction, torque), and heat generated by RF power deposition. Additionally, artifacts associated with the use of a surgical instrument or device can be particularly problematic if in the imaging area of interest during its intended use. To address these problems, various surgical instruments and devices have been developed that do not present a hazard or additional risk to the patient or MR healthcare practitioner in the interventional MR environment (see **Appendix I**).

246

REFERENCES

Hinks RS, Bronskill MJ, Kucharczyk W, Bernstein M, Collick BD, Henkelman RM. MR systems for image-guided therapy. J Magn Reson Imaging 1998;8:19-25.

Jolesz FA. Interventional and intraoperative MRI: a general overview of the field. J Magn Reson Imaging 1998;3-7.

Jolesz FA. Image-guided procedures and the operating room of the future. Radiology 1997; 204:601-612.

Jolesz FA, et al. Compatible instrumentation for intraoperative MRI: expanding resources. J Magn Reson Imaging 1998;8:8-11.

Shellock FG. Compatibility of an endoscope designed for use in interventional MR imaging procedures. AJR Amer J Roentgenol 1998;171:1297-1300.

Shellock FG. Metallic surgical instruments for interventional MRI procedures: evaluation of MR safety. J Magn Reson Imaging 2001;13:152-157.

Shellock FG. MRI safety of instruments designed for interventional MRI: assessment of ferromagnetism, heating, and artifacts. Workshop on New Insights into Safety and Compatibility Issues Affecting In Vivo MR, Syllabus, International Society of Magnetic Resonance in Medicine, Berkeley, 1998; pp. 39.

Shellock FG, Shellock VJ. Ceramic surgical instruments: Evaluation of MR-compatibility at 1.5 Tesla. J Magn Reson Imaging 1996;6:954-956.

SUTURES

A variety of materials, including nonmetallic and metallic materials, are used to make sutures. Various sutures with the needles removed have been testing at 1.5- and 3.0-Tesla because they have not been previously evaluated in association with the MR environment and there is confusion regarding the implications of these materials for patients undergoing MR procedures. At 1.5-Tesla, for the 13 different sutures evaluated, all were considered safe for patients.

MR Safety at 3.0-Tesla and Sutures. At 3.0-Tesla, most sutures displayed no magnetic field interactions, while two (Flexon suture and Steel suture, United States Surgical, North Haven, CT) showed minor deflection angles and torque. For these two sutures, the *in situ* application of these materials is likely to provide sufficient counter-forces to prevent movement or dislodgment. Therefore, in consideration of the intended *in vivo* use of these materials, all of the sutures with the needles removed that have been tested are regarded to be safe at 3.0-Tesla.

REFERENCES

American Society for Testing and Materials (ASTM) Designation: F 2052. Standard test method for measurement of magnetically induced displacement force on passive implants in the magnetic resonance environment. In: Annual Book of ASTM Standards, Section 13, Medical Devices and Services, Volume 13.01 Medical Devices; Emergency Medical Services. West Conshohocken, PA, 2002, pp. 1576-1580.

Shellock FG. Biomedical implants and devices: assessment of magnetic field interactions with a 3.0-Tesla MR system. J Magn Reson Imaging J Magn Reson Imaging 2002;16:721-732.

SYNCHROMED INFUSION SYSTEM

The SynchroMed® Infusion System (Medtronic, Minneapolis, MN) has two parts that are both implanted in the body during a surgical procedure: the catheter and the pump. The catheter is a small, soft tube. One end is connected to the pump, and the other is placed in the intrathecal space (where fluid flows around the spinal cord). The pump is a round metal device that stores and releases prescribed amounts of medication directly into the intrathecal space. It is about one inch (2.5-cm) thick, three inches (8.5-cm) in diameter, weighs about six ounces (205-grams) and is made of titanium.

The reservoir is the space inside the pump that holds the medication. There is a raised center portion of the pump through which the pump is refilled. A physician or nurse inserts a needle through the patient's skin and through the fill port to fill the pump. Some pumps have a side catheter access port that allows injection of other medications or sterile solutions directly into the catheter, bypassing the pump.

MRI Safety Information for the SynchroMed Infusion System

Exposure to the MR system and an MR procedure will temporarily stop the rotor of the pump motor due to the magnetic field of the MR system and suspend drug infusion for the duration of the MRI exposure. The pump should resume normal operation upon termination of the MRI exposure. Prior to MRI, the physician should determine if the patient can safely be deprived of drug delivery. If the patient cannot be safely deprived of drug delivery, alternative delivery methods for the drug can be utilized during the time required for the MRI scan.

If there is concern that depriving the patient of drug delivery may be unsafe for the patient during the MR procedure, medical supervision should be provided while the MRI is conducted. Prior to scheduling an MRI scan and upon completion of the MRI scan, or shortly thereafter, the pump status should be confirmed using the SynchroMed programmer.

In the unlikely event that any change to the pump status has occurred, a "pump memory error" message will be displayed and the pump will sound a Pump Memory Error Alarm (double tone). The pump should then be reprogrammed and Medtronic Technical Services notified at (800) 328-0810.

MRI Safety Testing

Testing conducted on the SynchroMed pump has established the following with regard to other MR safety issues:

Tissue Heating Adjacent to Implant During MRI Scans

Specific Absorption Rate (SAR): The presence of the pump can potentially cause a two-fold increase of the local temperature rise in tissues near the pump.

During a 20-minute pulse sequence in a 1.5-T (Tesla) GE Signa Scanner with a whole-body average SAR of 1.0 W/kg, a temperature rise of 1 degree Celsius in a static phantom was observed near the pump implanted in the "abdomen" of the phantom. The temperature rise in a static phantom represents a worst case for physiological temperature rise and the 20 minute scan time is representative of a typical imaging session.

FDA MRI guidance allows a physiological temperature rise of up to 2 degrees Celsius in the torso, therefore the local temperature rise in the phantom is considered by FDA guidance to be below the level of concern. Implanting the pump more lateral to the midline of the abdomen may result in higher temperature rises in tissues near the pump.

In the unlikely event that the patient experiences uncomfortable warmth near the pump, the MRI scan should be stopped and the scan parameters adjusted to reduce the SAR to comfortable levels.

Peripheral Nerve Stimulation During MRI Scans

Time-Varying Gradient Magnetic Fields: Presence of the pump may potentially cause a two-fold increase of the induced electric field in tissues near the pump. With the pump implanted in the abdomen, using pulse sequences that have dB/dt up to 20 T/s, the measured induced electric field near the pump is below the threshold necessary to cause stimulation.

In the unlikely event that the patient reports stimulation during the scan, the proper procedure is the same as for patients without implants: Stop the MRI scan and adjust the scan parameters to reduce the potential for nerve stimulation.

Static Magnetic Field: For magnetic fields up to 1.5-Tesla, the magnetic force and torque on the SynchroMed pump will be less than the force and torque due to gravity.

For magnetic fields of 2.0-T, the patient may experience a slight tugging sensation at the pump implant site. An elastic garment or wrap will prevent the pump from moving and reduce the sensation the patient may experience. SynchroMed pump performance has not been established in >2.0-Tesla MR scanners and it is not recommended that patients have MRI using these scanners.

Image Distortion. The SynchroMed pump contains ferromagnetic components that will cause image distortion and image dropout in areas around the pump. The severity of image artifact is dependent on the MR pulse sequence used. For spin echo pulse sequences, the area of significant image artifact may be 20 to 25-cm across. Images of the head or lower extremities should be largely unaffected.

Minimizing Image Distortion: MR image artifact may be minimized by careful choice of pulse sequence parameters and location of the angle and location of the imaging plane. However, the reduction in image distortion obtained by adjustment of pulse sequence parameters will usually be at a cost in signal to noise ratio. The following general principles should be followed:

(1) Use imaging sequences with stronger gradients for both slice and read encoding directions. Employ higher bandwidth for both RF pulse and data sampling.

(2) Choose an orientation for read-out axis that minimizes the appearance of in-plane distortion.

(3) Use spin echo (SE) or gradient echo (GE) MR imaging sequences with a relatively high data sampling bandwidth.

(4) Use shorter echo time (TE) for gradient echo techniques, whenever possible.

(5) Be aware that the actual imaging slice shape can be curved in space due to the presence of the field disturbance of the pump (as stated above, this is image distortion).

(6) Identify the location of the implant in the patient and when possible, orient all imaging slices away from the implanted pump.

[*The majority of this information was obtained with permission from the technical manuals for the SynchroMed Pump and SynchroMed EL Pump, Medtronic, Minneapolis, MN]

THERASEED RADIOACTIVE SEED IMPLANT

The TheraSeed radioactive seed implant (Theragenics Corporation, Buford, GA) is used to deliver low-level radiation to the prostate to treat cancer. Palladium-103 is the isotope contained in TheraSeed implant. Because the radiation is so low and the seeds are placed so precisely, virtually all the radiation is absorbed by the prostate. Treatments with the TheraSeed implant may involve placement in the prostate of from 80 to 120 of these devices. Importantly, these seeds are implanted with "spacers" between them and placed according to treatment plans, so that they do not contact one another during their intended "*in vivo*" use.

TheraSeed radioactive seed is relatively small implant comprised of a titanium tube with two graphite pellets and a lead marker inside. The outside dimensions of tube are approximately length 0.177" in length with a diameter of 0.032".

Magnetic resonance safety tests (magnetic field interactions, heating, induced currents, and artifacts) conducted on the TheraSeed revealed that this implant is safe for patients undergoing MR procedures at 1.5-Tesla or less.

REFERENCE

http://www.MRIsafety.com

TRANSDERMAL PATCHES

There have been several anecdotal reports pertaining to patients undergoing MR procedures who experienced heating in association with wearing drug delivery systems that involve transdermal "patch" techniques (e.g., for administration of nitroglycerine, nicotine, or other similar medication). Apparently, the metallic components of these devices can be heated in association with MR procedures.

In one reported case, a Deponit (nitroglycerin transdermal delivery system) patch, which contains an aluminum component, was worn by a patient during MR imaging. The metallic component of this patch is nonferromagnetic and, therefore, not attracted to the static magnetic field of the MR system. However, the patient wearing this patch received a second degree burn during MR imaging performed using standard pulse sequences (Personal Communication, Robert E. Mucha, Schwarz Pharma, Milwaukee, WI; 1995). This likely occurred due to the conductive qualities of the metallic component of the transdermal patch.

In consideration of the above, it is recommended that any patient using the Deponit or similar transdermal delivery system with a metallic component have the patch removed prior to the MR procedure. A new patch should be applied after the examination is completed (Personal Communication, Robert E. Mucha, Schwarz Pharma, Milwaukee, WI; 1995). The patient's physician should be consulted prior to removing the transdermal patch to obtain information related to the proper administration of any transdermal patch medication that is dispensed by a prescription.

REFERENCES

Shellock FG. Magnetic Resonance Procedures: Health Effects and Safety. CRC Press, LLC, Boca Raton, FL, 2001.

Shellock FG. Pocket Guide to MR Procedures and Metallic Objects: Update 2001, Seventh Edition, Lippincott Williams & Wilkins Healthcare, Philadelphia, 2001.

Shellock FG, Kanal E. Magnetic Resonance: Bioeffects, Safety, and Patient Management. Second Edition, Lippincott-Raven Press, New York, 1996.

VASCULAR ACCESS PORTS, INFUSION PUMPS*, AND CATHETERS

Vascular access ports, infusion pumps, and catheters are implants and devices that are commonly used to provide long-term vascular administration of chemotherapeutic agents, antibiotics, analgesics and other medications (also, see information pertaining to the **AccuRx Constant Flow Implantable Pump**, the **IsoMed Implantable Constant-Flow Infusion Pump**, and the **SynchroMed Infusion System**). Vascular access ports are implanted typically in a subcutaneous pocket over the upper chest wall with the catheters inserted either in the jugular, subclavian, or cephalic vein. Smaller vascular access ports, which are less obtrusive and tend to be tolerated better, have also been designed for implantation in the arms of children or adults, with vascular access via an antecubital vein.

Vascular access ports have a variety of inherent features (e.g., a reservoir, central septum, and catheter) and are constructed from different types of materials including stainless steel, titanium, silicone, and various forms of plastic. Because of the widespread use of vascular access ports and associated catheters and the high probability that patients with these devices may require MR procedures, it was important to determine MR-safety or MR-compatibility of these implants.

Three of the implantable vascular access ports and catheters evaluated for safety with MR procedures showed measurable magnetic field interactions during exposure to the MR systems (typically 1.5-Tesla) used for testing, but the interactions were considered to be minor relative to the *in vivo* applications of these implants. Therefore, an MR procedure is safe to perform using an MR system operating at 1.5-Tesla or less in a patient that has one of the vascular access ports or catheters listed in **The List**.

With respect to MR imaging and artifacts, in general, the vascular access ports that will produce the least amount of artifact in association with MR imaging are made entirely from nonmetallic materials. Conversely, the ones that produce the greatest amount of artifact are composed of metal(s) or have metal in an unusual shape (e.g., the OmegaPort Access devices).

Some manufacturers of vascular access ports have decided to make devices entirely from nonmetallic materials under the assumption that this is required for the device to be "MR-compatible." In fact, several manufacturers have produced brochures that state that their devices allow "distortion free imaging" or "will not obscure important structures" during MR imaging.

In one marketing brochure, an MR image is shown that is "color-enhanced" such that the artifact caused by a competitor's metallic vascular access port appears to be inordinately large, whereas the manufacturer's plastic vascular access port caused essentially no distortion of the image (Unpublished Observations, F.G. Shellock, 1994). This misrepresents the actual MR-compatibility issue and promotes a marketing claim that is without support from a diagnostic MR imaging standpoint.

Even the so-called "MR-compatible" or "MRI ports" made entirely from nonmetallic materials are, in fact, seen on the MR images because they contain silicone. The septum portion of each of the vascular access port typically is made from silicone. Using MR imaging, the Larmor precessional frequency of fat is close to that of silicone (i.e., 100 Hz at 1.5-T). Therefore, silicone used in the construction of a vascular access port may be observed on MR images with varying degrees of signal intensity depending on the pulse sequence used for imaging.

Manufacturers of nonmetallic vascular access ports have not addressed this finding during advertising and marketing of their products. On the contrary, vascular access ports made from nonmetallic materials are claimed to be "MR-compatible" and "invisible" on MR images. However, if a radiologist did not know that this type of vascular access port was present in a patient, the MR signal produced by the silicone component of the device could be considered an abnormality, or at the very least, present a confusing image. For example, this may present a diagnostic problem in a patient being evaluated for a rupture of a silicone breast implant, because silicone from the vascular access port may be misread as an "extracapsular silicone implant rupture."

In more general terms, it is improbable that an artifact produced by the presence of a metallic vascular access port will detract from the diagnostic capabilities of MR imaging. The extent of the artifact is relatively minor and, as such, is unlikely to obscure any important anatomical structures by its presence. MR imaging examinations of the chest, where most vascular access ports are typically implanted in a subcutaneous pocket, account for a relatively small percentage of diagnostic studies performed using this imaging modality.

Finally, an important issue related to the construction of a vascular access port should be discussed. Metal is typically used to make this type of implant in order to guard against piercing of the injection site by repetitive insertions of needles used to refill the reservoir. Of note is that repeated needle access of a plastic reservoir compared to a metal reservoir may perturb the functional integrity and long-term durability of the vascular access port (Unpublished Observations, 1994). This could result in embolization by fragmented plastic pieces or a reduced ability to properly flush the vascular access port. Therefore, a vascular access port with a reservoir made from metal or other similar hard material may, in fact, be more acceptable for use in a patient compared to one made from plastic.

MR Safety at 3.0-Tesla and Vascular Access Ports, Infusion Pumps, and Catheters. For the vascular access ports and infusion pumps assessed for magnetic field interactions at 3.0-Tesla, none of them exhibited magnetic field interactions and, therefore, will not move or dislodge in this MR environment. For the accessories, the infusion set and needles showed measurable ferromagnetism, with the PORT-A-CATH Needle (Deltec, Inc., St. Paul, MN) exceeding the recommended ASTM deflection angle safety guideline (i.e., greater than 45 degrees). However, during the actual use of this accessory, it is unlikely that it will present a problem in the 3.0-Tesla MR environment considering that the simple application of a small amount of adhesive tape effectively counterbalances the relatively minor ferromagnetism that was determined for this device (Unpublished 0bservations, F.G. Shellock, 2002).

[*Refer to additional information for infusion pumps and similar devices published in this textbook.]

REFERENCES

American Society for Testing and Materials (ASTM) Designation: F 2052. Standard test method for measurement of magnetically induced displacement force on passive implants in the magnetic resonance environment. In: Annual

Book of ASTM Standards, Section 13, Medical Devices and Services, Volume 13.01 Medical Devices; Emergency Medical Services. West Conshohocken, PA, 2002, pp. 1576-1580.

Shellock FG. Magnetic Resonance Procedures: Health Effects and Safety. CRC Press, LLC, Boca Raton, FL, 2001.

Shellock FG. Biomedical implants and devices: assessment of magnetic field interactions with a 3.0-Tesla MR system. J Magn Reson Imaging J Magn Reson Imaging 2002;16:721-732.

Shellock FG. Pocket Guide to MR Procedures and Metallic Objects: Update 2001, Seventh Edition, Lippincott Williams & Wilkins Healthcare, Philadelphia, 2001.

Shellock FG, Kanal E. Magnetic Resonance: Bioeffects, Safety, and Patient Management. Second Edition, Lippincott-Raven Press, New York, 1996.

Shellock FG, Nogueira M, Morisoli S. MR imaging and vascular access ports: *ex vivo* evaluation of ferromagnetism, heating, and artifacts at 1.5-T. J Magn Reson Imaging 1995;4:481-484.

Shellock FG, Shellock VJ. Vascular access ports and catheters tested for ferromagnetism, heating, and artifacts associated with MR imaging. Magn Reson Imaging 1996;14:443-447.

VOCARE BLADDER SYSTEM, IMPLANTABLE FUNCTIONAL NEUROMUSCULAR STIMULATION

The NeuroControl VOCARE Bladder System is a radiofrequency (RF) powered motor control neuroprosthesis or neurostimulation system that consists of both implanted and external components. The VOCARE Bladder System delivers low levels of electrical stimulation to a spinal cord injured patient's intact sacral spinal nerve roots in order to elicit functional contraction of the muscles innervated by them. The NeuroControl VOCARE Bladder System consists of the following subsystems:

- The **Implanted Components** include the Implantable Receiver-Stimulator and Extradural Electrodes.

- The **External Components** include the External Controller, External Transmitter, External Cable, Transmitter Tester, Battery Charger and Power Cord.

- The **Surgical Components** include the Surgical Stimulator, Intradural Surgical Probe, Extradural Surgical Probe, Surgical Test Cable, and Silicone Adhesive.

The NeuroControl VOCARE Bladder System is indicated for the treatment of patients who have clinically complete spinal cord lesions (ASIA Classification) with intact parasympathetic innervation of the bladder and are skeletally mature and neurologically stable, to provide urination on demand and to reduce post-void residual volumes of urine. Secondary intended use is to aid in bowel evacuation.

CONTRAINDICATIONS

The NeuroControl VOCARE Bladder System is contraindicated for patients with the following characteristics:

- Poor or inadequate bladder reflexes

- Active or recurrent pressure ulcers

- Active sepsis

- Implanted cardiac pacemaker

WARNINGS

The NeuroControl VOCARE Bladder System may only be prescribed, implanted, or adjusted by clinicians who have been trained and certified in its implementation and use.

PRECAUTIONS

- **X-rays, Diagnostic Ultrasound:** X-ray imaging, and diagnostic ultrasound have not been reported to affect the function of the Implantable Receiver-Stimulator or Extradural Electrodes. However, the implantable components may obscure the view of other anatomic structures.

- **MRI:** Testing of the VOCARE Bladder System in a 1.5-Tesla scanner with a maximum spatial gradient of 450 gauss/cm or less, exposed to an average Specific Absorption Rate (SAR) of 1.1 W/kg, for a 30 minute duration resulted in localized temperature rises up to 5.5°C in a gel phantom (without blood flow) and translational force less than that of a 12 gram mass and torque of 0.47 N-cm (less than that produced by the weight of the device). A patient with a VOCARE Bladder System may undergo an MR procedure using a shielded or unshielded MR system with a static magnetic field of 1.5-Tesla only and a maximum spatial gradient of 450 gauss/cm or less. The implantable components may obscure the view of other nearby anatomic structures. Artifact size is dependent on variety of factors including the type of pulse sequence used for imaging (e.g. larger for gradient echo pulse sequences and smaller for spin echo and fast spin echo pulse sequences), the direction of the frequency encoding, and the size of the field of view used for imaging. The use of non-standard scanning modes to minimize image artifact or improve visibility should be applied with caution and with the Specific Absorption Rate (SAR) not to exceed an average of 1.1 W/kg and gradient magnetic fields no greater than 20 Tesla/sec. The use of Transmit RF Coils other than the scanner's Body RF Coil or a Head RF Coil is prohibited. Testing of the function of each electrode should be conducted prior to MRI scanning to ensure no leads are broken. Do not expose patients to MRI if any lead is broken or if integrity cannot be established as excessive heating may result in a broken lead. Patients should be advised to empty their bladder or bowel prior to MRI scanning as a precaution. The external components of the VOCARE Bladder System must be removed prior to MRI scanning. Patients must be continuously observed during the MRI procedure and instructed to report any unusual sensations (e.g., warming, burning, or neuromuscular stimulation). Scanning must be discontinued immediately if any unusual sensation occurs. Contact NeuroControl Corporation for additional information.

- **Therapeutic Ultrasound, Therapeutic Diathermy, and Microwave Therapy:** Therapeutic ultrasound, therapeutic diathermy, and microwave therapy should not be performed over the area of the Implantable Receiver-Stimulator or Extradural Electrodes as it may damage the VOCARE Bladder System.

- **Electrocautery:** Do not touch the Implantable Components of the VOCARE Bladder System with electrocautery instruments. Do not use electrocautery within 1 cm of the metal electrode contacts.

- **Antibiotic Prophylaxis:** Standard antibiotic prophylaxis for patients with an implant should be utilized to protect the patient when invasive procedures (e.g., oral surgery) are performed.

- **Drug Interactions:** Anticholinergic medications, or other medications which reduce the contraction of smooth muscle, may reduce the strength of bladder contraction achieved using the VOCARE Bladder System. Anticholinergic medications should be discontinued at least three days prior to evaluating patients for the VOCARE Bladder System and prior to implantation surgery so that bladder reflexes and response to electrical stimulation can be accurately evaluated. In addition, long-acting neuromuscular blocking agents must not be used during surgery.

- **Prior procedures** (such as bladder neck surgery or bladder augmentation) or conditions (such as severe urethral damage, stricture, or erosion) may affect patient suitability for the VOCARE Bladder System or clinical outcome. Patients with bladder augmentation may not be candidates for this procedure unless they can still achieve appropriate bladder pressures through reflex contractions. Patients should be thoroughly evaluated and counseled regarding the effect of any prior procedures or conditions.

- **Post-operative incontinence** may occur following posterior rhizotomy, which is typically performed in conjunction with implantation of the VOCARE Bladder System. While rhizotomy generally abolishes reflex incontinence, some patients may still experience stress incontinence. Patients should be evaluated for open bladder neck pre-operatively and counseled regarding the factors that may increase the risk of stress incontinence.

- **Bowel motility** may be affected by the rhizotomy procedure and by use of the VOCARE Bladder System. Patients should be advised that the rhizotomy may decrease the response to suppositories and digital stimulation of the rectum. Conversely, use of the VOCARE Bladder System may increase bowel motility. Patients may need to adjust the frequency and/or method of their bowel management routine postoperatively.

- The **rhizotomy** procedure typically performed in conjunction with implantation of the VOCARE Bladder System may cause loss of erectile function and ejaculation in men who had these responses before surgery.

- **Spinal instability** may result from the laminectomies required during implantation and rhizotomy surgery. Patients should be evaluated carefully for added risk factors, such as significant osteoporosis or scoliosis.

- **Studies have not been conducted** on the use of the NeuroControl VOCARE Bladder System in pregnant women.

- **Post-operatively**, the patient should be advised to check the condition of his or her skin over the VOCARE Bladder System Receiver-Stimulator and leads daily for signs of redness, swelling, or breakdown. If skin breakdown becomes apparent, patients should contact their clinician immediately. The clinician should treat the infection aggressively, taking into consideration the extra risk presented by the presence of the implanted materials.

- **Unintended Stimulation:** While there have been no reports of VOCARE Bladder System activation or malfunction due to electromagnetic interference (such as from retail anti-theft detectors, airport metal detectors, or other electronic devices) testing has not been conducted to rule out the possibility of this occurring. Patients should be advised to notify their clinician if they experience unintended stimulation when the VOCARE Bladder System is not in use. If possible, patients should note when and where the stimulation occurred.

- **Keep it dry:** The user should avoid getting the external components, cables, and attachments of the VOCARE Bladder System wet.

- The patient and caregiver should be advised to **inspect the external cables and connectors** daily for fraying or damage and replace components when necessary.

- To avoid possible interference, patients with electric wheelchairs should be advised to **turn off their wheelchair controller** prior to turning on the VOCARE Bladder System External Controller.

- **External Defibrillation:** The effect of external defibrillation on the VOCARE Bladder System is unknown.

- Patients should be advised to turn off the VOCARE Bladder System External Controller when not in use. **The External Transmitter can become hot** if the VOCARE Bladder System is left on for extended periods of time.

[Excerpted with permission from VOCARE Bladder System, Implantable Functional Neuromuscular Stimulator, NeuroControl Corporation, Valley View, OH]

THE LIST - INFORMATION AND TERMINOLOGY

The List contains MR-safety or MR-compatibility information for over 1,200 implants, devices, materials, and other products. The objects in **The List** are divided into general categories to facilitate access and review of pertinent information.

To properly utilize **The List**, particular attention must be given to the information indicated for the highest static magnetic **Field Strength** used for testing and the **Status** information indicated for a given **Object** (Note: These specific terms correspond to the column headings for information compiled in **The List**). In addition, for certain objects, it may also be necessary to refer to specific recommendations or guidelines in **SECTION II** of this textbook.

The relevant terminology for **The List** is, as follows:

Object: This is the implant, device, material, or product that underwent evaluation for MR-safety or MR-compatibility. Information is also provided for the material(s) used to make the object and the manufacturer of the object, if known. The term "SS" refers to stainless steel.

Status: This information pertains to the results of the tests conducted for the object. MR safety tests typically included an assessment of magnetic field interactions (i.e., deflection and/or torque) and heating. In some cases, medical products were assessed for induced electrical currents, as well.

MR-safety information for each object has been specifically categorized using a **Status** designation, which indicates the object to be **Safe, Conditional, or Unsafe**, as follows:

Safe - The object is considered to be safe for the patient or individual in the MR environment, with special reference to the highest static magnetic field strength that was used for the MR safety test. The object has undergone testing to demonstrate that it is safe or it is made from material(s) considered to be safe with regard to the MR environment (e.g., plastic, silicone, etc.). Refer to additional information for the particular object indicated in the text of **SECTION II** of this book (see **Table of Contents**).

Conditional - The object may or may not be safe for the patient or individual in the MR environment, depending on the specific MR conditions that are present for the MR procedure. This information has been sub-categorized to indicate specific recommendations for the particular object, as follows:

Conditional 1 - The object is considered safe for the patient or individual in the MR environment, despite the fact that it showed positive findings for magnetic field interaction during testing. Notably, the object is considered to be only "weakly" ferromagnetic.

In general, the object is safe because the magnetic field interactions were characterized as "mild" relative to the *in vivo* forces present for the object. For example, certain prosthetic heart valve prostheses and annuloplasty rings showed measurable magnetic field interactions during exposure to the MR systems used for testing, but the magnetic field interactions were less than the forces exerted on the implants by the beating heart.

Additionally, there may be substantial "retentive" or counter forces provided by the presence of sutures or other means of fixation, tissue ingrowth, scarring, or granulation that serve to prevent the object from presenting a risk or hazard to the patient or individual in the MR environment.

For a device or product that is used for an MR-guided procedure (e.g., laryngoscope, endoscope, etc.), there may be minor magnetic field interactions in association with the MR system. However, the device or product is considered to be MR-safe and/or MR-compatible if it is used in its "intended" manner, as indicated by the manufacturer. Special attention should be given to the strength of the static magnetic field used for testing the device or product. Additionally, specific recommendations for the use of the device or product in the MR environment or during an MR procedure (i.e., typically presented in the Product Insert) should be followed carefully.

Conditional 2 - These "weakly" ferromagnetic intravascular coils, filters, stents, and cardiac occluders, or other implants typically become firmly incorporated into the tissue six to eight weeks following placement. Therefore, it is unlikely that these objects will be moved or dislodged by interactions with the magnetic fields of MR systems operating at the static magnetic field strength used for testing. Furthermore, to date, there has been no report of an injury to a patient or individual in association with an MR procedure for these coils, stents, filters, cardiac occluders or other implants.

Of note is that if the implant is made from a nonmagnetic material (e.g., Phynox, Elgiloy, titanium, titanium alloy, MP35N, Nitinol, etc.), it is unnecessary to wait six to eight weeks before performing and MR procedure using an MR system operating at 1.5-Tesla or less. In fact, MR-guided procedures are now being used to implant certain stents using MR systems operating at 1.5-Tesla.

Special Note: If there is any concern regarding the integrity of the implant or the integrity of the tissue with regard to its ability to retain the object in place during an MR procedure or during exposure to the MR environment, the patient or individual should not be allowed into the MR environment.

Conditional 3 - The Deponit (nitroglycerin transdermal delivery system) although not attracted to an MR system, has been found to heat excessively during an MR procedure. This excessive heating may produce discomfort or burn a patient or individual wearing this patch. Therefore, it is recommended that the patch be removed prior to the MR procedure. A new patch should be applied immediately after the examination. This procedure should only be done in consultation with the patient's or individual's personal physician. Other similar transdermal delivery systems that contain metallic materials should be considered in a similar manner.

Conditional 4 - This halo vest or cervical fixation device is known to have ferromagnetic components, however, the magnetic field interactions have not been determined. Nevertheless, there has been no report of patient injury in association with the presence of this device in the MR environment at the static magnetic field strength used for MR safety testing.

Conditional 5 - This object is considered safe for a patient or individual in the MR environment as long as highly specific guidelines or recommendations are followed (see specific information for a given object in **SECTION II**). Please refer to the specific criteria for performing a safe MR procedure by reviewing the information for the object in the text of this book (i.e., refer to the **Table of Contents** to find the medical device or product).

Unsafe 1 - The object is considered to pose a potential or realistic risk or hazard to a patient or individual in the MR environment primarily as the result of movement or dislodgment of the object. Other hazards may also exist. Therefore, in general, the presence of this object is considered to be a contraindication for an MR procedure and/or for an individual to enter the MR environment. Note that the "default" static magnetic field strength for an unsafe implant or device is 1.5-Tesla.

Unsafe 2 - This object displays only minor magnetic field interactions which, in consideration of the *in vivo* application of this object, is unlikely to pose a hazard or risk in association with movement or dislodgment. Nevertheless, the presence of this object is considered to be a contraindication for an MR procedure or for an individual in the MR environment. Potential risks of performing an MR procedure in a patient or individual with this object are related to possible induced currents, excessive heating, or other potentially hazardous conditions. Therefore, it is inadvisable to perform an MR procedure in a patient or individual with this object.

For example, although certain cardiovascular catheters and accessories typically do not exhibit magnetic field interactions, there are other mechanisms whereby these devices may pose a hazard to the patient or individual or in the MR environment (e.g., induced currents or heating).

The Swan-Ganz thermodilution catheter (and other similar catheters) displays no attraction to the MR system. However, there has been a report of a Swan-Ganz catheter that "melted" in a patient during an MR procedure. Therefore, the presence of this cardiovascular catheter and any other similar device is considered to be a contraindication for a patient undergoing an MR procedure.

Field Strength - This is the highest strength of the static magnetic field of the MR system that was used for safety testing of the object. In most cases, a 1.5-Tesla MR system was used for testing. However, there are some objects that were tested at field strengths lower (e.g., 0.15-Tesla) or higher (e.g., 3.0-Tesla) than 1.5-Tesla. Note that the "default" field strength for an unsafe implant or device is 1.5-Tesla.

There are MR systems with static magnetic field strengths that exceed 2.0-Tesla (i.e., as high as 8.0-Tesla). Several objects have been assessed to determine the relative amount of magnetic field interactions in association with these very high-field-strength MR systems, namely 3.0-Tesla MR systems.

Important Note: An object that exhibits only "mild" or "weak" magnetic field interactions in association with exposure to a 1.5-Tesla MR system may be attracted with sufficient force to a higher field strength scanner (e.g., 3.0-Tesla), potentially posing a risk to a patient or individual. Therefore, careful consideration must be given to each object relative to the static magnetic field strength of the MR system used for testing as well as the conditions that present for the patient or individual under consideration prior to exposure to the MR environment. Please refer to **3.0-Tesla MR Safety Information for Implants and Devices** in this book for additional guidance and recommendations.

Reference - This is the peer-reviewed publication or other documentation used for the MR safety information indicated for a particular object.

The List

Object	Status	Field Strength (T)	Reference
Aneurysm Clips			
Aesculap AVM clip curved Phynox Aesculap Inc. Center Valley, PA	Safe	1.5, 3	
Aesculap AVM clip straight Phynox Aesculap Inc. Center Valley, PA	Safe	1.5, 3	
Downs Multi-Positional, Aneurysm Clip 17-7PH	Unsafe 1	1.39	1
Drake Aneurysm Clip 301 SS Edward Weck Triangle Park, NJ	Unsafe 1	1.5	2
Drake (DR 14, DR 21) Aneurysm Clip Edward Weck Triangle Park, NJ	Unsafe 1	1.39	1
Drake (DR 16) Aneurysm Clip Edward Weck Triangle Park, NJ	Unsafe 1	0.147	1
Heifetz Aneurysm Clip 17-7PH Edward Weck Triangle Park, NJ	Unsafe 1	1.89	4
Heifetz Aneurysm Clip Elgiloy Edward Weck Triangle Park, NJ	Safe	1.89	2
Housepian Aneurysm Clip	Unsafe 1	0.147	1
Kapp Aneurysm Clip 405 SS V. Mueller	Unsafe 1	1.89	2

Object	Status	Field Strength (T)	Reference

Aneurysm Clips (continued)

Object	Status	Field Strength (T)	Reference
Kapp, Curved Aneurysm Clip 404 SS V. Mueller	Unsafe 1	1.39	1
Kapp, Straight Aneurysm Clip 404 SS V. Mueller	Unsafe 1	1.39	1
Mayfield Aneurysm Clip 301 SS Codman Randolf, MA	Unsafe 1	1.5	3
Mayfield Aneurysm Clip 304 SS Codman Randolf, MA	Unsafe 1	1.89	5
McFadden Aneurysm Clip 301 SS Codman Randolf, MA	Unsafe 1	1.5	2
McFadden Vari-Angle, Aneurysm Clip micro clip, fenestrated 9 mm straight blade MP35N Codman Johnson & Johnson Professional, Inc. Raynham, MA	Safe	1.5	
McFadden Vari-Angle, Aneurysm Clip micro clip 9 mm straight blade MP35N Codman Johnson & Johnson Professional, Inc. Raynham, MA	Safe	1.5	
Olivercrona, Aneurysm Clip	Safe	1.39	1
Perneczky Aneurysm Clip straight, 2 mm blade SS alloy Zeppelin Chirugishe Instrumente Pullach, Germany	Safe	1.5, 3	82

Object	Status	Field Strength (T)	Reference
Perneczky Aneurysm Clip straight, 6 mm blade SS alloy Zeppelin Chirugishe Instrumente Pullach, Germany	Safe	1.5, 3	82
Perneczky Aneurysm Clip straight, 7 mm blade SS alloy Zeppelin Chirugishe Instrumente Pullach, Germany	Safe	1.5, 3	82
Perneczky Aneurysm Clip curved, 20 mm Zeppelin Chirurgische Instrumente Germany	Safe	1.5	
Perneczky Aneurysm Clip straight, 3 mm Zeppelin Chirurgische Instumente Germany	Safe	1.5	
Perneczky Aneurysm Clip curved, 9 mm Zeppelin Chirurgische Instrumente Germany	Safe	1.5	
Perneczky Aneurysm Clip straight, 9 mm Zeppelin Chirurgische Instrumente Germany	Safe	1.5	
Pivot Aneurysm Clip 17-7PH	Unsafe 1	1.89	5
R. Spetzler Titanium Aneurysm Clip for Permanent Occlusion 5-mm Fenestrated Bayonet 11-mm blade length (M-9248) Ti6Al4V (Titanium alloy) Allegiance Healthcare Corporation V. Mueller Neuro/Spine San Carlos, CA	Safe	1.5, 3	

| | | Field | |
Object	Status	Strength (T)	Reference

Aneurysm Clips (continued)

Object	Status	Field Strength (T)	Reference

R. Spetzler Titanium Aneurysm Clip
    for Permanent Occlusion
    5-mm Fenestrated Straight
    12-mm blade length (M-9227)
    Ti6Al4V (Titanium alloy)
    Allegiance Healthcare Corporation
    V. Mueller Neuro/Spine
    San Carlos, CA
	Safe	1.5, 3	

R. Spetzler Titanium Aneurysm Clip
    for Permanent Occlusion
    Mini 40 Degree Side
    5-mm blade length (M-9450)
    Ti6Al4V (Titanium alloy)
    Allegiance Healthcare Corporation
    V. Mueller Neuro/Spine
    San Carlos, CA
	Safe	1.5, 3	

R. Spetzler Titanium Aneurysm Clip
    for Permanent Occlusion
    Standard Bayonet
    12-mm blade length (M-9152)
    Ti6Al4V (Titanium alloy)
    Allegiance Healthcare Corporation
    V. Mueller Neuro/Spine
    San Carlos, CA
	Safe	1.5, 3	

R. Spetzler Titanium Aneurysm Clip
    for Permanent Occlusion
    Standard Straight
    15-mm blade length (M-9113)
    Ti6Al4V (Titanium alloy)
    Allegiance Healthcare Corporation
    V. Mueller Neuro/Spine
    San Carlos, CA
	Safe	1.5, 3	

Scoville Aneurysm Clip
    EN58J
    Downs Surgical, Inc.
    Decatur, GA
	Safe	1.89	2

Object	Status	Field Strength (T)	Reference
Spetzler Pure Titanium Aneurysm Clip 　　Model C-2200 　　straight, 5 mm blade 　　C.P. Titanium 　　NMT Neurosciences 　　Duluth, Georgia	Safe	1.5, 3	82
Spetzler Pure Titanium Aneurysm Clip 　　Model C-2212 　　curved, 7 mm blade 　　C.P. Titanium 　　NMT Neurosciences 　　Duluth, Georgia	Safe	1.5, 3	82
Spetzler Pure Titanium Aneurysm Clip 　　straight, 9 mm blade 　　C. P. Titanium 　　Elekta Instruments 　　Atlanta, GA	Safe	1.5, 3	82
Spetzler Pure Titanium Aneurysm Clip 　　Model C-2214 　　curved, 11 mm blade 　　C.P. Titanium 　　NMT Neurosciences 　　Duluth, Georgia	Safe	1.5, 3	82
Spetzler Pure Titanium Aneurysm Clip 　　Model C-2203 　　straight, 11 mm blade 　　C.P. Titanium 　　NMT Neurosciences 　　Duluth, Georgia	Safe	1.5, 3	82
Spetzler Pure Titanium Aneurysm Clip 　　Model C-2526 　　straight, 11 mm blade 　　C.P. Titanium 　　NMT Neurosciences 　　Duluth, Georgia	Safe	1.5, 3	82
Spetzler Pure Titanium Aneurysm Clip 　　Model C-2224 　　straight, 11 mm/3.5 mm 　　fenestrated blade 　　C.P. Titanium 　　NMT Neurosciences 　　Duluth, Georgia	Safe	1.5, 3	82

Object	Status	Field Strength (T)	Reference

Aneurysm Clips (continued)

Object	Status	Field Strength (T)	Reference

Spetzler Titanium Aneurysm Clip
    straight, 13 mm blade, double turn
    C.P. Titanium
    Elekta Instruments, Inc.
    Atlanta, GA
	Safe	1.5	68

Spetzler Titanium Aneurysm Clip
    straight, 13 mm blade, single turn
    C.P. Titanium
    Elekta Instruments, Inc.
    Atlanta, GA
	Safe	1.5, 3	68, 82

Spetzler Titanium Aneurysm Clip
    straight, 9 mm blade, single turn
    C.P. Titanium
    Elekta Instruments, Inc.
    Atlanta, GA
	Safe	1.5, 3	68, 82

Spetzler Titanium Aneurysm Clip
    straight, 9 mm blade, double turn
    C.P. Titanium
    Elekta Instruments, Inc.
    Atlanta, GA
	Safe	1.5	68

Stevens Aneurysm Clip silver alloy	Safe	0.15	6

Sugita Aneurysm Clip
    Elgiloy
    Downs Surgical, Inc.
    Decatur, GA
	Safe	1.89	2

Sugita, AVM Micro Clip
    Aneurysm Clip
    Elgiloy
    Mizuho America, Inc.
    Beverly, MA
	Safe	1.5	

Sugita
    Fenestrated large, bent
    Aneurysm Clip
    for permanent occlusion
    Elgiloy
    Mizuho America, Inc.
    Beverly, MA
	Safe	1.5	

Object	Status	Field Strength (T)	Reference
Sugita Aneurysm Clip Fenestrated large, bent 7.5 mm Elgiloy Mizuho America, Inc. Beverly, MA	Safe	1.5, 3	82
Sugita Fenestrated Large Fujita Blade Deflected Type Aneurysm Clip for Permanent Occlusion angled, 10 mm serrated blade Elgiloy Mizuho America, Inc. Beverly, MA	Safe	1.5, 3	82
Sugita Large Aneurysm Clip for Permanent Occlusion straight, 21 mm serrated blade Elgiloy Mizuho America, Inc. Beverly, MA	Safe	1.5, 3	82
Sugita Long Aneurysm Clip for Permanent Occlusion straight, 19 mm nonserration blade Elgiloy Mizuho America, Inc. Beverly, MA	Safe	1.5, 3	82
Sugita Standard bent Aneurysm Clip 8 mm blade Elgiloy Mizuho America, Inc. Beverly, MA	Safe	1.5, 3	82

Object	Status	Field Strength (T)	Reference

Aneurysm Clips (continued)

Object	Status	Field Strength (T)	Reference
Sugita Standard curved Aneurysm Clip 6 mm blade Elgiloy Mizuho America, Inc. Beverly, MA	Safe	1.5, 3	82
Sugita Temporary Mini Aneurysm Clip bent, 7 mm blade Elgiloy Mizuho America, Inc. Beverly, MA	Safe	1.5, 3	82
Sugita Temporary standard straight Aneurysm Clip 7 mm blade Elgiloy Mizuho America, Inc. Beverly, MA	Safe	1.5, 3	82
Sugita Titanium Aneurysm Clip mini temporary straight slim, 6 mm blade Titanium alloy Mizuho America Beverly, MA	Safe	1.5	
Sugita Titanium Aneurysm Clip mini temporary angled, 5 mm blade Titanium alloy Mizuho America, Inc. Beverly, MA	Safe	1.5	
Sugita Titanium Aneurysm Clip mini temporary straight large, 6 mm blade Titanium alloy Mizuho America Beverly, MA	Safe	1.5	

Object	Status	Field Strength (T)	Reference
Sugita Titanium Aneurysm Clip mini, curved slim, 5.2 mm blade Titanium alloy Mizuho America, Inc. Beverly, MA	Safe	1.5	
Sugita Titanium Aneurysm Clip mini, curved slim, 8.2 mm blade Titanium alloy Mizuho America, Inc. Beverly, MA	Safe	1.5	
Sugita Titanium Aneurysm Clip mini, curved slim, 6.7 mm blade Titanium alloy Mizuho America, Inc. Beverly, MA	Safe	1.5	
Sugita Titanium Aneurysm Clip mini, straight slim, 6 mm blade Titanium alloy Mizuho America, Inc. Beverly, MA	Safe	1.5	
Sugita Titanium Aneurysm Clip mini, straight slim, 4 mm blade Titanium alloy Mizuho America, Inc. Beverly, MA	Safe	1.5	
Sugita Titanium Aneurysm Clip mini, bayonet, 5 mm blade Titanium alloy Mizuho America, Inc. Beverly, MA	Safe	1.5	
Sugita Titanium Aneurysm Clip mini, angled, 7 mm blade Titanium alloy Mizuho America, Inc. Beverly, MA	Safe	1.5	
Sugita Titanium Aneurysm Clip mini, curved, 5.5 mm blade Titanium alloy Mizuho America, Inc. Beverly, MA	Safe	1.5	

Object	Status	Field Strength (T)	Reference

Aneurysm Clips (continued)

Object	Status	Field Strength (T)	Reference

Sugita Titanium Aneurysm Clip
  mini, side curved, 7 mm blade
  Titanium alloy
  Mizuho America, Inc.
  Beverly, MA
	Safe	1.5	

Sugita Titanium Aneurysm Clip
  mini, curved, 4.5 mm blade
  Titanium alloy
  Mizuho America, Inc.
  Beverly, MA
	Safe	1.5	

Sugita Titanium Aneurysm Clip
  mini, angled, 5 mm blade
  Titanium alloy
  Mizuho America, Inc.
  Beverly, MA
	Safe	1.5	

Sugita Titanium Aneurysm Clip
  mini, curved, 5 mm blade
  Titanium alloy
  Mizuho America, Inc.
  Beverly, MA
	Safe	1.5	

Sugita Titanium Aneurysm Clip
  mini, straight large, 6 mm blade
  Titanium alloy
  Mizuho America, Inc.
  Beverly, MA
	Safe	1.5	

Sugita Titanium Aneurysm Clip
  mini, straight small, 4 mm blade
  Titanium alloy
  Mizuho America, Inc.
  Beverly, MA
	Safe	1.5	

Sugita Titanium Aneurysm Clip
  temporary
  slightly curved, 11 mm blade
  Titanium alloy
  Mizuho America
  Beverly, MA
	Safe	1.5	

Object	Status	Field Strength (T)	Reference

Sugita Titanium Aneurysm Clip
    temporary
    curved, 8 mm blade
    Titanium alloy
    Mizuho America, Inc.
    Beverly, MA
	Safe	1.5	

Sugita Titanium Aneurysm Clip
    temporary
    bayonet, 7 mm blade
    Titanium alloy
    Mizuho America, Inc.
    Beverly, MA
	Safe	1.5	

Sugita Titanium Aneurysm Clip
    temporary
    straight large, 10 mm blade
    Titanium alloy
    Mizuho America, Inc.
    Beverly, MA
	Safe	1.5	

Sugita Titanium Aneurysm Clip
    temporary
    straight small, 7 mm blade
    Titanium alloy
    Mizuho America, Inc.
    Beverly, MA
	Safe	1.5	

Sugita Titanium Aneurysm Clip
    standard
    1/4 curved, 11 mm blade
    Titanium alloy
    Mizuho America, Inc.
    Beverly, MA
	Safe	1.5	

Sugita Titanium Aneurysm Clip
    standard
    curved (160 g), 9 mm blade
    Titanium alloy
    Mizuho America, Inc.
    Beverly, MA
	Safe	1.5	

Sugita Titanium Aneurysm Clip
    standard
    slightly curved (170), 11 mm blade
    Titanium alloy
    Mizuho America
    Beverly, MA
	Safe	1.5	

Object	Status	Field Strength (T)	Reference

Aneurysm Clips (continued)

Object	Status	Field Strength (T)	Reference
Sugita Titanium Aneurysm Clip standard slightly curved (165), 11 mm blade Titanium alloy Mizuho America Beverly, MA	Safe	1.5	
Sugita Titanium Aneurysm Clip standard 1/4 curved, large, 8.6 mm blade Titanium alloy Mizuho America Beverly, MA	Safe	1.5	
Sugita Titanium Aneurysm Clip standard 1/4 curved, small, 7.6 mm blade Titanium alloy Mizuho America Beverly, MA	Safe	1.5	
Sugita Titanium Aneurysm Clip standard L-curved, large, 10 mm blade Titanium alloy Mizuho America, Inc. Beverly, MA	Safe	1.5	
Sugita Titanium Aneurysm Clip standard L-curved, small, 7.5 mm blade Titanium alloy Mizuho America, Inc. Beverly, MA	Safe	1.5	
Sugita Titanium Aneurysm Clip standard slightly curved, small, 14 mm blade Titanium alloy Mizuho America Beverly, MA	Safe	1.5	

Object	Status	Field Strength (T)	Reference
Sugita Titanium Aneurysm Clip standard for Permanent Occlusion 45 degree angled, 19 mm serrated blade Titanium alloy Mizuho America, Inc. Beverly, MA	Safe	1.5, 3	82
Sugita Titanium Aneurysm Clip standard 45° curved, large, 10 mm blade Titanium alloy Mizuho America, Inc. Beverly, MA	Safe	1.5	
Sugita Titanium Aneurysm Clip standard 45° curved, medium, 7.5 mm blade Titanium alloy Mizuho America Beverly, MA	Safe	1.5	
Sugita Titanium Aneurysm Clip standard 45° curved, small, 5 mm blade Titanium alloy Mizuho America, Inc. Beverly, MA	Safe	1.5	
Sugita Titanium Aneurysm Clip standard J-angled, large, 9 mm blade Titanium alloy Mizuho America, Inc. Beverly, MA	Safe	1.5	
Sugita Titanium Aneurysm Clip standard J-angled, medium, 7.5 mm blade Titanium alloy Mizuho America Beverly, MA	Safe	1.5	

Object	Status	Field Strength (T)	Reference

Aneurysm Clips (continued)

Object	Status	Field Strength (T)	Reference

Sugita Titanium Aneurysm Clip
    standard
    J-angled, small, 6 mm blade
    Titanium alloy
    Mizuho America, Inc.
    Beverly, MA
	Safe	1.5	

Sugita Titanium Aneurysm Clip
    standard
    L-angled, large, 10 mm blade
    Titanium alloy
    Mizuho America, Inc.
    Beverly, MA
	Safe	1.5	

Sugita Titanium Aneurysm Clip
    standard
    L-angled, medium, 7.5 mm blade
    Titanium alloy
    Mizuho America
    Beverly, MA
	Safe	1.5	

Sugita Titanium Aneurysm Clip
    standard
    L-angled, small, 5 mm blade
    Titanium alloy
    Mizuho America, Inc.
    Beverly, MA
	Safe	1.5	

Sugita Titanium Aneurysm Clip
    standard
    straight large, 18 mm blade
    Titanium alloy
    Mizuho America, Inc.
    Beverly, MA
	Safe	1.5	

Sugita Titanium Aneurysm Clip
    standard
    slightly curved, 18 mm blade
    Titanium alloy
    Mizuho America, Inc.
    Beverly, MA
	Safe	1.5	

Object	Status	Field Strength (T)	Reference
Sugita Titanium Aneurysm Clip standard slight curved bayonet, 12 mm blade Titanium alloy Mizuho America Beverly, MA	Safe	1.5	
Sugita Titanium Aneurysm Clip standard straight medium, 15 mm blade Titanium alloy Mizuho America, Inc. Beverly, MA	Safe	1.5	
Sugita Titanium Aneurysm Clip standard bayonet large, 12 mm blade Titanium alloy Mizuho America, Inc. Beverly, MA	Safe	1.5	
Sugita Titanium Aneurysm Clip standard bayonet small, 10 mm blade Titanium alloy Mizuho America, Inc. Beverly, MA	Safe	1.5	
Sugita Titanium Aneurysm Clip standard angled, 12 mm blade Titanium alloy Mizuho America, Inc. Beverly, MA	Safe	1.5	
Sugita Titanium Aneurysm Clip standard side angled, 12 mm blade Titanium alloy Mizuho America, Inc. Beverly, MA	Safe	1.5	
Sugita Titanium Aneurysm Clip standard staight (small), 12 mm blade Titanium alloy Mizuho America, Inc. Beverly, MA	Safe	1.5	

Object	Status	Field Strength (T)	Reference

Aneurysm Clips (continued)

Object	Status	Field Strength (T)	Reference

Sugita Titanium Aneurysm Clip
  standard
  small curved, 9 mm blade
  Titanium alloy
  Mizuho America, Inc.
  Beverly, MA
	Safe	1.5	

Sugita Titanium Aneurysm Clip
  standard
  slightly curved, 9 mm blade
  Titanium alloy
  Mizuho America, Inc.
  Beverly, MA
	Safe	1.5	

Sugita Titanium Aneurysm Clip
  standard
  curved (155 g), 9 mm blade
  Titanium alloy
  Mizuho America, Inc.
  Beverly, MA
	Safe	1.5	

Sugita Titanium Aneurysm Clip
  standard
  side curved bayonet, 8.5 mm blade
  Titanium alloy
  Mizuho America
  Beverly, MA
	Safe	1.5	

Sugita Titanium Aneurysm Clip
  standard
  side curved, 9 mm blade
  Titanium alloy
  Mizuho America, Inc.
  Beverly, MA
	Safe	1.5	

Sugita Titanium Aneurysm Clip
  standard
  side angled, 8 mm blade
  Titanium alloy
  Mizuho America, Inc.
  Beverly, MA
	Safe	1.5	

Object	Status	Field Strength (T)	Reference

Sugita Titanium Aneurysm Clip
    standard
    slightly angled, 8 mm blade
    Titanium alloy
    Mizuho America, Inc.
    Beverly, MA
	Safe	1.5	

Sugita Titanium Aneurysm Clip
    standard
    angled, 8 mm blade
    Titanium alloy
    Mizuho America, Inc.
    Beverly, MA
	Safe	1.5	

Sugita Titanium Aneurysm Clip
    standard
    bayonet, 7 mm blade
    Titanium alloy
    Mizuho America, Inc.
    Beverly, MA
	Safe	1.5	

Sugita Titanium Aneurysm Clip
    standard
    straight, large, 10 mm blade
    Titanium alloy
    Mizuho America, Inc.
    Beverly, MA
	Safe	1.5	

Sugita Titanium Aneurysm Clip
    standard
    straight, small, 7 mm blade
    Titanium alloy
    Mizuho America, Inc.
    Beverly, MA
	Safe	1.5	

Sugita, bent, mini aneurysm clip
    for temporary occlusion
    Elgiloy
    Mizuho America, Inc.
    Beverly, MA
	Safe	1.5	

Sugita, bent, standard aneurysm clip
    for temporary occlusion
    Elgiloy
    Mizuho America, Inc.
    Beverly, MA
	Safe	1.5	

Object	Status	Field Strength (T)	Reference

Aneurysm Clips (continued)

Object	Status	Field Strength (T)	Reference
Sugita, sideward CVD bayonet, standard aneurysm clip for permanent occlusion Mizuho America, Inc. Beverly, MA	Safe	1.5	
Sugita, straight, large aneurysm clip for permanent occlusion Elgiloy Mizuho America, Inc. Beverly, MA	Safe	1.5	
Sundt AVM, Micro Clip aneurysm clip MP35N Codman Johnson & Johnson Professional, Inc. Raynham, MA	Safe	1.5	
SUNDT SLIM-LINE ANEURYSM CLIP # 6, Bayonet MP35N Codman Codman & Shurtleff, Inc. Johnson and Johnson Company Raynham, MA	Safe	1.5, 3	
SUNDT SLIM-LINE ANEURYSM CLIP # 6, Forward Angle MP35N Codman Codman & Shurtleff, Inc. Johnson and Johnson Company Raynham, MA	Safe	1.5, 3	
SUNDT SLIM-LINE ANEURYSM CLIP Fenestrated Straight 4-mm opening, 24-mm blade MP35N Codman Codman & Shurtleff, Inc. Johnson and Johnson Company Raynham, MA	Safe	1.5, 3	

Object	Status	Field Strength (T)	Reference
SLIM-LINE ANEURYSM CLIP Fenestrated Straight 4-mm opening 6-mm blade MP35N Codman Codman & Shurtleff, Inc. Johnson and Johnson Company Raynham, MA	Safe	1.5, 3	
SUNDT SLIM-LINE ANEURYSM CLIP Fenestrated Bayonet Straight, 6-mm opening 15-mm blade MP35N Codman Codman & Shurtleff, Inc. Johnson and Johnson Company Raynham, MA	Safe	1.5, 3	
SUNDT SLIM-LINE ANEURYSM CLIP Sharp Right Angle 12-mm blade MP35N Codman Codman & Shurtleff, Inc. Johnson and Johnson Company Raynham, MA	Safe	1.5, 3	
SUNDT SLIM-LINE MINI ANEURYSM CLIP # 7, Right Angle MP35N Codman Codman & Shurtleff, Inc. Johnson and Johnson Company Raynham, MA	Safe	1.5, 3	
SUNDT SLIM-LINE MINI ANEURYSM CLIP # 6, Right Angle MP35N Codman Codman & Shurtleff, Inc. Johnson and Johnson Company Raynham, MA	Safe	1.5, 3	

Object	Status	Field Strength (T)	Reference

Aneurysm Clips (continued)

Object	Status	Field Strength (T)	Reference
SUNDT SLIM-LINE MINI ANEURYSM CLIP # 6, Straight MP35N Codman Codman & Shurtleff, Inc. Johnson and Johnson Company Raynham, MA	Safe	1.5, 3	
SUNDT AVM MICRO CLIP # 3 MP35N Codman Codman & Shurtleff, Inc. Johnson and Johnson Company Raynham, MA	Safe	1.5, 3	
SUNDT AVM MICRO CLIP # 4 MP35N Codman Codman & Shurtleff, Inc. Johnson and Johnson Company Raynham, MA	Safe	1.5, 3	
SUNDT AVM MICRO CLIP # 5 MP35N Codman Codman & Shurtleff, Inc. Johnson and Johnson Company Raynham, MA	Safe	1.5, 3	
Sundt Slim-Line, Graft Clip Aneurysm Clip MP35N Codman Johnson & Johnson Professional, Inc. Raynham, MA	Safe	1.5	

Object	Status	Field Strength (T)	Reference
Sundt Slim-Line, Temporary Aneurysm Clip 10 mm blade MP35N Codman Johnson & Johnson Professional, Inc. Raynham, MA	Safe	1.5	
Sundt-Kees Multi-Angle Aneurysm Clip 17-7PH Downs Surgical, Inc. Decatur, GA	Unsafe 1	1.89	2
Sundt-Kees Slim-Line fenestrated, Aneurysm Clip 9 mm blade MP35N Codman Johnson & Johnson Professional, Inc. Raynham, MA	Safe	1.5	
Sundt-Kees, Slim-Line 9 mm blade MP35N Aneurysm Clip Codman Johnson & Johnson Professional, Inc. Raynham, MA	Safe	1.5	
Vari-Angle Aneurysm Clip 17-7PH Codman Randolf, MA	Unsafe 1	1.89	5
Vari-Angle McFadden Aneurysm Clip MP35N Codman Randolf, MA	Safe	1.89	2
Vari-Angle Micro Aneurysm Clip 17-7PH Codman Randolf, MA	Unsafe 1	0.15	2

Object	Status	Field Strength (T)	Reference

Aneurysm Clips (continued)

Object	Status	Field Strength (T)	Reference
Vari-Angle Spring Aneurysm Clip 17-7PH Codman Randolf, MA	Unsafe 1	0.15	2
Yasargil Aneurysm Clip 316L SS Aesculap, Inc. Center Valley, PA	Safe	1.89	5
Yasargil Mini Clip, Titanium Aneurysm Clip Model FT728T bayonet 7 mm blade Titanium alloy Aesculap, Inc. Center Valley, PA	Safe	1.5, 3	82
Yasargil Aneurysm Clip Model FD Aesculap, Inc. Center Valley, PA	Unsafe 1	1.5	
Yasargil Aneurysm Clip Model FE 720T mini, permanent, 7 mm blade, Titanium alloy Aesculap, Inc. Center Valley, PA	Safe	1.5	
Yasargil Aneurysm Clip Model FE 740T standard, permanent, 7 mm blade, Titanium alloy Aesculap, Inc. Center Valley, PA	Safe	1.5	

Object	Status	Field Strength (T)	Reference
Yasargil Aneurysm Clip Model FE 748 standard, 9 mm blade, bayonet Phynox Aesculap, Inc. Center Valley, PA	Safe	1.5	
Yasargil Aneurysm Clip Model FE 750 9 mm blade, straight Phynox Aesculap, Inc. Center Valley, CA	Safe	1.5	
Yasargil Aneurysm Clip Model FE 750T standard, permanent, 9 mm blade Titanium alloy Aesculap, Inc. Center Valley, PA	Safe	1.5	
Yasargil Aneurysm Clip Model FE Aesculap, Inc. Center Valley, PA	Safe	1.5	
Yasargil Standard Aneurysm Clip Model FE750 straight, 9 mm blade Phynox Aesculap, Inc. Center Valley, PA	Safe	1.5, 3	82
Yasargil Standard Aneurysm Clip Model FE780 straight, 14 mm blade Phynox Aesculap, Inc. Center Valley, PA	Safe	1.5, 3	82

Object	Status	Field Strength (T)	Reference

Aneurysm Clips (continued)

Object	Status	Field Strength (T)	Reference
Yasargil Standard Aneurysm Clip Model FE786 curved, 15.3 mm blade Phynox Aesculap, Inc. Center Valley, PA	Safe	1.5, 3	82
Yasargil Standard Aneurysm Clip Model FE790K straight, 20 mm blade Phynox Aesculap, Inc. Center Valley, PA	Safe	1.5, 3	82
Yasargil Standard Aneurysm Clip Model FE798 bayonet, 20 mm blade Phynox Aesculap, Inc. Center Valley, PA	Safe	1.5, 3	82
Yasargil Standard Aneurysm Clip Model FE798 bayonet, 20 mm blade Phynox Aesculap, Inc. Center Valley, PA	Safe	1.5, 3	82
Yasargil Standard Aneurysm Clip Model FE887 7 mm blade Phynox Aesculap, Inc. Center Valley, PA	Safe	1.5, 3	82

Object	Status	Field Strength (T)	Reference
Yasargil Standard Aneurysm Clip Titanium Model FT740T straight, 7 mm blade Titanium alloy Aesculap, Inc. Center Valley, PA	Safe	1.5, 3	82
Yasargil Standard Aneurysm Clip Titanium Model FT758T bayonet, 12 mm blade Titanium alloy Aesculap, Inc. Center Valley, PA	Safe	1.5, 3	82
Yasargil Standard Aneurysm Clip Titanium Model FT760T 11 mm blade Titanium alloy Aesculap, Inc. Center Valley, PA	Safe	1.5, 3	82
Yasargil Standard Aneurysm Clip Titanium Model FT790T straight, 20 mm blade Titanium alloy Aesculap, Inc. Center Valley, PA	Safe	1.5, 3	82

Biopsy Needles, Markers, and Devices

Object	Status	Field Strength (T)	Reference
Adjustable, Automated Aspiration Biopsy Gun 10, 15, and 20 mm 304 SS MD Tech Watertown, MA	Unsafe 1	1.5	7
Adjustable, Automated Biopsy Gun 6, 13, and 19 mm 304 SS MD Tech Watertown, MA	Unsafe 1	1.5	7

Object	Status	Field Strength (T)	Reference

Biopsy Needles, Markers, and Devices (continued)

Object	Status	Field Strength (T)	Reference
ASAP 16, Automatic 16 G Core Biopsy System 19 cm length 304 SS	Unsafe 1	1.5	7
AspirationNeedle MRI MRI Devices Corporation Waukesha, WI	Safe	1.5	
Automatic Cutting Needle with Depth Markings 14 G, 10 cm length 304 SS biopsy needle Manan Northbrook, IL	Unsafe 1	1.5	7
Automatic Cutting Needle with Ultrasound Tip & Depth Markings 18 G, 16 cm length 304 SS biopsy needle Manan Northbrook, IL	Unsafe 1	1.5	7
Automatic Cutting Needle with Ultrasound Tip & Depth Markings 18 G, 20 cm length 304 SS biopsy needle Manan Northbrook, IL	Unsafe 1	1.5	7
Basic II Hookwire Breast Localization Needle 304 SS MD Tech Watertown, MA	Unsafe 1	1.5	7
Beaded Breast Localization Wire Set 19 G, 3-1/2 inch needle with 7-7/8 inch wire 304 SS Inrad Grand Rapids, MI	Unsafe 1	1.5	7

Object	Status	Field Strength (T)	Reference
Beaded Breast Localization Wire Set	Unsafe 1	1.5	7
20 G, 2 inch needle with 5-7/8 inch wire			
304 SS			
Inrad			
Grand Rapids, MI			
Biopsy Gun	Unsafe 1	1.5	7
13 mm, biopsy needle			
Meadox			
Oakland, NJ			
Biopsy Gun	Unsafe 1	1.5	7
25 mm, biopsy needle			
Meadox			
Oakland, NJ			
Biopsy Needle	Unsafe 1	1.5	7
17 G, 10 cm length			
Meadox			
Oakland, NJ			
Biopsy Needle	Unsafe 1	1.5	7
20 G, 15 cm length			
Meadox			
Oakland, NJ			
Biopsy Needle	Unsafe 1	1.5	7
22 G, 15 cm length			
Cook, Inc.			
Bloomington, IN			
Biopsy Needle	Unsafe 1	1.5	7
22 G, 15 cm length			
Meadox			
Oakland, NJ			
Biopty-Cut Biopsy Needle	Unsafe 1	1.5	7
14 G, 10 cm length			
304 SS			
C.R. Bard, Inc.			
Covington, GA			
Biopty-Cut Biopsy Needle	Unsafe 1	1.5	7
16 G, 16 cm length			
304 SS			
C.R. Bard, Inc.			
Covington, GA			

Object	Status	Field Strength (T)	Reference

Biopsy Needles, Markers, and Devices (continued)

Object	Status	Field Strength (T)	Reference
Biopty-Cut Biopsy Needle 18 G, 18 cm length 304 SS C.R. Bard, Inc. Covington, GA	Unsafe 1	1.5	7
Biopty-Cut Biopsy Needle 18 G, 20 cm length 304 SS C.R. Bard, Inc. Covington, GA	Unsafe 1	1.5	7
BoneBiopsy Set MRI manual version, includes trocar, stylet, drill and ejector MRI Devices Corporation Waukesha, WI	Safe	1.5	
Breast Localization Needle 20 G, 5 cm length 304 SS Manan Northbrook, IL	Unsafe 1	1.5	7
Breast Localization Needle 20 G, 7 cm length 304 SS Manan Northbrook, IL	Unsafe 1	1.5	7
Chiba Needle and HiLiter Ultrasound Enhancement 22 G, 3-7/8 inch biopsy needle 304 SS Inrad Grand Rapids, MI	Unsafe 1	1.5	7
Coaxial Needle Set Chiba-type 22 G, 5-7/8 inch biopsy needle 304 SS Inrad Grand Rapids, MI	Unsafe 1	1.5	7

Object	Status	Field Strength (T)	Reference
Coaxial Needle Set Introducer 19G, 2-15/16 inch biopsy needle 304 SS Inrad Grand Rapids, MI	Unsafe 1	1.5	7
CoaxNeedle MRI MRI Devices Corporation Waukesha, WI	Safe	1.5	
Cutting Needle & Gun 18 G, 155 mm length, biopsy needle Meadox Oakland, NJ	Unsafe 1	1.5	7
Cutting Needle 14 G, 9 cm length biopsy needle West Coast Medical Laguna Beach, CA	Unsafe 1	1.5	7
Cutting Needle 16 G, 17 mm length 304 SS biopsy needle BIP USA, Inc. Niagara Falls, NY	Unsafe 1	1.5	7
Cutting Needle 16 G, 19 mm length 304 SS biopsy needle BIP USA, Inc. Niagara Falls, NY	Unsafe 1	1.5	7
Cutting Needle 18 G, 100 mm length biopsy needle Meadox Oakland, NJ	Unsafe 1	1.5	7
Cutting Needle 18 G, 15 cm length biopsy needle West Coast Medical Laguna Beach, CA	Unsafe 1	1.5	7

Object	Status	Field Strength (T)	Reference

Biopsy Needles, Markers, and Devices (continued)

Object	Status	Field Strength (T)	Reference
Cutting Needle 18 G, 150 mm length biopsy needle Meadox Oakland, NJ	Unsafe 1	1.5	7
Cutting Needle 18 G, 9 cm length biopsy needle West Coast Medical Laguna Beach, CA	Unsafe 1	1.5	7
Cutting Needle 19 G, 15 cm length biopsy needle West Coast Medical Laguna Beach, CA	Unsafe 1	1.5	7
Cutting Needle 19 G, 6 cm length biopsy needle West Coast Medical Laguna Beach, CA	Unsafe 1	1.5	7
Cutting Needle 19 G, 9 cm length biopsy needle West Coast Medical Laguna Beach, CA	Unsafe 1	1.5	7
Cutting Needle 20 G, 15 cm length biopsy needle West Coast Medical Laguna Beach, CA	Unsafe 1	1.5	7
Cutting Needle 20 G, 20 cm length biopsy needle West Coast Medical Laguna Beach, CA	Unsafe 1	1.5	7
Cutting Needle 20 G, 9 cm length biopsy needle West Coast Medical Laguna Beach, CA	Unsafe 1	1.5	7

Object	Status	Field Strength (T)	Reference

Fully Automatic BiopsyGun MRI
    MRI Devices Corporation
    Waukesha, WI
	Safe	1.5	

Hawkins Blunt Needle
    304 SS
    biopsy needle
    MD Tech
    Watertown, MA
	Unsafe 1	1.5	7

Hawkins III Breast Localization
    Needle
    MD Tech
    Watertown, MA
	Unsafe 1	1.5	7

LeLoc MRI
    Tumor Localizer
    MRI Devices Corporation
    Waukesha, WI
	Safe	1.5	

Lufkin Aspiration Cytology Needle
    20 G, 5 cm length, biopsy needle
    high nickel alloy
    E-Z-Em, Inc.
    Westbury, NY
	Safe	1.5	9

Lufkin Biopsy Needle
    18 G, 15 cm length
    high nickel alloy
    E-Z-Em, Inc.
    Westbury, NY
	Safe	1.5	8

Lufkin Biopsy Needle
    18 G, 5 cm length
    high nickel alloy
    E-Z-Em, Inc.
    Westbury, NY
	Safe	1.5	8

Lufkin Biopsy Needle
    22 G, 10 cm length
    high nickel alloy
    E-Z-Em, Inc.
    Westbury, NY
	Safe	1.5	8

Lufkin Biopsy Needle
    22 G, 15 cm length
    high nickel alloy
    E-Z-Em, Inc.
    Westbury, NY
	Safe	1.5	8

	Status	Field Strength (T)	Reference

Biopsy Needles, Markers, and Devices (continued)

Object	Status	Field Strength (T)	Reference
Lufkin Biopsy Needle 22 G, 5 cm length high nickel alloy E-Z-Em, Inc. Westbury, NY	Safe	1.5	8
Micromark Clip, marking clip 316L SS Biopsys Medical Irvine, CA	Safe	1.5	61
MicroMark II Clip 316L SS Ethicon Endosurgery Cincinnati, Ohio	Safe	1.5, 3	
MReye Chiba Biopsy Needle William Cook Europe A/S Bjaeverskov, Denmark	Safe	1.5	
MReye Franseen Lung Biopsy Needle William Cook Europe A/S Bjaeverskov, Denmark	Safe	1.5	
MReye Interventional Needle biopsy needle William Cook Europe A/S Bjaeverskov, Denmark	Safe	1.5	
MReye Kopans Breast Lesion Localization Needles (21, 20, 19 gauges; 5.0, 9.0, 15.0 lengths) William Cook Europe A/S Bjaeverskov, Denmark	Safe	1.5	
MRI BioGun 18 G, 10 cm length high nickel alloy biopsy needle E-Z-Em, Inc. Westbury, NY	Safe	1.5	8
MRI BiopsyKit MRI Devices Corporation Waukesha, WI	Safe	1.5	

Object	Status	Field Strength (T)	Reference

MRI Histology Needle
  18 G, 15 cm length
  high nickel alloy
  biopsy needle
  E-Z-Em, Inc.
  Westbury, NY
	Safe	1.5	7

MRI Histology Needle
  18 G, 5 cm length
  high nickel alloy
  biopsy needle
  E-Z-Em, Inc.
  Westbury, NY
	Safe	1.5	8

MRI Histology Needle
  20 G, 10 cm length
  high nickel alloy
  biopsy needle
  E-Z-Em, Inc.
  Westbury, NY
	Safe	1.5	8

MRI Histology Needle
  20 G, 15 cm length
  high nickel alloy
  biopsy needle
  E-Z-Em, Inc.
  Westbury, NY
	Safe	1.5	8

MRI Histology Needle
  20 G, 5 cm length
  high nickel alloy
  biopsy needle
  E-Z-Em, Inc.
  Westbury, NY
	Safe	1.5	7

MRI Histology Needle
  20 G, 7.5 cm length
  high nickel alloy
  biopsy needle
  E-Z-Em, Inc.
  Westbury, NY
	Safe	1.5	8

MRI Lesion Marking System
  20 G, 7.5 cm length
  high nickel alloy
  E-Z-Em, Inc.
  Westbury, NY
	Safe	1.5	8

Object	Status	Field Strength (T)	Reference

Biopsy Needles, Markers, and Devices (continued)

Object	Status	Field Strength (T)	Reference
MRI Needle 　　surgical grade SS 　　biopsy needle 　　Cook, Inc. 　　Bloomington, IN	Safe	1.5	7
mrt Biopsy Needle 　　all sizes 　　Titanium alloy 　　Daum Medical 　　Baltimore, MD and 　　Schwerin,Germany	Safe	1.5	
NeuroCut Needle MRI 　　MRI Devices Corporation 　　Waukesha, WI	Safe	1.5	
NeuroGate Set MRI 　　MRI Devices Corporation 　　Waukesha, WI	Safe	1.5	
NeuroPunctureNeedle MRI 　　MRI Devices Corporation 　　Waukesha, WI	Safe	1.5	
Percucut Biopsy Needle and Stylet 　　19.5 gauge x 10 cm 　　316L SS 　　E-Z-Em, Inc. 　　Westbury, NY	Unsafe 1	1.5	7
Percucut Biopsy Needle and Stylet 　　21 gauge x 10 cm 　　316L SS 　　E-Z-Em, Inc. 　　Westbury, NY	Unsafe 1	1.5	7
PunctureNeedle MRI 　　MRI Devices Corporation 　　Waukesha, WI	Safe	1.5	
Sadowsky Breast Marking System 　　20 G, 5 cm length needle and 　　7 inch hook wire 　　316 L SS 　　Ranfac Corporation 　　Avon, MA	Unsafe 1	1.5	7

Object	Status	Field Strength (T)	Reference

Semi-Automatic BiopsyGun MRI
    MRI Devices Corporation
    Waukesha, WI
	Safe	1.5	

SmartGuide CT/MRI
    MRI Devices Corporation
    Waukesha, WI
	Safe	1.5	

Soft Tissue Biopsy Needle Gun
    & biopsy needle
    304 SS
    Anchor Procducts Co.
    Addison, IL
	Unsafe 1	1.5	7

TargoGrid
    positioning grid for
    biopsies and punctures
    MRI Devices Corporation
    Waukesha, WI
	Safe	1.5	

Trocar Needle
    304 SS
    biopsy needle
    BIP USA, Inc.
    Niagara Falls, NY
	Unsafe 1	1.5	7

Trocar Needle, Disposable
    SS
    biopsy needle
    Cook, Inc.
    Bloomington, IN
	Unsafe 1	1.5	7

TumoLoc MRI
    Tumor Localizer
    MRI Devices Corporation
    Waukesha, WI
	Safe	1.5	

Ultra-Core, biopsy needle
    16 G, 16 cm length
    304 SS
	Unsafe 1	1.5	7

Breast Tissue Expanders and Implants

Object	Status	Field Strength (T)	Reference

Becker Expander/Mammary
    Mentor H/S
    Prosthesis
    316L SS
    breast implant
    Santa Barbara, CA
	Safe	1.5	10

Object	Status	Field Strength (T)	Reference

Breast Tissue Expanders and Implants (continued)

Object	Status	Field Strength (T)	Reference
Breast Tissue Expander Style 133 FV with MAGNA-SITE Injection Site magnetic port McGhan Medical/ INAMED Aesthetics Santa Barbara, CA	Unsafe 1	1.5	86
Breast Tissue Expander Style 133 LV with MAGNA-SITE Injection Site magnetic port McGhan Medical/ INAMED Aesthetics Santa Barbara, CA	Unsafe 1	1.5	86
Breast Tissue Expander Style 133 MV with MAGNA-SITE Injection Site magnetic port McGhan Medical/INAMED Aesthetics Santa Barbara, CA	Unsafe 1	1.5	86
Contour Profile Tissue Expander Breast Tissue Expander Mentor Santa Barbara, CA	Unsafe 1	1.5	
Double Chamber Breast Tissue Expander Model 20739-400 SILIMED, Inc. Dallas, TX	Unsafe 1	1.5	
Infall, breast implant (inflatable with magnetic port) 3101198 Model, breast implant Heyerschultzz	Unsafe 1	1.5	
Radovan Tissue Expander 316L SS Mentor H/S Santa Barbara, CA	Safe	1.5	10

Object	Status	Field Strength (T)	Reference
Siltex Spectrum Post-Operatively Adjustable Saline-Filled Mammary Prosthesis 316L SS Mentor H/S Santa Barbara, CA	Safe	1.5	10
Tissue expander with magnetic port breast implant McGhan Medical Corporation Santa Barbara, CA	Unsafe 1	1.5	

Cardiac Pacemakers and ICDs

Object	Status	Field Strength (T)	Reference
Cosmos Model 283-01 Pacemaker Intermedics, Inc. Freeport, TX	Unsafe 1	3	85
Cosmos II Model 283-03 Pacemaker Intermedics, Inc. Freeport, TX	Unsafe 2	3	85
Cosmos II Model 284-05 Pacemaker Intermedics, Inc. Freeport, TX	Unsafe 1	3	85
Delta TRS Type DDD Model 0937 Pacemaker Cardiac Pacemakers, Inc. St. Paul, MN	Unsafe 1	3	85
GEM DR 7271 Dual Chamber Implantable Cardioverter Defibrillator Medtronic, Inc. Minneapolis, MN	Unsafe 1	3	85

Object	Status	Field Strength (T)	Reference

Cardiac Pacemakers and ICDs (continued)

Object	Status	Field Strength (T)	Reference
GEM II DR 7273 Dual Chamber Implantable Cardioverter Defibrillator Medtronic, Inc. Minneapolis, MN	Unsafe 1	3	85
KAPPA DR403 Dual Chamber Rate Responsive Pacemaker Medtronic, Inc. Minneapolis, MN	Unsafe 2	3	85
KAPPA DR706 Dual Chamber Rate Responsive Pacemaker Medtronic, Inc. Minneapolis, MN	Unsafe 1	3	85
MARQUIS DR 7274 Implantable Cardioverter Defibrillator Medtronic, Inc. Minneapolis, MN	Unsafe 1	3	85
MICRO JEWEL II 7223CX Implantable Cardioverter Defibrillator Medtronic, Inc. Minneapolis, MN	Unsafe 1	3	85
Nova Model 281-01 Intermedics, Inc. Freeport, TX	Unsafe 1	3	85
Nova II Model 281-05 Pacemaker Intermedics, Inc. Freeport, TX	Unsafe 2	3	85
Nova II Model 282-04 Pacemaker Intermedics, Inc. Freeport, TX	Unsafe 2	3	85

Object	Status	Field Strength (T)	Reference
Quantum	Unsafe 1	3	85
Model 253-19			
Pacemaker			
Intermedics, Inc.			
Freeport, TX			
Relay	Unsafe 1	3	85
Model 294-03			
Pacemaker			
Intermedics, Inc.			
Freeport, TX			
Res-Q ACE	Unsafe 1	3	85
Model 101-01			
Pacemaker			
Intermedics, Inc.			
Freeport, TX			
SIGMA SDR306	Unsafe 1	3	85
Dual Chamber Rate			
Responsive Pacemaker			
Medtronic, Inc.			
Minneapolis, MN			
THERA VDD 8968I	Unsafe 2	3	85
Dual Chamber Atrial Sensing			
Ventricular Sensing			
Pacing Pacemaker			
Medtronic, Inc.			
Minneapolis, MN			

Cardiovascular Catheters and Accessories

Object	Status	Field Strength (T)	Reference
Opti-Q SvO2/CCO	Unsafe 2	1.5	60
catheter			
Abbott Laboratories			
Morgan Hill, CA			
Opticath Catheter, Model U400,	Unsafe 2	1.5	60
catheter			
Abbott Laboratories			
Morgan Hill, CA			
Opticath PA Catheter with extra port	Unsafe 2	1.5	60
Abbott Laboratories			
Morgan Hill, CA			

Object	Status	Field Strength (T)	Reference

Cardiovascular Catheters and Accessories (continued)

Object	Status	Field Strength (T)	Reference
Opticath PA Catheter with RV Pacing Port Abbott Laboratories Morgan Hill, CA	Unsafe 2	1.5	60
Oximetric 3, SO2 Optical Module Abbott Laboratories Morgan Hill, CA	Unsafe 2	1.5	60
RV Pacing Lead Abbott Laboratories Morgan Hill, CA	Unsafe 2	1.5	60
Swan-Ganz thermodilution Catheter American Edwards Laboratories Irvine, CA	Unsafe 2	1.5	57
Swan-Ganz triple-lumen thermodilution catheter American Edwards Laboratories Irvine, CA	Unsafe 2	1.5	60
TD Thermodilution Catheter Flow-directed thermodilution pulmonary artery catheter Abbott Laboratories Morgan Hill, CA	Unsafe 2	1.5	60
TDQ CCO Catheter Flow-directed thermodilution continuous cardiac output pulmonary artery catheter Abbott Laboratories Morgan Hill, CA	Unsafe 2	1.5	60
Torque-Line Flow-directed thermodilution pulmonary artery catheter Abbott Laboratories Morgan Hill, CA	Unsafe 2	1.5	60
Transpac IV Abbott Laboratories Morgan Hill, CA	Safe	1.5	60

Object	Status	Field Strength (T)	Reference

Carotid Artery Vascular Clamps

Object	Status	Field Strength (T)	Reference
Crutchfield SS carotid artery vascular clamp Codman Randolf, MA	Conditional 1	1.5	11
Kindt SS carotid artery vascular clamp V. Mueller	Conditional 1	1.5	11
Poppen-Blaylock SS carotid artery vascular clamp Codman Randolf, MA	Unsafe 1	1.5	11
Salibi SS carotid artery vascular clamp Codman Randolf, MA	Conditional 1	1.5	11
Selverstone SS carotid artery vascular clamp Codman Randolf, MA	Conditional 1	1.5	11

Coils, Stents, and Filters

Object	Status	Field Strength (T)	Reference
Abdominal Aortic Aneurysm (AAA) Bifurcated Stent Graft Nitinol TriVascular, Inc. Santa Rosa, CA	Safe	1.5	
ACS Multi-Link RX Duet 18 x 3.0 mm stent 316L SS Guidant Austin, TX	Safe	1.5	72

Object	Status	Field Strength (T)	Reference

Coils, Stents, and Filters (continued)

Object	Status	Field Strength (T)	Reference
ACS Multi-Link Rx Duet 8 x 3.5 mm stent 316L SS Guidant Austin, TX	Safe	1.5	72
ACS RX Multi-Link 15 x 3.0 mm stent 316L SS Guidant Austin, TX	Safe	1.5	72
ACS RX Multi-Link 15 x 4.0 mm stent 316L SS Guidant Austin, TX	Safe	1.5	72
ACS RX Multi-Link 25 x 3.0 mm stent 316L SS Guidant Austin, TX	Safe	1.5	72
ACS Stent 316L SS Guidant Menlo Park, CA	Safe	1.5	
Amplatz IVC filter Cook, Inc. Bloomington, IN	Safe	4.7	21
AneuRX Graft Stent Medtronic AneuRx Sunnyvale, CA	Safe	1.5	
AneuRx Stent Graft Medtronic AneuRx Sunnyvale, CA	Safe	1.5	70

Object	Status	Field Strength (T)	Reference
Angiomed Memotherm Femoral stent 4 mm x 120 mm Nitinol C.R. Bard, Inc. Billerica, MA	Safe	1.5	
Angiomed Memotherm Femoral stent 5 mm x 20 mm Nitinol C.R. Bard, Inc. Billerica, MA	Safe	1.5	
Angiomed Memotherm Iliac stent 12 mm x 110 mm Nitinol C.R. Bard, Inc. Billerica, MA	Safe	1.5	
Angiomed Memotherm Iliac stent 8 mm x 20 mm Nitinol C.R. Bard, Inc. Billerica, MA	Safe	1.5	
AngioStent stent 15 mm platinum, iridium Angiodynamics Queensbury, NY	Safe	1.5	
AngioStent stent	Safe	1.5	
ASpire Covered Stent 100 x 14 mm Nitinol, nickel, Titanium Vascular Architects Portola, CA	Safe	1.5	
AVE GFX Stent 316L SS Arterial Vascular Engineering Santa Rosa, CA	Safe	1.5	69

Object	Status	Field Strength (T)	Reference

Coils, Stents, and Filters (continued)

Object	Status	Field Strength (T)	Reference
AVE Micro I Stent 316L SS Arterial Vascular Engineering Santa Rosa, CA	Safe	1.5	69
AVE Micro II Stent 316L SS Arterial Vascular Engineering Santa Rosa, CA	Safe	1.5	69
BARD CONFORMEXX Biliary Stent 12 x 120 mm Nitinol C. R. Bard Angiomed GmbH & Co. Medizintechnik KG Karlsruhe, Germany	Safe	1.5, 3	83
BARD LUMINEXX Biliary and Vascular Stent 12 x 120 mm Nitinol, tantalum C. R. Bard Angiomed GmbH & Co. Medizintechnik KG Karlsruhe, Germany	Safe	1.5, 3	83
Bard LUMINEXX 3 Biliary and Vascular Stent Nitinol, tantalum 14-mm x 120-mm C. R. Bard Angiomed GmbH & Co. Medizintechnik KG Karlsruhe, Germany	Safe	1.5, 3	
Bard Stent, bifurcated 36 x 20 mm Nitinol	Safe	1.5	70
Bard XT Stent 15 x 3.0 mm 316 LVM Bard Interventional Products Billerica, MA	Safe	1.5	72

Object	Status	Field Strength (T)	Reference
Bard XT Stent 316 LVM Bard Limited Ireland	Safe	1.5	69
BeStent MR-safe SS stent Medtronic Minneapolis, MN	Safe	1.5	69
BeStent 15 x 3.0 mm 316L SS stent Medtronic AVE Minneapolis, MN	Safe	1.5	72
BeStent 25 x 3.0 mm 316L SS stent Medtronic AVE Minneapolis, MN	Safe	1.5	72
BGRE Stent 5.0 mm x 45 mm Nitinol, gold markers W.L. Gore and Associates Flagstaff, AZ	Safe	1.5, 3	
Bifurcated EXCLUDER Endoprosthesis Nitinol stent W.L. Gore and Associates, Inc. Sunnyvale, CA	Safe	1.5	
BX Stent 316L SS Cordis Miami, FL	Safe	1.5	69
BX Velocity Balloon-Expander Intracranial Intravascular stent 4.0 mm x 8 mm Cordis	Safe	1.5, 3	83

Object	Status	Field Strength (T)	Reference

Coils, Stents, and Filters (continued)

Object	Status	Field Strength (T)	Reference
Cook occluding spring embolization coil MWCE- 338-5-10 Cook, Inc. Bloomington, IN	Conditional 2	1.5	
Cook-Z Stent Gianturco-Rosch Biliary Design 10 mm x 3 cm Cook, Inc. Bloomington, IN	Conditional 2	1.5	
Cook-Z Stent Gianturco-Rosch Tracheobronchial Design 20 mm x 5 cm Cook, Inc. Bloomington, IN	Conditional 2	1.5	
CoRectCoil Nitinol stent Intratherapeutics, Inc. St. Paul, MN	Safe	1.5	
Corvita Endoluminal Graft for Abdominal Aortic Aneurysm 27 x 120 Schneider (USA) Inc. Pfizer Medical Technology Group Minneapolis, MN	Safe	1.5	64
Covered UltraFlex Esophageal Stent 23 mm diameter x 12 cm length Nitinol stent Boston Scientific Corporation Natick, MA	Safe	1.5	
Cragg Nitinol spiral filter	Safe	4.7	21
Crossflex Stent 316L SS Cordis Miami, FL	Safe	1.5	69

Object	Status	Field Strength (T)	Reference
Crown Stent	Safe	1.5	69
316L SS			
Cordis			
Miami, FL			
Diamond Colonic Stent	Safe	1.5	
26 mm diameter x 10 cm length			
Nitinol stent with			
platinum/iridium markers			
Boston Scientific Corporation			
Natick, MA			
Diverter Stent	Safe	1.5	
8-mm x 32-mm			
Elgiloy (Conichrome), Tantalum			
MindGuard, Ltd.			
Israel			
Driver Stent	Safe	3	
MP35N			
Medtronic AVE			
Santa Rosa, CA			
Dynalink Stent	Safe	1.5, 3	
5-mm x 100-mm			
Nickel/Titanium			
Guidant Corporation			
Santa Clara, CA			
Dynalink Stent	Safe	1.5, 3	
10-mm x 100-mm			
Nickel/Titanium			
Guidant Corporation			
Santa Clara, CA			
Dynalink Stent	Safe	1.5	
5 to 10-mm (OTW)			
Nitinol			
ACS/Guidant			
Santa Clara, CA			
Endeavor Stent	Safe	1.5	
MP35N			
Medtronic AVE			
Santa Rosa, CA			

Object	Status	Field Strength (T)	Reference

Coils, Stents, and Filters (continued)

Object	Status	Field Strength (T)	Reference

EndoCoil
    Nitinol
    stent
    Intratherapeutics, Inc.
    St. Paul, MN
	Safe	1.5	

EndoCoil-T
    Nitinol
    stent
    Intratherapeutics, Inc.
    St. Paul, MN
	Safe	1.5	

EndoFit Endoluminal Stent Graft
    Cuff(C)
    ENDOMED, Inc.
    Phoenix, AZ
	Safe	1.5	

EndoFit Endoluminal Stent Graft
    Extender (E)
    ENDOMED, Inc.
    Phoenix, AZ
	Safe	1.5	

EndoFit Endoluminal Stent Graft
    Occluder (O)
    ENDOMED, Inc.
    Phoenix, AZ
	Safe	1.5	

EndoFit Endoluminal Stent Graft
    Tapered Aortomonoiliac (A)
    ENDOMED, Inc.
    Phoenix, AZ
	Safe	1.5	

EndoFit Endoluminal Stent Graft
    Thoracic (T)
    ENDOMED, Inc.
    Phoenix, AZ
	Safe	1.5	

Enforcer Stent
    316L SS
    Cardiovascular Dynamics, Inc.
	Safe	1.5	

EsophaCoil-SR
    Nitinol
    stent
    Intratherapeutics, Inc.
    St. Paul, MN
	Safe	1.5	

Object	Status	Field Strength (T)	Reference
Fanelli Laparoscopic Endobiliary Stent Stent Cook Surgical Products Bloomington, IN	Safe	1.5	
Filiform Double Pigtail Silicone Stent Cook Urological, Inc. Spencer, IN	Safe	1.5	
Flamingo Wallstent Esophageal Endoprosthesis Elgiloy Schneider GmbH Bulach, Switzerland	Safe	1.5	
Flower embolization microcoil platinum Target Therapeutics San Jose, CA	Safe	1.5	22
FLUENCY Vascular and Tracheobronchial Stent Grafts 10-mm x 80-mm Nitinol C. R. Bard Angiomed GmbH & Co. Medizintechnik KG Karlsruhe, Germany	Safe	1.5	
GDC 3D Shape various sizes platinum Boston Scientific/Target Wayne, NJ	Safe	1.5	
GDC SR Coil stretch resistant various sizes platinum Boston Scientific/Target Wayne, NJ	Safe	1.5	
GDC TriSpan 14 mm Nitinol, platinum coil Boston Scientific Corporation Fremont, CA	Safe	1.5, 3	

Object	Status	Field Strength (T)	Reference

Coils, Stents, and Filters (continued)

Object	Status	Field Strength (T)	Reference
GDC TriSpan 16 mm Nitinol, platinum coil Boston Scientific Corporation Fremont, CA	Safe	1.5, 3	
Gianturco bird nest IVC filter Cook, Inc. Bloomington, IN	Conditional 2	1.5	21
Gianturco embolization coil Cook, Inc. Bloomington, IN	Conditional 2	1.5	21
Gianturco zig-zag stent Cook, Inc. Bloomington, IN	Conditional 2	1.5	21
Gianturco-Roubin II 20 x 3.0 mm stent 316L SS Cook Bloomington, IN	Safe	1.5	72
Gianturco-Roubin Stent 316L SS Cook Bloomington, IN	Safe	1.5	69
Gore TIPS Endoprosthesis Nitinol W.L. Gore and Associates, Inc. Flagstaff, AZ	Safe	1.5	
GRAFTcath 316L SS, Titanium, silicone GRAFTcath, Inc. Coon Rapids, MN	Safe	1.5	
Greenfield vena cava filter SS MD Tech Watertown, MA	Conditional 2	1.5	21

Object	Status	Field Strength (T)	Reference

Greenfield vena cava filter
    Titanium alloy
    Ormco
    Glendora, CA
	Safe	1.5	21

Guglielmi detachable coil
    platinum
    Target Therapeutics
    San Jose, CA
	Safe	1.5	25

Gunther IVC filter
    William Cook
    Europe
	Conditional 2	1.5	21

Gunther Tulip Vena Cava
    MReye Filter
    Conichrome
    Cook Incorporated
    Bloomington, IN
	Safe	1.5	

Gunther Tulip Vena Cava
    MReye Filter
    Conichrome
    Cook Incorporated
    Bloomington, IN
	Conditional 2	3	

HERCULINK
    Stent
    4.0 to 7.0 mm (RX)
    316L SS
    Guidant
    Santa Clara, CA
	Safe	1.5	

Hilal embolization microcoil
    Cook, Inc.
    Bloomington, IN
	Safe	1.5	23

Horizon Prostatic Stent
    Nitinol
    Endocare
    Irvine, CA
	Safe	1.5	

Iliac Wallgraft Endoprosthesis
    12 x 90
    Schneider (USA) Inc.
    Pfizer Medical Technology Group
    Minneapolis, MN
	Safe	1.5	64

Object	Status	Field Strength (T)	Reference

Coils, Stents, and Filters (continued)

Object	Status	Field Strength (T)	Reference
Iliac Wallstent Endoprosthesis 12 x 90 Schneider (USA) Inc. Pfizer Medical Technology Group Minneapolis, MN	Safe	1.5	64
Iliac Wallstent Endoprosthesis 5 x 80 Schneider (USA) Inc. Pfizer Medical Technology Group Minneapolis, MN	Safe	1.5	64
Iliac Wallstent Endoprosthesis 6 X 90 Schneider (USA) Inc. Pfizer Medical Technology Group Minneapolis, MN	Safe	1.5	64
Inflow 15 x 3.0 mm 316L SS stent Inflow Dynamics Munich, Germany	Safe	1.5	72
Inflow Gold 9 x 3.0 mm 316L SS, gold plate stent Inflow Dynamics Munich, Germany	Safe	1.5	72
Inflow Gold 15 x 3.0 mm 316L SS, gold plate stent Inflow Dynamics Munich, Germany	Safe	1.5	72
Inflow Stent 316L SS stent Inflow Dynamics Munich, Germany	Safe	1.5	69

Object	Status	Field Strength (T)	Reference

IntraCoil Bronchial Endoprosthesis
    Nitinol
    stent
    Intratherapeutics, Inc.
    St. Paul, MN
	Safe	1.5	

IntraStent Biliary Endoprosthesis
    stent
    Intratherapeutics, Inc.
    St. Paul, MN
	Safe	1.5	

IntraStent DoubleStrut Biliary
    Endoprosthesis
    stent
    Intratherapeutics, Inc.
    St. Paul, MN
	Safe	1.5	

IntraStent LP Biliary Endoprosthesis
    stent
    Intratherapeutics, Inc.
    St. Paul, MN
	Safe	1.5	

IVC venous clip
    Teflon
    Pilling Weck Co.
	Safe	1.5	

Jostent
    316L SS
    Jomed
    Helsingborg, Sweden
	Safe	1.5	69

LGM IVC filter
    Phynox
    B. Braun Vena Tech
    Evanston, IL
	Safe	1.5	26

LPS Stent, bifurcated
    36 x 20 mm
    Nitinol
    World Medical Manufacturing Corp.
    Sunrise, FL
	Safe	1.5	70

LPS Thoracic Stent
    46 mm
    Nitinol
    World Medical Manufacturing Corp.
    Sunrise, FL
	Safe	1.5	70

Object	Status	Field Strength (T)	Reference

Coils, Stents, and Filters (continued)

Object	Status	Field Strength (T)	Reference
MAC-Stent 　　17 x 3 mm 　　316L SS 　　stent 　　AMG 　　Munich, Germany	Safe	1.5	72
Maas helical endovascular stent 　　Medinvent 　　Lausanne, Switzerland	Safe	4.7	21
Maas helical IVC filter 　　Medinvent 　　Lausanne, Switzerland	Safe	4.7	21
Magic Wallstent 　　platinum and cobalt-alloy 　　stent 　　Schneider 　　Bulach, Switzerland	Safe	1.5	69
Medtronic AVE Stent 　　316 L SS, gold 　　Medtronic AVE 　　Ireland	Safe	1.5	70
MEGALINK Stent 　　6.0 to 10.0 mm (RX and OTW) 　　316L SS 　　Guidant 　　Santa Clara, CA	Safe	1.5	
Micro Stent II 　　24 x 3.0 mm 　　316L SS 　　stent 　　Medtronic AVE 　　Minneapolis, MN	Safe	1.5	72
Mini-Crown Stent 　　316L SS 　　Cordis 　　Miami, FL	Safe	1.5	69
Mobin-Uddin IVC/umbrella filter 　　American Edwards 　　Santa Ana, CA	Safe	4.7	21

Object	Status	Field Strength (T)	Reference

MReye Embolization Coil
   William Cook Europe A/S
   Bjaeverskov, Denmark
	Safe	1.5	

Multi-Link Stent
   316L SS
   Guidant
   Santa Clara, CA
	Safe	1.5	69

MULTI-LINK
   Stent
   2.5 to 4.0 mm (RX and OTW)
   316L SS
   ACS/Guidant
   Santa Clara, CA
	Safe	1.5	

MULTI-LINK DUET
   Stent
   2.5 to 4.0 mm (RX and OTW)
   316L SS
   ACS/Guidant
   Santa Clara, CA
	Safe	1.5	

MULTI-LINK PENTA
   Stent
   2.5 to 3.0 mm (RX and OTW)
   316L SS
   Guidant
   Santa Clara, CA
	Safe	1.5	

MULTI-LINK PENTA
   Stent
   3.5 to 4.0 mm (RX and OTW)
   316L SS
   Guidant
   Santa Clara, CA
	Safe	1.5	

MULTI-LINK PIXEL
   Stent
   2.0 to 2.5 mm (RX and OTW)
   316L SS
   Guidant
   Santa Clara, CA
	Safe	1.5	

Object	Status	Field Strength (T)	Reference

Coils, Stents, and Filters (continued)

Object	Status	Field Strength (T)	Reference
MULTI-LINK TETRA Stent 2.5 to 4.0 mm (RX and OTW) 316L SS Guidant Santa Clara, CA	Safe	1.5	
MULTI-LINK TRISTAR Stent 2.5 to 4.0 mm (RX and OTW) 316L SS ACS/Guidant Santa Clara, CA	Safe	1.5	
MULTI-LINK ULTRA Stent 3.5 to 5.0 mm (RX and OTW) 316L SS Guidant Santa Clara, CA	Safe	1.5	
New retrievable IVC filter Thomas Jefferson University Philadelphia, PA	Conditional 2	1.5	21
NIR Stent MR-safe metal Medinol Ltd. Jerusalem, Israel	Safe	1.5	69
Palmaz endovascular stent Ethicon	Conditional 2	1.5	21
Palmaz endovascular stent Johnson & Johnson Interventional Warren, NJ	Safe	1.5	
Palmaz-Schatz Stent 15 x 3.0 mm 316L SS Johnson and Johnson Miami Lakes, FL	Safe	1.5	72
Palmaz-Schatz Stent P-S 153 316L SS Cordis Miami, FL	Safe	1.5	69

Object	Status	Field Strength (T)	Reference

Palmaz-Schatz Stent
    P-S 154
    316L SS
    Cordis
    Miami, FL
	Safe	1.5	69

Palmaz-Shatz balloon-
    expandable stent
    Johnson & Johnson Interventional
    Warren, NJ
	Conditional 2	1.5	

Passager Stent
    tantalum
    10 mm x 30 mm
    Meadox Surgimed
    Oakland, NJ
	Safe	1.5	

Passager Stent
    tantalum
    4 mm x 30 mm
    Meadox Surgimed
    Oakland, NJ
	Safe	1.5	

Percuflex Plus
    Stent Graft with Suprarenal
    Ureteral Stent
    4.8 Fr. (1.6 mm x 220 mm)
    Microvasive
    Boston Scientific Corporation
    Watertown, MA
	Safe	1.5, 3	83

PowerLink
    cobalt-alloy
    Endologix, Inc.
    Irvine, CA
	Safe	1.5	70

PowerWeb, Model 1
    cobalt-alloy
    Endologix, Inc.
    Irvine, CA
	Safe	1.5	70

PowerWeb
    cobalt-alloy
    Endologix, Inc.
    Irvine, CA
	Safe	1.5	70

Precedent Stent
    Nitinol, platinum
    Boston Scientific
    Wayne, NJ
	Safe	1.5	70

Object	Status	Field Strength (T)	Reference

Coils, Stents, and Filters (continued)

Object	Status	Field Strength (T)	Reference
Precise-TM Microvascular Anastomotic Device (MACD) 316 L SS	Safe	1.5	71
R Stent 316 LVM SS Spectranetics Corporation Colorado Springs, CO	Safe	1.5	
R Stent 316 L SS Orbus Medical Technologies Fort Lauderdale, FL	Safe	1.5	
Radius Stent Nitinol Scimed Maple Grove, MN	Safe	1.5	69
Radius Stent 4.0 mm x 31 mm Nitinol Boston Scientific Natick, MA	Safe	1.5, 3	
Recovery Nitinol Filter Nitinol C.R. Bard Bard Peripheral Vascular Tempe, AZ	Safe	1.5, 3	
Seaquence Stent 15 x 3.5 316 L SS Nycomed Amersham Princeton, NJ	Safe	1.5	72
Strecker stent tantalum MD Tech Watertown, MA	Safe	1.5	27

Object	Status	Field Strength (T)	Reference
Talent Graft Stent bare spring model 16 x 8 mm Nitinol World Medical Manufacturing Corp. Sunrise, FL	Safe	1.5	
Talent Graft Stent bare spring model 36 x 20 mm Nitinol World Medical Manufacturing Corp. Sunrise, FL	Safe	1.5	
Talent Graft Stent open web model 16 x 8 mm Nitinol World Medical Manufacturing Corp. Sunrise, FL	Safe	1.5	
Talent Graft Stent open web model 36 x 20 mm Nitinol World Medical Manufacturing Corp. Sunrise, FL	Safe	1.5	
Tracheobronchial Wallstent Endoprosthesis 14 x 80, stent Schneider (USA) Inc. Pfizer Medical Technology Group Minneapolis, MN	Safe	1.5	64
Tracheobronchial Wallstent Endoprosthesis 24 x 70, stent Schneider (USA) Inc. Pfizer Medical Technology Group Minneapolis, MN	Safe	1.5	64
TRUFILL Pushable Coils Vascular Occlusion System platinum Cordis Endovascular Systems, Inc. Miami Lakes, FL	Safe	1.5	

Object	Status	Field Strength (T)	Reference

Coils, Stents, and Filters (continued)

Object	Status	Field Strength (T)	Reference

Ultraflex
    stent
    Titanium alloy
	Safe	1.5	73

Ureteral stent	Safe	1.5	

Vanguard Stent
    Nitinol, platinum
    Boston Scientific
    Wayne, NJ
	Safe	1.5	70

VIABAHN Endoprosthesis
    9-mm x 15-cm
    Nitinol
    W. L. Gore & Associates, Inc.
    Flagstaff, AZ
	Safe	1.5	

VIABAHN Endoprosthesis
    13-mm x 10-cm
    Nitinol
    W. L. Gore & Associates, Inc.
    Flagstaff, AZ
	Safe	1.5	

VIACARDIA Stent
    5-mm x 45-mm
    Nitinol, gold end markers
    W. L. Gore and Associates
    Flagstaff, AZ
	Safe	1.5	

VIATORR Endoprosthesis
    Stent
    Nitinol
    W.L. Gore and Associates, Inc.
    Flagstaff, AZ
	Safe	1.5	

Wallstent biliary endoprosthesis
    stent
    high nickle stainless steel
    Schneider USA
    Plymouth, MN
	Safe	1.5	28

Wallstent Endoprosthesis
    Magic Wallstent
    3.5 x 25, stent
    Schneider (USA) Inc.
    Pfizer Medical Technology Group
    Minneapolis, MN
	Safe	1.5	64

Object	Status	Field Strength (T)	Reference

Wallstent Endoprosthesis
    With Permalume covering
    8 x 80, stent
    Schneider (USA) Inc.
    Pfizer Medical Technology Group
    Minneapolis, MN
	Safe	1.5	64

Wallstent Esophageal II Endoprosthesis
    20 x 130, stent
    Schneider (USA) Inc.
    Pfizer Medical Technology Group
    Minneapolis, MN
	Safe	1.5	64

Wallstent
    platinum and cobalt-alloy
    Schneider
    Bulach, Switzerland
	Safe	1.5	69

Wiktor coronary artery stent
    Medtronic Inverventional
    Vascular, Inc.
	Safe	1.5	

Wiktor GX
    tantalum
    stent
    Medtronic AVE
    Minneapolis, MN
	Safe	1.5	72

Wilson-Cook Pancreatic Stent
    Wilson-Cook Medical, Inc.
    Winston-Salem, NC
	Safe	1.5	

Wilson-Cook Pancreatic Wedge Stent
    Wilson-Cook Medical, Inc.
    Winston-Salem, NC
	Safe	1.5	

X-Trode, 3 segment
    stent
    316 SS
    C.R. Bard, Inc.
    Billerica, MA
	Safe	1.5	

X-Trode, 9 segment
    stent
    316 SS
    C.R. Bard, Inc.
    Billerica, MA
	Safe	1.5	

Object	Status	Field Strength (T)	Reference

Coils, Stents, and Filters (continued)

Object	Status	Field Strength (T)	Reference
Zilver Stent 10 mm x 80 mm Cook Incorporated Bloomington, IN	Safe	1.5	

Dental Implants, Devices, and Materials

Object	Status	Field Strength (T)	Reference
Brace band SS dental American Dental Missoula, MT	Conditional 1	1.5	3
Brace wire chrome alloy dental Ormco Corp. San Marcos, CA	Conditional 1	1.5	3
Castable alloy, dental Golden Dental Products, Inc. Golden, CO	Conditional 1	1.5	12
Cement-in keeper, dental Solid State Innovations, Inc. Mt. Airy, NC	Conditional 1	1.5	12
Dental amalgam, dental	Safe	1.39	1
GDP Direct Keeper Pre-formed post, dental Golden Dental Products, Inc. Golden, CO	Conditional 1	1.5	12
Gutta Percha Points, dental	Safe	1.5	
Indian Head Real Silver Points, dental Union Broach Co., Inc. New York, NY	Safe	1.5	
Keeper, pre-formed post, dental Parkell Products, Inc. Farmingdale, NY	Conditional 1	1.5	1
Magna-Dent, large indirect keeper, dental Dental Ventures of America Yorba Linda, CA	Conditional 1	1.5	1

Object	Status	Field Strength (T)	Reference

Palladium clad magnet, dental
    Parkell Products, Inc.
    Farmingdale, NY
	Unsafe 1	1.5	13

Palladium/palladium keeper, dental
    Parkell Products, Inc.
    Farmingdale, NY
	Conditional 1	1.5	13

Palladium/platinum casting alloy,
    dental
    Parkell Products, Inc.
    Farmingdale, NY
	Conditional 1	1.5	13

Permanent crown
    amalgam
    dental
    Ormco Corp.
	Safe	1.5	3

Silver point, dental
    Union Broach Co., Inc.
    New York, NY
	Safe	1.5	3

Stainless steel clad magnet, dental
    Parkell Products, Inc.
    Farmingdale, NY
	Unsafe 1	1.5	13

Stainless steel keeper, dental
    Parkell Products, Inc.
    Farmingdale, NY
	Conditional 1	1.5	13

Titanium clad magnet, dental
    Parkell Products, Inc.
    Farmingdale, NY
	Unsafe 1	1.5	13

ECG Electrodes

Object	Status	Field Strength (T)	Reference

Accutac, ECG electrode
    ConMed Corp.
    Utica, NY
	Safe	1.5	

Accutac, ECG electrode
    Diaphoretic
    ConMed Corp.
    Utica, NY
	Safe	1.5	

Adult Cloth, ECG electrode
    ConMed Corp.
    Utica, NY
	Safe	1.5	

Object	Status	Field Strength (T)	Reference

ECG Electrodes (continued)

Object	Status	Field Strength (T)	Reference

Adult ECG, ECG electrode
    Electrode 3-Pack
    ConMed Corp.
    Utica, NY
	Safe	1.5	

Adult Foam, ECG electrode
    ConMed Corp.
    Utica, NY
	Safe	1.5	

Cleartrace 2, ECG electrode
    ConMed Corp.
    Utica, NY
	Safe	1.5	

Dyna/Trace Diagnostic
    ECG Electrode
    ConMed Corp.
    Utica, NY
	Safe	1.5	

Dyna/Trace Mini, ECG electrode
    ConMed Corp.
    Utica, NY
	Safe	1.5	

Dyna/Trace Stress, ECG electrode
    ConMed Corp.
    Utica, NY
	Safe	1.5	

Dyna/Trace, ECG electrode
    ConMed Corp.
    Utica, NY
	Safe	1.5	

High Demand, ECG electrode
    ConMed Corp.
    Utica, NY
	Safe	1.5	

Holtrode, ECG electrode
    ConMed Corp.
    Utica, NY
	Safe	1.5	

HP M2202A Radio-lucent,
    ECG electrode
    Monitoring Electrode, Ag/AgCL
    Hewlett-Packard
    Medical Supplies
    Andover, MA
	Safe	1.5	

Invisatrace Adult, ECG electrode
    ConMed Corp.
    Utica, NY
	Safe	1.5	

Object	Status	Field Strength (T)	Reference
Pediatric Foam, ECG electrode ConMed Corp. Utica, NY	Safe	1.5	
Plia Cell Diagnostic, ECG electrode ConMed Corp. Utica, NY	Safe	1.5	
Plia-Cell Diaphoretic, ECG electrode ConMed Corp. Utica, NY	Safe	1.5	
Plia-Cell, ECG electrode ConMed Corp. Utica, NY	Safe	1.5	
Quadtrode MRI, ECG electrode InVivo Research, Inc. Orlando, FL	Safe	1.5	
Silvon Adult ECG Electrode ConMed Corp. Utica, NY	Safe	1.5	
Silvon Diaphoretic, ECG electrode ConMed Corp. Utica,NY	Safe	1.5	
Silvon Stress, ECG electrode ConMed Corp. Utica, NY	Safe	1.5	
Silvon, ECG electrode ConMed Corp. Utica, NY	Safe	1.5	
Snaptrace, ECG electrode ConMed Corp. Utica, NY	Safe	1.5	
SSE Radiotransparent ECG Electrode ConMed Corp. Utica, NY	Safe	1.5	
SSE, ECG electrode ConMed Corp. Utica, NY	Safe	1.5	

Object	Status	Field Strength (T)	Reference

Foley Catheters With and Without Temperature Sensors

Object	Status	Field Strength (T)	Reference
Bardex I.C. Foley Catheter with silver and hydrogel coating,16 Fr. Bard Medical Division Covington, GA	Conditional 5	1.5	
Bardex I.C. Temp. Sensing Foley Catheter with a 6 foot cable, 16 Fr. Bard Medical Division Covington, GA	Conditional 5	1.5	
Bardex Lubricath Temp. Sensing Urotrack Plus Foley Catheter with a 6 foot cable, 16 Fr. Bard Medical Division Covington, GA	Conditional 5	1.5	
Bardex Pediatric Temp. Sensing 400-Series Urotrack Foley Catheter with a detachable cable, 12 Fr. Bard Medical Division Covington, GA	Conditional 5	1.5	
Extension cable for Foley Catheter with temperature sensor, 10 feet RSP Respiratory Support Products, Inc. SIMS Smiths Industries Irvine, CA	Conditional 5	1.5	
Foley catheter Bard Medical Division Covington, GA	Safe	1.5	

Object	Status	Field Strength (T)	Reference
Foley Catheter with temperature sensor, 10 Fr. RSP Respiratory Support Products, Inc. SIMS Smiths Industries Irvine, CA	Conditional 5	1.5	
Foley Catheter with temperature sensor, 18 Fr. RSP Respiratory Support Products, Inc. SIMS Smiths Industries Irvine, CA	Conditional 5	1.5	

Halo Vests and Cervical Fixation Devices

Object	Status	Field Strength (T)	Reference
Ambulatory Halo System halo and cervical fixation	Conditional 4	1.5	14
Bremer standard halo crown and vest halo and cervical fixation Bremmer Medical Co. Jacksonville, FL	Safe	1.0	15
Bremmer halo system MR-compatible halo and cervical fixation Bremmer Medical Co. Jacksonville, FL	Safe	1.0	15
Closed-back halo Titanium halo and cervical fixation DePuy ACE Medical Co. El Segundo, CA	Safe	1.5	16
EXO adjustable coller halo and cervical fixation Florida Manufacturing Co. Daytona, FL	Conditional 4	1.0	15
Guilford cervical orthosis, modified halo and cervical fixation Guilford & Son, Ltd. Cleveland, OH	Safe	1.0	15

Object	Status	Field Strength (T)	Reference

Halo Vests and Cervical Fixation Devices (continued)

Object	Status	Field Strength (T)	Reference

Guilford cervical orthosis
    halo and cervical fixation
    Guilford & Son, Ltd.
    Cleveland, OH
	Conditional 4	1.0	15

Mark III halo vest
    aluminum superstructure
    stainless steel rivets, Titanium bolts
    DePuy ACE Medical Co.
    El Segundo, CA
	Safe	1.5	16

Mark IV halo vest
    aluminum superstructure and
    Titanium bolts
    DePuy ACE Medical Co.
    El Segundo, CA
	Safe	1.5	16

MR-compatible
    halo vest and cervical fixation
    Lerman & Son Co.
    Beverly Hills, CA
	Safe	1.5	

Open-back halo
    aluminum
    DePuy ACE Medical Co.
    El Segundo, CA
	Safe	1.5	16

Open-back halo
    with Delrin inserts for skull pins
    aluminum and Delrin
    DePuy ACE Medical Co.
    El Segundo, CA
	Safe	1.5	16

Philadelphia coller
    Philadelphia Coller Co.
    Westville, NJ
	Safe	1.0	15

PMT halo cervical orthosis
    PMT Corp.
    Chanhassen, MN
	Safe	1.0	15

PMT halo cervical orthosis
    with graphite rods and halo ring
    PMT Corp.
    Chanhassen, MN
	Safe	1.0	15

S.O.M.I. cervical orthosis
    U.S. Manufacturing Co.
    Pasadena, CA
	Conditional 4	1.0	15

Object	Status	Field Strength (T)	Reference

Trippi-Wells tong
  Titanium
  DePuy ACE Medical Co.
  El Segundo, CA
	Safe	1.5	16

Heart Valve Prostheses and Annuloplasty Rings

Object	Status	Field Strength (T)	Reference

Annuloflex
  Annuloplasty Ring
  Size 26 mm
  L001285A 26M
  Model AF800
  Sulzer Carbomedics, Inc.
  Austin, TX
	Safe	1.5, 3	

Annuloflex
  Annuloplasty Ring
  Size 36 mm
  L004032A 36M
  Sulzer Carbomedics, Inc.
  Austin, TX
	Safe	1.5, 3	

Annuloflo
  Annuloplasty Ring
  Size 26 mm
  S011896A 26M
  Sulzer Carbomedics, Inc.
  Austin, TX
	Safe	1.5, 3	

Annuloflo
  Annuloplasty Ring
  Size 36 mm
  S013460A 36M
  Sulzer Carbomedics, Inc.
  Austin, TX
	Safe	1.5, 3	

AnnuloFlo
  Mitral Annuloplasty Device
  Size 36
  Model AR-736
  Titanium
  Sulzer Carbomedics, Inc.
  Austin, TX
	Safe	1.5, 3	83

Object	Status	Field Strength (T)	Reference

Heart Valve Prostheses and Annuloplasty Rings (continued)

Object	Status	Field Strength (T)	Reference
Annuloplasty ring Titanium Sulzer Medica and Sulzer Carbomedics	Safe	1.5	
AorTech Aortic Model 3800 Titanium heart valve Aortech Ltd. Strathclyde, U.K.	Conditional 1	1.5	75
AorTech Mitral Model 4800 Titanium heart valve Aortech Ltd. Strathclyde, U.K.	Conditional 1	1.5	75
Aortic Mitroflow Synergy PC Aortic Pericardial Heart Valve Size 19 mm Sulzer Carbomedics, Inc. Austin, TX	Safe	1.5, 3	
Aortic Mitroflow Synergy PC Aortic Pericardial Heart Valve Size 29 mm Sulzer Carbomedics, Inc. Austin, TX	Safe	1.5, 3	
Aortic SJM Regent Valve Mechanical Heart Valve Size 27 mm 27AGN-751 Rotatable Aortic Standard Cuff-Polyester, AGN St. Jude Medical St. Paul, MN	Safe	1.5, 3	

Object	Status	Field Strength (T)	Reference

Aortic Valve
    Size 16 mm
    A419529D 16A
    Sulzer Carbomedics, Inc.
    Austin, TX
	Safe	1.5, 3	

Apical Connector
    Model 174A
    heart valve
    Medtronic Heart Valve Division
    Minneapolis, MN
	Safe	1.5, 3	84

ATS Medical Open Pivot
    BiLeaflet Heart Valve
    Mitral, Model 500DM29
    Standard Valve
    pyrolytic carbon
    heart valve
    ATS Medical
    Minneapolis, MN
	Conditional 1	1.5	75

ATS Medical Open Pivot
    BiLeaflet Heart Valve
    Aortic, Model 501DA18
    Advanced Performance (AP)
    pyrolytic carbon
    heart valve
    ATS Medical
    Minneapolis, MN
	Conditional 1	1.5	75

Autogenics Autologous
    Model APHV
    Eligoy
    heart valve
    Autogenics Europe Ltd
    Glasgow, Scotland
	Safe	1.5	75

Autogenics Autologous
    Model ATCV
    Eligoy
    heart valve
    Autogenics Europe Ltd
    Glasgow, Scotland
	Safe	1.5	75

Beall
    heart valve
    Coratomic Inc.
    Indiana, PA
	Conditional 1	2.35	17

Object	Status	Field Strength (T)	Reference

Heart Valve Prostheses and Annuloplasty Rings (continued)

Object	Status	Field Strength (T)	Reference
Beall Mitral pyrolitic carbon heart valve Coratomic Inc. Indianapolis, IN	Conditional 1	1.5	75
Bileaflet Model A7760, 29 mm heart valve Medtronic Heart Valve Division Minneapolis, MN	Safe	1.5	
Biocor Aortic Model H3636 heart valve St. Jude Medical St. Paul, MN	Safe	1.5	75
Bjork-Shiley (convexo/concave) heart valve Shiley Inc. Irvine, CA	Safe	1.5	3
Bjork-Shiley (universal/spherical) heart valve Shiley Inc. Irvine, CA	Conditional 1	1.5	3
Bjork-Shiley, Model 22 MBRC 11030 heart valve Shiley Inc. Irvine, CA	Conditional 1	2.35	18
Bjork-Shiley, Model MBC heart valve Shiley Inc. Irvine, CA	Conditional 1	2.35	18

Object	Status	Field Strength (T)	Reference
Bjork Shiley Monostrut Aortic Model ABMS chromium cobalt alloy heart valve Pfizer, Inc. Cincinnati, OH	Safe	1.5	75
Bjork Shiley Monostrut Mitral Model MBRMS chromium cobalt alloy heart valve Pfizer, Inc. Cincinnati, OH	Safe	1.5	75
Bjork Shiley Monostrut Mitral Model MBUM chromium cobalt alloy heart valve Pfizer, Inc. Cincinnati, OH	Conditional 1	1.5	75
Bjork Shiley Pyrolitic Carbon Conical Disc Mitral Model MBRP chromium cobalt alloy heart valve Pfizer, Inc. Cincinnati, OH	Conditional 1	1.5	75
Bjork Shiley Pyrolitic Carbon Conical Disc Mitral Model MBUP chromium cobalt alloy heart valve Pfizer, Inc. Cincinnati, OH	Safe	1.5	75
Carbomedics Heart Valve Prosthesis Annuloflo Annuloplasty Ring, Size 26 Carbomedics Austin, TX	Safe	1.5	

Object	Status	Field Strength (T)	Reference

Heart Valve Prostheses and Annuloplasty Rings (continued)

Object	Status	Field Strength (T)	Reference
Carbomedics Heart Valve Prosthesis Annuloflo Annuloplasty Ring, Size 36 Carbomedics Austin, TX	Safe	1.5	
CarboMedics Heart Valve Prosthesis Aortic Reduced, Model R500, Size 19 CarboMedics Austin, TX	Safe	1.5	19
CarboMedics Heart Valve Prosthesis Aortic Reduced, Model R500, Size 21 CarboMedics Austin, TX	Safe	1.5	19
CarboMedics Heart Valve Prosthesis Aortic Reduced, Model R500, Size 23 CarboMedics Austin, TX	Safe	1.5	19
CarboMedics Heart Valve Prosthesis Aortic Reduced, Model R500, Size 25 CarboMedics Austin, TX	Safe	1.5	19
CarboMedics Heart Valve Prosthesis Aortic Reduced, Model R500, Size 27 CarboMedics Austin, TX	Safe	1.5	19
CarboMedics Heart Valve Prosthesis Aortic Reduced, Model R500, Size 29 CarboMedics Austin, TX	Safe	1.5	19

Object	Status	Field Strength (T)	Reference
CarboMedics Heart Valve Prosthesis Aortic Standard, Model 500, Size 31 CarboMedics Austin, TX	Safe	1.5	19
Carbomedics Heart Valve Prosthesis Aortic Valve, Size 16 Carbomedics Austin, TX	Safe	1.5	
Carbomedics Heart Valve Prosthesis Carboseal Size 31 Carbomedics Austin, TX	Safe	1.5	
CarboMedics Heart Valve Prosthesis Mitral Standard, Model 700, Size 23 CarboMedics Austin, TX	Safe	1.5	19
CarboMedics Heart Valve Prosthesis Mitral Standard, Model 700, Size 25 CarboMedics Austin, TX	Safe	1.5	19
CarboMedics Heart Valve Prosthesis Mitral Standard, Model 700, Size 27 CarboMedics Austin, TX	Safe	1.5	19
CarboMedics Heart Valve Prosthesis Mitral Standard, Model 700, Size 29 CarboMedics Austin, TX	Safe	1.5	19
CarboMedics Heart Valve Prosthesis Mitral Standard, Model 700, Size 31 CarboMedics Austin, TX	Safe	1.5	19

Object	Status	Field Strength (T)	Reference

Heart Valve Prostheses and Annuloplasty Rings (continued)

Object	Status	Field Strength (T)	Reference
CarboMedics Heart Valve Prosthesis Mitral Standard, Model 700 Size 33 CarboMedics Austin, TX	Safe	1.5	19
Carbomedics Heart Valve Prosthesis Mitral Valve, Size 33 Carbomedics Austin, TX	Safe	1.5	
Carboseal Ascending Aortic Prosthesis Size 33 mm, Model AP 33 S/N C489492-H Sulzer Carbomedics, Inc. Austin, TX	Safe	1.5, 3	
Carboseal Ascending Aortic Valve Conduit Size 33 mm Model AP-033 Nitinol Sulzer Carbomedics, Inc. Austin, TX	Safe	1.5, 3	83
Carboseal Ascending Aortic Valve Conduit Size 33 mm Model AP-033 Titanium Sulzer Carbomedics, Inc. Austin, TX	Safe	1.5, 3	83
Carpentier Edwards BioPhysio Heart Valve Prosthesis Model 3100 29M Nitinol, silicone Edwards Lifesciences Irvine, CA	Safe	1.5, 3, 8	

Object	Status	Field Strength (T)	Reference

Carpentier-Edwards
    Classic Annuloplasty Ring
    Mitral Model 4400, size 40 mm
    Edwards Lifesciences
    Irvine, CA
	Safe	1.5, 3	77, 83

Carpentier-Edwards (porcine)
    heart valve
    American Edwards Laboratories
    Santa Ana, CA
	Conditional 1	2.35	18

Carpentier-Edwards Annuloplasty Ring
    Model 4400
    Baxter Healthcare Corporation
    Santa Ana, CA
	Safe	1.5	

Carpentier-Edwards Annuloplasty Ring
    Model 4500
    Baxter Healthcare Corporation
    Santa Ana, CA
	Safe	1.5	

Carpentier-Edwards Annuloplasty Ring
    Model 4600
    Baxter Healthcare Corporation
    Santa Ana, CA
	Safe	1.5	

Carpentier-Edwards Bioprosthesis
    Model 2625, heart valve
    Baxter Healthcare Corporation
    Santa Ana, CA
	Safe	1.5	

Carpentier-Edwards Bioprosthesis
    Model 6625, heart valve
    Baxter Healthcare Corporation
    Santa Ana, CA
	Safe	1.5	

Carpentier-Edwards
    Low Pressure Bioprosthesis
    Porcine, Mitral Model 6625-LP
    size 35 mm
    heart valve
    Edwards Lifesciences
    Irvine, CA
	Safe	1.5, 3	77, 83

Carpentier-Edwards Pericardial
    Bioprosthesis
    Model 2700, heart valve
    Baxter Healthcare Corporation
    Santa Ana, CA
	Safe	1.5	

Object	Status	Field Strength (T)	Reference

Heart Valve Prostheses and Annuloplasty Rings (continued)

Object	Status	Field Strength (T)	Reference
Carpentier-Edwards PERIMOUNT Pericardial Bioprosthesis Mitral Model 6900 size 33 mm heart valve Edwards Lifesciences Irvine, CA	Conditional 1	1.5, 3	77, 83
Carpentier-Edwards Physio Annuloplasty Ring Mitral Model 4450, size 40 mm Edwards Lifesciences Irvine, CA	Conditional 1	1.5, 3	77, 83
Carpentier-Edwards Physio Annuloplasty Ring Model 4450 Baxter Healthcare Corporation Santa Ana, CA	Safe	1.5	77
Carpentier-Edwards Model 2650, heart valve American Edwards Laboratories Santa Ana, CA	Conditional 1	2.35	18
Colvin-Galloway Future Band 638B annuloplasty device Medtronic Heart Valve Division Minneapolis, MN	Safe	1.5, 3	84
Contegra 200 heart valve Medtronic Heart Valve Division Minneapolis, MN	Safe	1.5, 3	84
Contegra 200S heart valve Medtronic Heart Valve Division Minneapolis, MN	Safe	1.5, 3	84
Cosgrove-Edwards Annuloplasty Ring Model 4600 Baxter Healthcare Corporation Santa Ana, CA	Safe	1.5	

Object	Status	Field Strength (T)	Reference
CPHV annuloplasty ring pyrolytic carbon, Titanium Sulzer Medica and Sulzer Carbomedics	Safe	1.5	
Durafic Aortic, Model AD size 33 mm heart valve	Conditional 1	1.5	75
Durafic Model MD, Mitral heart valve	Conditional 1	1.5	75
Duraflex Low Pressure Bioprosthesis Model 6625E6R-LP heart valve Baxter Healthcare Corporation Santa Ana, CA	Safe	1.5	
Duraflex Low Pressure Bioprosthesis Model 6625LP heart valve Baxter Healthcare Corporation Santa Ana, CA	Safe	1.5	
Duran Ring Model H601H annuloplasty device Medtronic Heart Valve Division Minneapolis, MN	Safe	1.5, 3	84
Duran Ring Model H608H annuloplasty device Medtronic Heart Valve Division Minneapolis, MN	Safe	1.5, 3	84
Duran Ring Model 610R annuloplasty device Medtronic Heart Valve Division Minneapolis, MN	Safe	1.5, 3	84

Object	Status	Field Strength (T)	Reference

Heart Valve Prostheses and Annuloplasty Rings (continued)

Object	Status	Field Strength (T)	Reference
Duran Annuloplasty Ring Model H601H, 35 mm Medtronic Heart Valve Division Minneapolis, MN	Safe	1.5	
Edwards MIRA Mechanical Valve Mitral, Model 9600 Size 27 mm heart valve Edwards Lifesciences Irvine, CA	Safe	1.5, 3	83
Edwards MIRA Mechanical Valve Mitral, Model 9600 size 33 mm heart valve Edwards Lifesciences Irvine, CA	Safe	1.5	
Edwards Prima Plus Stentless Bioprosthesis Model 2500P aortic heart valve porcine, polyester cloth Edwards Lifesciences Irvine, CA	Safe	1.5	
Edwards TEKNA Bileaflet Valve Model 3200 heart valve Baxter Healthcare Corporation Santa Ana, CA	Safe	1.5	
Edwards TEKNA Bileaflet Valve Model 9200 heart valve Baxter Healthcare Corporation Santa Ana, CA	Safe	1.5	

Object	Status	Field Strength (T)	Reference

Edwards-Duromedics Bileaflet Valve
    Model 3160
    heart valve
    Baxter Healthcare Corporation
    Santa Ana, CA
	Safe	1.5	

Edwards-Duromedics Bileaflet Valve
    Model 9120
    heart valve
    Baxter Healthcare Corporation
    Santa Ana, CA
	Safe	1.5	

Freestyle
    Aortic Root
    Model 995
    heart valve
    Medtronic Heart Valve Division
    Minneapolis, MN
	Safe	1.5, 3	84

Freestyle
    Model 995
    27 mm
    heart valve
    Medtronic Heart Valve Division
    Minneapolis, MN
	Safe	1.5	

Hall-Kaster, Model A7700
    heart valve
    Medtronic Heart Valve Division
    Minneapolis, MN
	Conditional 1	1.5	3

Hancock
    Model 100
    Pulmonic Conduit
    heart valve
    Medtronic Heart Valve Division
    Minneapolis, MN
	Safe	1.5, 3	84

Hancock
    Model 105
    Low Porosity Conduit
    heart valve
    Medtronic Heart Valve Division
    Minneapolis, MN
	Safe	1.5, 3	84

Object	Status	Field Strength (T)	Reference

Heart Valve Prostheses and Annuloplasty Rings (continued)

Object	Status	Field Strength (T)	Reference
Hancock Model 150 Pulmonic Conduit heart valve Medtronic Heart Valve Division Minneapolis, MN	Safe	1.5, 3	84
Hancock Model 242 Aortic Valve heart valve Medtronic Heart Valve Division Minneapolis, MN	Safe	1.5, 3	84
Hancock Model 342 Mitral Valve heart valve Medtronic Heart Valve Division Minneapolis, MN	Safe	1.5, 3	84
Hancock 342 35 mm, Model 342 heart valve Medtronic Heart Valve Division Minneapolis, MN	Safe	1.5	
Hancock Conduit Model 100, 30 mm heart valve Medtronic Heart Valve Division Minneapolis, MN	Safe	1.5	
Hancock extracorporeal Model 242R heart valve Johnson & Johnson Anaheim, CA	Conditional 1	2.35	19
Hancock extracorporeal Model M 4365-33 heart valve Johnson & Johnson Anaheim, CA	Conditional 1	2.35	19

Object	Status	Field Strength (T)	Reference
Hancock I (porcine) heart valve Johnson & Johnson Anaheim, CA	Conditional 1	1.5	3
Hancock II (porcine) heart valve Johnson & Johnson Anaheim, CA	Conditional 1	1.5	3
Hancock II Model T505 Aortic Valve heart valve Medtronic Heart Valve Division Minneapolis, MN	Safe	1.5, 3	84
Hancock II Model T510 Mitral Valve heart valve Medtronic Heart Valve Division Minneapolis, MN	Safe	1.5, 3	84
Hancock II Model T510, 33 mm heart valve Medtronic Heart Valve Division Minneapolis, MN	Safe	1.5	
Hancock MO II Model 250 Aortic Valve heart valve Medtronic Heart Valve Division Minneapolis, MN	Safe	1.5, 3	84
Hancock MO II Model 250B Aortic Valve heart valve Medtronic Heart Valve Division Minneapolis, MN	Safe	1.5, 3	84

Object	Status	Field Strength (T)	Reference

Heart Valve Prostheses and Annuloplasty Rings (continued)

Object	Status	Field Strength (T)	Reference
Hancock MO II Model 250C Aortic Valve heart valve Medtronic Heart Valve Division Minneapolis, MN	Safe	1.5, 3	84
Hancock MO II Model 250D Aortic Valve heart valve Medtronic Heart Valve Division Minneapolis, MN	Safe	1.5, 3	84
Hancock MO II Model 250E Aortic Valve heart valve Medtronic Heart Valve Division Minneapolis, MN	Safe	1.5, 3	84
Hancock MO II Model 250H Aortic Valve heart valve Medtronic Heart Valve Division Minneapolis, MN	Safe	1.5, 3	84
Hancock Pericardial Mitral Model T410 Haynes alloy heart valve Medtronic Inc. Minneapolis, MN	Safe	1.5	75
Hancock Vascor, Model 505 heart valve Johnson & Johnson Anaheim, CA	Safe	2.35	19
Inonescu-Shiley, Universal ISM heart valve	Conditional 1	2.35	19

Object	Status	Field Strength (T)	Reference
Intact Aortic, Model A805 size 19 mm heart valve Medtronic Inc. Minneapolis, MN	Safe	1.5	75
Intact Model 705 Mitral Valve heart valve Medtronic Heart Valve Division Minneapolis, MN	Safe	1.5, 3	84
Intact Model 805 Aortic Valve heart valve Medtronic Heart Valve Division Minneapolis, MN	Safe	1.5, 3	84
Intact Mitral, Model M705 size 25 mm heart valve Medtronic Inc. Minneapolis, MN	Safe	1.5	75
Jyros Aortic, Model J1A carbon alloy heart valve Axion Medical Ltd.	Safe	1.5	75
Jyros Mitral, Model J1M carbon alloy heart valve Axion Medical Ltd.	Safe	1.5	75
Lillehi-Kaster Model 300S heart valve Medical Inc. Inver Grove Heights, MN	Conditional 1	2.35	17

Object	Status	Field Strength (T)	Reference

Heart Valve Prostheses and Annuloplasty Rings (continued)

Object	Status	Field Strength (T)	Reference
Lillehi-Kaster Model 5009 heart valve Medical Inc. Inver Grove Heights, MN	Conditional 1	2.35	19
Liotta Aortic, Model MA783 Delrin heart valve St. Jude Medical St. Paul, MN	Safe	1.5	75
Med Hall Conduit Model R7700, 33 mm heart valve Medtronic Heart Valve Division Minneapolis, MN	Safe	1.5	
Medtronic Advantage Model A7760 Aortic Valve heart valve Medtronic Heart Valve Division Minneapolis, MN	Safe	1.5, 3	84
Medtronic Advantage Model M7760 Mitral Valve heart valve Medtronic Heart Valve Division Minneapolis, MN	Safe	1.5, 3	84
Medtronic Hall heart valve Medtronic Heart Valve Division Minneapolis, MN	Conditional 1	2.35	18
Medtronic Hall Model 7700, 33 mm heart valve Medtronic Heart Valve Division Minneapolis, MN	Safe	1.5	

Object	Status	Field Strength (T)	Reference
Medtronic Hall Model A7700 Aortic Valve heart valve Medtronic Heart Valve Division Minneapolis, MN	Safe	1.5, 3	84
Medtronic Hall Model A7700-D-16 heart valve Medtronic Heart Valve Division Minneapolis, MN	Conditional 1	2.35	18
Medtronic Hall Model C7700 Valved Conduit heart valve Medtronic Heart Valve Division Minneapolis, MN	Safe	1.5, 3	84
Medtronic Hall Model M7700 Mitral Valve heart valve Medtronic Heart Valve Division Minneapolis, MN	Safe	1.5, 3	84
Medtronic Hall Model R7700 Low Porosity Valved Conduit heart valve Medtronic Heart Valve Division Minneapolis, MN	Safe	1.5, 3	84
Medtronic Hall Model Z7700 Low Porosity Valved Conduit heart valve Medtronic Heart Valve Division Minneapolis, MN	Safe	1.5, 3	84
Mitral Prosthetic Heart Valve Model 2100 heart valve TRI Technologies Brazil	Safe	1.5	70

Object	Status	Field Strength (T)	Reference

Heart Valve Prostheses and Annuloplasty Rings (continued)

Object	Status	Field Strength (T)	Reference

Mitral Valve
    Size 33 mm
    A307504F
    Sulzer Carbomedics, Inc.
    Austin, TX
	Safe	1.5, 3	

Mitroflow
    Aortic, Model 11A
    Delrin
    heart valve
    Mitroflow Sulzer CarboMedics
    U.K.
	Safe	1.5	75

Mitroflow
    Aortic, Model 14A
    Delrin
    heart valve
    Mitroflow Sulzer CarboMedics
    U.K.
	Safe	1.5	75

Mitroflow
    Mitral, Model 11M
    Delrin
    heart valve
    Mitroflow Sulzer CarboMedics
    U.K.
	Safe	1.5	75

Mitroflow Pericardial Heart Valve
    Model 12
    Sulzer-Medica and
    Mitroflow International
    Richmond, B.C.
    Canada
	Safe	1.5	70

Mosaic
    Model 305
    Aortic Valve
    heart valve
    Medtronic Heart Valve Division
    Minneapolis, MN
	Safe	1.5, 3	84

Object	Status	Field Strength (T)	Reference
Mosaic Model 310 Mitral Valve heart valve Medtronic Heart Valve Division Minneapolis, MN	Safe	1.5, 3	84
Mosaic Model 310, 33 mm heart valve Medtronic Heart Valve Division Minneapolis, MN	Safe	1.5	
Omnicarbon Model 35231029 heart valve Medical Inc. Inver Grove Heights, MN	Conditional 1	2.35	18
Omniscience Model 6522 heart valve Medical Inc. Inver Grove Heights, MN	Conditional 1	2.35	18
On-X Valve Model 6816 heart valve Medical Carbon Research Institute Austin, TX	Safe	1.5	
Percutaneous Heart Valve 316LMV SS Percutaneous Valve Technologies, Ltd. Israel	Safe	1.5, 3	
Porcine Synergy ST Aortic, Size 19 mm Sulzer Carbomedics, Inc. Austin, TX	Safe	1.5, 3	
Porcine Synergy ST Mitral, Size 33 mm Sulzer Carbomedics, Inc. Austin, TX	Safe	1.5, 3	

Object	Status	Field Strength (T)	Reference

Heart Valve Prostheses and Annuloplasty Rings (continued)

Object	Status	Field Strength (T)	Reference

Posterior Annuloplasty Band
    annuloplasty device
    Model H607
    Medtronic Heart Valve Division
    Minneapolis, MN
	Safe	1.5, 3	84

Posterior Annuloplasty Band
    Model 610B
    annuloplasty device
    Medtronic Heart Valve Division
    Minneapolis, MN
	Safe	1.5, 3	84

Reduced Aortic CPHV
    Carbomedics Prosthetic
    Model R5-029
    Size 29 mm
    heart valve
    Nitinol
    Sulzer Carbomedics, Inc.
    Austin, TX
	Safe	1.5, 3	83

Reduced Aortic CPHV
    Carbomedics Prosthetic
    Model R5-029
    Size 29 mm
    heart valve
    Titanium
    Sulzer Carbomedics, Inc.
    Austin, TX
	Safe	1.5, 3	83

Sculptor Ring
    Model 605M
    annuloplasty device
    Medtronic Heart Valve Division
    Minneapolis, MN
	Safe	1.5, 3	84

Sculptor Ring
    Model 605T
    annuloplasty device
    Medtronic Heart Valve Division
    Minneapolis, MN
	Safe	1.5, 3	84

Object	Status	Field Strength (T)	Reference
Sculptor Annuloplasty Ring Model 605M, 35 mm Medtronic Heart Valve Division Minneapolis, MN	Safe	1.5	
SJM Regent Valve Mechanical Heart Valve Rotatable, Aortic Standard Cuff-Polyester Model 25AGN-751-IDE St. Jude Medical St. Paul, MN	Conditional 1	1.5	75
Smelloff Cutter Aortic Titanium heart valve Sorin Biomedica Italy	Conditional 1	1.5	75
Smeloff-Cutter heart valve Cutter Laboratories Berkeley, CA	Conditional 1	2.35	18
Sorin Allcarbon, AS Model MTR-29AS, 29 mm pyrolitic carbon heart valve Sorin Biomedica Cardio S.p.A. Saluggia, Italy	Conditional 1	1.5	75
Sorin, No. 23 heart valve	Conditional 1	1.5	20
Sorin Pericarbon (stented) Mitral pyrolytic carbon heart valve Sorin Biomedica Italy	Conditional 1	1.5	75

Object	Status	Field Strength (T)	Reference

Heart Valve Prostheses and Annuloplasty Rings (continued)

Object	Status	Field Strength (T)	Reference
St. Jude Medical Mechanical Heart Valve SJM Masters Series Rotatable, Aortic Model 25AJ-501 heart valve St. Jude Medical St. Paul, MN	Conditional 1	1.5	75
St. Jude, Model A 100 heart valve St. Jude Medical Inc. St. Paul, MN	Conditional 1	2.35	19
St. Jude, Model M 101 heart valve St. Jude Medical Inc. St. Paul, MN	Conditional 1	2.35	19
St. Jude heart valve St. Jude Medical Inc. St. Paul, MN	Safe	1.5	3
Standard Mitral CPHV Carbomedics Prosthetic Model R5-029 Size 29 mm heart valve Nitinol Sulzer Carbomedics, Inc. Austin, TX	Safe	1.5, 3	83
Standard Mitral CPHV Carbomedics Prosthetic Model M7-033 Size 33 mm Heart Valve Titanium Sulzer Carbomedics, Inc. Austin, TX	Safe	1.5, 3	83

Object	Status	Field Strength (T)	Reference
Starr-Edwards, Model 1000 heart valve Baxter Healthcare Corporation Santa Ana, CA	Conditional 1	1.5	
Starr-Edwards, Model 1200 heart valve Baxter Healthcare Corporation Santa Ana, CA	Conditional 1	1.5	
Starr-Edwards, Model 1260 heart valve American Edwards Laboratories Baxter Healthcare Corporation Santa Ana, CA	Conditional 1	2.35	17
Starr-Edwards, Model 2300 heart valve Baxter Healthcare Corporation Santa Ana, CA	Conditional 1	1.5	
Starr-Edwards, Model 2310 heart valve Baxter Healthcare Corporation Santa Ana, CA	Conditional 1	1.5	
Starr-Edwards, Model 2320 heart valve American Edwards Laboratories Baxter Healthcare Corporation Santa Ana, CA	Conditional 1	2.35	17
Starr-Edwards, Model 2400 heart valve American Edwards Laboratories Baxter Healthcare Corporation Santa Ana, CA	Safe	1.5	3
Starr-Edwards, Model 6000 heart valve Baxter Healthcare Corporation Santa Ana, CA	Conditional 1	1.5	
Starr-Edwards, Model 6120 heart valve Baxter Healthcare Corporation Santa Ana, CA	Conditional 1	1.5	

Object	Status	Field Strength (T)	Reference

Heart Valve Prostheses and Annuloplasty Rings (continued)

Object	Status	Field Strength (T)	Reference
Starr-Edwards, Model 6300 heart valve Baxter Healthcare Corporation Santa Ana, CA	Conditional 1	1.5	
Starr-Edwards, Model 6310 heart valve Baxter Healthcare Corporation Santa Ana, CA	Conditional 1	1.5	
Starr-Edwards, Model 6320 heart valve Baxter Healthcare Corporation Santa Ana, CA	Conditional 1	1.5	
Starr-Edwards, Model 6400 heart valve Baxter Healthcare Corporation Santa Ana, CA	Conditional 1	1.5	
Starr-Edwards, Model 6520 heart valve Baxter Healthcare Corporation Santa Ana, CA	Conditional 1	2.35	19
Starr-Edwards, Model Pre 6000 heart valve American Edwards Laboratories Baxter Healthcare Corporation Santa Ana, CA	Conditional 1	2.35	17
Sulzer/Carbomedics Synergy PC Pericardial Heart Valve Sulzer-Medica and Mitroflow International Richmond, B.C. Canada	Safe	1.5	70
Tascon Aortic Elgiloy heart valve Medtronic Inc. Minneapolis, MN	Safe	1.5	75

Object	Status	Field Strength (T)	Reference

Toronto SPV Valve
    Stentless Porcine
    Aortic Model SPA-101-25
    heart valve
    St. Jude Medical
    St. Paul, MN
	Safe	1.5	75

Wessex
    Aortic, Model WAV10
    heart valve
    Sorin Biomedica
    Italy
	Safe	1.5	75

Wessex
    Mitral, Model WMV20
    heart valve
    Sorin Biomedica
    Italy
	Safe	1.5	75

Xenofic
    Aortic, Model AP80, size 23
    SS
    heart valve
	Safe	1.5	75

Hemostatic Clips, Other Clips, Fasteners, and Staples

Object	Status	Field Strength (T)	Reference

Absolok Plus
    large
    hemostatic clip
    polydioxanone
    Ethicon, Endo-Surgery
    Cincinnati, OH
	Safe	1.5	71

Absolok Plus
    medium
    hemostatic clip
    polydioxanone
    Ethicon, Endo-Surgery
    Cincinnati, OH
	Safe	1.5	71

Absolok Plus
    small
    hemostatic clip
    polydioxanone
    Ethicon, Endo-Surgery
    Cincinnati, OH
	Safe	1.5	71

Object	Status	Field Strength (T)	Reference

Hemostatic Clips, Other Clips, Fasteners, and Staples (continued)

Object	Status	Field Strength (T)	Reference
AcuClip C.P. Titanium hemostatic clip Origin Medsystems Menlo Park, CA	Safe	1.5	71
Endostaple Surgical Fastener MP35N MedSource Technologies Newton, MA	Safe	1.5, 3	83
Endostaple Surgical Fastener Nitinol MedSource Technologies Newton, MA	Safe	1.5, 3	83
Fascia staple 316L SS United States Surgical North Haven, CT	Safe	1.5, 3	83
Filshie Clip Avalon Medical Corporation Williston, VT	Safe	1.5	
Gastrointestinal anastomosis clip Auto Suture SGIA, SS hemostatic clip United States Surgical Corp. Norwalk, CT	Safe	1.5	3
GIA 4.8 staple Titanium United States Surgical North Haven, CT	Safe	1.5, 3	83
Hemoclip, #10 316L SS hemostatic clip Edward Weck Triangle Park, NJ	Safe	1.5	3

Object	Status	Field Strength (T)	Reference

Hemoclip
    tantalum
    hemostatic clip
    Edward Weck
    Triangle Park, NJ
	Safe	1.5	3

Ligaclip Extra LT 100
    C.P. Titanium
    hemostatic clip
    Ethicon, Endo-Surgery
    Cincinnati, OH
	Safe	1.5	71

Ligaclip Extra LT 200
    C.P. Titanium
    hemostatic clip
    Ethicon, Endo-Surgery
    Cincinnati, OH
	Safe	1.5	71

Ligaclip Extra LT 300
    C.P. Titanium
    hemostatic clip
    Ethicon, Endo-Surgery
    Cincinnati, OH
	Safe	1.5	71

Ligaclip Extra LT 400
    C.P. Titanium
    hemostatic clip
    Ethicon, Endo-Surgery
    Cincinnati, OH
	Safe	1.5	71

Ligaclip
    tantalum
    hemostatic clip
    Ethicon, Inc.
    Sommerville, NJ
	Safe	1.5	3

Ligaclip, #6
    316L SS
    hemostatic clip
    Ethicon, Inc.
    Sommerville, NJ
	Safe	1.5	3

Ligating clip, small
    C.P. Titanium
    hemostatic clip
    Horizon Surgical
    Evergreen, CO
	Safe	1.5	71

Object	Status	Field Strength (T)	Reference

Hemostatic Clips, Other Clips, Fasteners, and Staples (continued)

Object	Status	Field Strength (T)	Reference
Ligating clip, medium C.P. Titanium hemostatic clip Horizon Surgical Evergreen, CO	Safe	1.5	71
MCM 20 C.P. Titanium hemostatic clip Ethicon, Endo-Surgery Cincinnati, OH	Safe	1.5	71
MSM 20 C.P. Titanium hemostatic clip Ethicon, Endo-Surgery Cincinnati, OH	Safe	1.5	71
MCS 20 C.P. Titanium hemostatic clip Ethicon, Endo-Surgery Cincinnati, OH	Safe	1.5	71
Micro SurgiClips Titanium U.S. Surgical Corporation North Haven, CT	Safe	1.5	
MultApplier clip Titanium United States Surgical North Haven, CT	Safe	1.5, 3	83
Ogden Suture Anchor Titanium United States Surgical North Haven, CT	Safe	1.5, 3	83
Premium SurgiClip, L-13 C.P. Titanium hemostatic clip United States Surgical Corp. Norwalk, CT	Safe	1.5	71

Object	Status	Field Strength (T)	Reference

Premium SurgiClip, M-11
    C.P. Titanium
    hemostatic clip
    United States Surgical Corp.
    Norwalk, CT
	Safe	1.5	71

Premium SurgiClip, S-9
    C.P. Titanium
    hemostatic clip
    United States Surgical Corp.
    Norwalk, CT
	Safe	1.5	71

Royal staple
    316L SS
    United States Surgical
    North Haven, CT
	Safe	1.5, 3	83

Surgiclip, M-9.5
    C.P. Titanium
    hemostatic clip
    United States Surgical Corp.
    Norwalk, CT
	Safe	1.5	71

Surgiclip, M-11
    C.P. Titanium
    hemostatic clip
    United States Surgical Corp.
    Norwalk, CT
	Safe	1.5	71

Surgiclip, Auto Suture M-9.5
    SS
    hemostatic clip
    United States Surgical Corp.
    Norwalk, CT
	Safe	1.5	3

Surgiclip spring
    carbon steel
    United States Surgical
    North Haven, CT
	Safe	3	83

TA 90-4.8 directional staples
    Titanium
    United States Surgical
    North Haven, CT
	Safe	1.5, 3	83

Tacker helical fastener
    Titanium
    United States Surgical
    North Haven, CT
	Safe	1.5, 3	83

Object	Status	Field Strength (T)	Reference

Hemostatic Clips, Other Clips, Fasteners, and Staples (continued)

Object	Status	Field Strength (T)	Reference
Ultraclip 316 stainless steel INRAD, Inc. Grand Rapids, MI	Safe	1.5	
Ultraclip Titanium alloy INRAD, Inc. Grand Rapids, MI	Safe	1.5	

Miscellaneous

Object	Status	Field Strength (T)	Reference
357 Magnum Revolver Model 66-3 Smith and Wesson Springfield, MA	Unsafe 1	1.5	46
Accura II Adult Integral Shunt System Phoenix Biomedical Corporation Valley Forge, PA	Safe	1.5	
Accura II Pediatric Integral Shunt System Phoenix Biomedical Corporation Valley Forge, PA	Safe	1.5	
Accusite pH Enteral Feeding System pH Site Locator 10 Fr. Zinetics Medical Salt Lake City, UT	Unsafe 2	1.5	
Adapter used for ICP Aesculap Inc. Center Valley, PA	Safe	1.5, 3	
Adson Tissue Forcep Ti6Al4V Johnson & Johnson Professional, Inc. Raynham, MA	Safe	1.5	62

Object	Status	Field Strength (T)	Reference

AMS Acticon Neosphincter Prosthesis
    American Medical Systems
    Minnetonka, MN
	Safe	1.5, 3	

AMS Artificial Urinary Sphincter 791
    AMS
    American Medical Systems
    Minnetonka, MN
	Safe	1.5, 3	

AMS Mainstay Urologic Soft-Tissue
    Anchor
    AMS
    American Medical Systems
    Minnetonka, MN
	Safe	1.5, 3	

AMS Sphincter 800
    Urinary Control System
    AMS
    American Medical Systems
    Minnetonka, MN
	Safe	1.5, 3	3

Atrostim Phrenic Nerve Stimulator
    ATROTECH OY
    Tampere, Finland
	Unsafe 1	1.5	

Battery, lithium, 3.9 Volt
    304 SS and 316L SS, nickle
    Greatbatch Scientific
    Clarence, NY
	Conditional 1	1.5	

B-D PosiFlow
    needleless IV access connector
    Becton Dickinson
    Sandy, Utah
	Safe	1.5	

Bipolar Coagulation Forceps
    For use in intraoperative
    MRI systems
    Aesculap AG & CO.KG
    Tuttlingen, Germany
	Safe	1.5, 3	83

BioClip
    neurosurgery fixation
    Codman
    Johnson and Johnson
    Berkshire, UK
	Safe	1.5	

Object	Status	Field Strength (T)	Reference

Miscellaneous (continued)

Object	Status	Field Strength (T)	Reference
BioMesh neurosurgery fixation C.P. Titanium Johnson and Johnson Berkshire, UK	Safe	1.5	
Biosearch endo-feeding tube	Safe	1.5	
Blood Collection Set SS United States Surgical North Haven, CT	Safe	1.5, 3	83
Burr Hole Cover neurosurgery fixation C.P. Titanium Johnson and Johnson Berkshire, UK	Safe	1.5	
Cannon Catheter II Chronic Hemodialysis Catheter polyurethane, 304 stainless steel Arrow International, Inc. Reading, PA	Safe	1.5, 3	
Cerebral ventricular shunt tube connector (type unknown)	Unsafe 1	0.147	1
Cerebral ventricular shunt tube connector, Accu-flow right angle Codman Randolf, MA	Safe	1.5	3
Cerebral ventricular shunt tube connector, Accu-Flow, straight Codman Randolf, MA	Safe	1.5	3
Cerebral ventricular shunt tube connector, Accu-flow, T-connector Codman Randolf, MA	Safe	1.5	3

Object	Status	Field Strength (T)	Reference

Codman EDS CSF
    External Drainage System
    Codman
    Johnson and Johnson Company
    Raynham, MA
	Safe	1.5, 3	

Codman-Medos
    Programmable Valve
    Medos S.A.,
    LeLocle, Switzerland
	Conditional 5	1.5	67

Codman MicroSensor Skull Bolt
    Codman
    Johnson and Johnson Company
    Raynham, MA
	Safe	1.5, 3	

Contraceptive diaphragm
    All Flex
    Ortho Pharmaceutical
    Raritan, NJ
	Conditional 1	1.5	3

Contraceptive diaphragm
    Flat Spring
    Ortho Pharmaceutical
    Raritan, NJ
	Conditional 1	1.5	3

Contraceptive diaphragm Gyne T	Safe	1.5	47

Contraceptive diaphragm
    Koroflex
    Young Drug Products
    Piscataway, NJ
	Conditional 1	1.5	3

Contraceptive IUD
    Multiload Cu375
    copper, silver
	Safe	1.5	47

Contraceptive IUD
    Nova T
    copper, silver
	Safe	1.5	47

Contraceptive IUD/IUS
    MIRENA
    intrauterine system (IUS)
    polyethylene, barium sulfate,
    silicone
    Berlex Laboratories
    Montville, NJ
	Safe	1.5, 3	

Object	Status	Field Strength (T)	Reference

Miscellaneous (continued)

Object	Status	Field Strength (T)	Reference

Cranial Ceramic Drill bit
  ceramic
  MicroSurgical Techniques Inc.
  Fort Collins, CO
	Safe	1.5	48

CranioFix
  bone flap fixation system
  Titanium alloy
  Aesculap, Inc.
  Center Valley, PA
	Safe	1.5	50

CranioFix
  Burr Hole Clamp
  FF100T, 11 mm
  Titanium
  Aesculap, Inc.
  Center Valley, PA
	Safe	1.5, 3	83

CranioFix
  Burr Hole Clamp
  FF101T, 16 mm
  Titanium
  Aesculap, Inc.
  Center Valley, PA
	Safe	1.5, 3	83

CranioFix
  Burr Hole Clamp
  FF0997, 20 mm
  Titanium
  Aesculap, Inc.
  Center Valley, PA
	Safe	1.5, 3	83

Cranial Screw
  neurosurgery fixation
  size, 1.5 x 4.5 mm
  Titanium alloy
  Johnson and Johnson
  Berkshire, UK
	Safe	1.5	

Cranial Screw
  neurosurgery fixation
  size 1.9 x 4.5 mm
  Titanium alloy
  Johnson and Johnson
  Berkshire, UK
	Safe	1.5	

Object	Status	Field Strength (T)	Reference
Deponit, nitroglycerin transdermal delivery system aluminized plastic Schwarz Pharma Milwaukee, WI	Conditional 3	1.5	
Disetronic Pumps Disetronic Medical Systems, Inc. St. Paul, MN	Unsafe 1	1.5	
EEG electrodes, Adult E-6-GH gold plated silver Grass Co. Quincy, MA	Safe	0.3	51
EEG electrodes, Pediatric E-5-GH gold plated silver Grass Co. Quincy, MA	Safe	0.3	51
Endoscope, rigid, 2.7 mm (Sinuscope) Greatbatch Scientific Clarence, NY	Safe	1.5	59
Endoscope, rigid, 8.0 mm (Laryngoscope) Greatbatch Scientific Clarence, NY	Safe	1.5	59
Endotracheal (ET) Tube polyurethane coating with silver C.R. Bard, Inc. Covington, GA	Safe	1.5	
Endotracheal tube with metal ring marker Trachmate	Safe	1.5	
ESSURE Device 316L SS, platinum ridium, nickel, Titanium Conceptus San Marcos, CA	Safe	1.5, 3	76, 83
Eyelid weight gold	Safe	1.5	52
FREEHAND System Implantable Functional Neurostimulator (FNS) NeuroControl Corporation Cleveland, OH	Conditional 5	1.5	

Object	Status	Field Strength (T)	Reference
Miscellaneous (continued)			

Fiber-Optic Cardiac Pacing Lead
    photonic device
    Biophan Technologies, Inc.
    Rochester, NY
	Safe	1.5, 3	81

Fiber-optic Intubating
    Laryngoscope Blade
    Greatbatch Scientific
    Clarence, NY
	Safe	1.5	

Fiber-optic Intubating
    Laryngoscope Handle
    Greatbatch Scientific
    Clarence, NY
	Safe	1.5	

Firestar
    9-mm semiautomatic
    Star Bonifacio Echeverria
    Eibar, Spain
	Unsafe 1	1.5	46

Flex-tip Plus Epidural Catheter
    304V SS
    Arrow International Inc.
    Reading, PA
	Unsafe 2	1.5	

Forceps, ceramic
    MicroSurgical Techniques Inc.
    Fort Collins, CO
	Safe	1.5	48

Forceps, Titanium	Safe	1.39	1
Hakim valve and pump	Safe	1.39	1

In-Fast Sling System
    AMS
    American Medical Systems
    Minnetonka, MN
	Safe	1.5, 3	

In-Fast Ultra Sling System
    AMS
    American Medical Systems
    Minnetonka, MN
	Safe	1.5, 3	

Infusion Set
    MMT-11X
    Medtronic MiniMed Inc.
    Northridge, CA
	Safe	1.5	

Object	Status	Field Strength (T)	Reference
Infusion Set MMT-31X Medtronic MiniMed Inc. Northridge, CA	Safe	1.5	
Infusion Set MMT-32X Medtronic MiniMed Inc. Northridge, CA	Safe	1.5	
Infusion Set MMT-37X Medtronic MiniMed Inc. Northridge, CA	Safe	1.5	
Infusion Set MMT-39X Medtronic MiniMed Inc. Northridge, CA	Safe	1.5	
Infusion Set Polyfin MMT-106 Medtronic MiniMed Inc. Northridge, CA	Unsafe 1	1.5	
Infusion Set Polyfin MMT-107 Medtronic MiniMed Inc. Northridge, CA	Unsafe 1	1.5	
Infusion Set Polyfin MMT-16X Medtronic MiniMed Inc. Northridge, CA	Unsafe 1	1.5	
Infusion Set Polyfin MMT-30X Medtronic MiniMed Inc. Northridge, CA	Unsafe 1	1.5	
Infusion Set Polyfin MMT-36X Medtronic MiniMed Inc. Northridge, CA	Unsafe 1	1.5	

Object	Status	Field Strength (T)	Reference

Miscellaneous (continued)

Object	Status	Field Strength (T)	Reference
Implantable Spinal Fusion Stimulator	Conditional 5	1.5	63
Bone Fusion Stimulator			
Electro-Biology, Inc. (EBI)			
Parsippany, NJ			
Implantable Spinal Fusion Stimulator	Conditional 5	1.5, 3	83
Bone Fusion Stimulator			
Model SpF-100			
Electro-Biology, Inc.			
Parsippany, NJ			
Intracranial depth electrodes	Safe	1.5	53
for EEG recordings			
nickle- chromium alloy			
Superior Tube Company			
Norristown, NY			
Intraflex Feeding Tube	Safe	1.5	
tungstun weight, plastic			
Intrauterine contraceptive	Safe	1.5	54
device (IUD), Copper T			
copper			
Searle Pharmaceuticals			
Chicago, IL			
Intrauterine contraceptive device (IUD)	Safe	1.5	
Lippey loop, plastic			
Intrauterine contraceptive device (IUD)	Safe	1.5	
Perigard			
Gyne Pharmaceuticals			
Intima-II I.V. Catheter System	Safe	1.5	
BD Medical Systems			
Becton Dickinson			
Sandy, Utah			
InVance Male Sling System	Safe	1.5, 3	
AMS			
American Medical Systems			
Minnetonka, MN			
Langenbeck Periosteal Elevator	Safe	1.5	62
304 SS			
Johnson & Johnson Professional, Inc.			
Raynham, MA			

Object	Status	Field Strength (T)	Reference
Laparoscopic Graspers Greatbatch Scientific Clarence, NY	Safe	1.5	
Large Flexible Burr Hole Cover neurosurgery fixation C.P. Titanium Johnson and Johnson Berkshire, UK	Safe	1.5	
Large Rigid Burr Hole Cover neurosurgery fixation C.P. Titanium Johnson and Johnson Berkshire, UK	Safe	1.5	
LMA Fastrach Endotracheal Tube size 8 mm endotracheal tube LMA North America, Inc. San Diego, CA	Conditional 5	1.5	
LMA-Classic size 5, large adult laryngeal mask airway LMA North America, Inc. San Diego, CA	Conditional 5	1.5	
LMA-Flexible size 2 larygeal mask airway LMA North America, Inc. San Diego, CA	Conditional 5	1.5	
LMA-ProSeal endotracheal tube LMA North America, Inc. San Diego, CA	Conditional 5	1.5	
Low Magnetic Signature Lithium Battery (C size) Greatbatch Scientific Clarence, NY	Safe	1.5	
May Hegar Needle Holder Ti6Al-4V Johnson & Johnson Professional, Inc. Raynham, MA	Safe	1.5	62

Object	Status	Field Strength (T)	Reference

Miscellaneous (continued)

Object	Status	Field Strength (T)	Reference

Medex 3000 Series MRI Syringe
    Infusion Pump
    Medex
    Dublin, OH
	Conditional 5	1.5	

Mercury Duotube-feeding, feeding tube	Safe	1.5	

Micro Needle Holder
    Greatbatch Scientific
    Clarence, NY
	Safe	1.5	

Micro Round Handled Scissors
    Greatbatch Scientific
    Clarence, NY
	Safe	1.5	

Micro Tissue Forceps
    Greatbatch Scientific
    Clarence, NY
	Safe	1.5	

Micro Tying Forceps
    Greatbatch Scientific
    Clarence, NY
	Safe	1.5	

MiniMed 2007
    Implantable Insulin Pump
    Medtronic MiniMed Inc.
    Northridge, CA
	Unsafe 1	1.5	

MiniMed 407C
    Infusion Pump
    Medtronic MiniMed Inc.
    Northridge, CA
	Unsafe 1	1.5	

MiniMed 508
    Insulin Pump
    Medtronic MiniMed Inc.
    Northridge, CA
	Unsafe 1	1.5	

Mitek anchor
    Miteck Products
    Westood, MA
	Safe	1.5	

MR Bone Punch
    Dismantalable
    Aesculap, Inc.
    Center Valley, PA
	Safe	1.5	74

MR-Brain Spatula with Silicone
    Model FF408K
    Aesculap AG & CO.KG
    Tuttlingen, Germany
	Safe	1.5, 3	83

Object	Status	Field Strength (T)	Reference
MR Currette Curved, 90° Blunt Ring Aesculap, Inc. Center Valley, PA	Safe	1.5	74
MRI FastSystem Retractor System Omni-Tract Surgical Minneapolis, MN	Safe	1.5	83
MR-Kocher-Langenbeck Retractor 70 X 14 mm Aesculap, Inc. Center Valley, PA	Safe	1.5	74
MR-Mallet with Alloy Handle 50 grams Aesculap, Inc. Center Valley, PA	Safe	1.5	74
MR-Septum Speculum, 75 X 7 Resp. 60 X 7 mm Aesculap, Inc. Center Valley, PA	Safe	1.5	74
MR Suction Cannula with Suction Stop Aesculap, Inc. Center Valley, PA	Safe	1.5	74
MR-Weil-Blakesley Ethmoid Forceps Aesculap, Inc. Center Valley, PA	Safe	1.5	74
Neurostimulation System, Medtronic Activa System includes Model 7426 Soletra and Model 7424 Itrel II neurostimulators, Model 3387 and Model 3389 DBS leads Medtronic Minneapolis, MN	Conditional 5	1.5	79, 80
Neurostimulation System InterStim Therapy Sacral Nerve Stimulation for Urinary Control Medtronic Minneapolis, MN	Unsafe 1	1.5	

Object	Status	Field Strength (T)	Reference

Miscellaneous (continued)

Object	Status	Field Strength (T)	Reference
Optistar MR Contrast Delivery System Mallinckrodt St. Louis, MO	Safe	1.5	
Penfield Dissector 304 SS Johnson & Johnson Professional, Inc. Raynham, MA	Safe	1.5	62
Pericardial Patch Hancock Model 710 Medtronic Heart Valve Division Minneapolis, MN	Safe	1.5, 3	84
Pericardial Patch Hancock Model 710L Medtronic Heart Valve Division Minneapolis, MN	Safe	1.5, 3	84
Peripheral Nerve Stimulator MR-STIM, Model GN-013 Greatbatch Scientific Clarence, NY	Safe	1.5	
Straight-In Sacral Colpopexy System AMS American Medical Systems Minnetonka, MN	Safe	1.5, 3	
Scalpel, SS	Unsafe 1	1.5	
Scalpel, Microsharp Ceramic Scalpels, sizes #10, #11, #11c, #15 ceramic MicroSurgical Techniques, Inc. Fort Collins, CO	Safe	1.5	48
Scissors, Ceramic prototype, ceramic Microsurgical Techniques, Inc. Fort Collins, CO	Safe	1.5	48
Shunt valve, Holter-Hausner type Holter-Hausner, Inc. Bridgeport, PA	Safe	1.5	55

Object	Status	Field Strength (T)	Reference
Shunt valve, Holtertype The Holter Co. Bridgeport, PA	Unsafe 1	1.5	55
Sophy adjustable pressure valve	Conditional 5	1.5	56
Sophy programmable pressure valve Model SM8 Sophysa Orsay, France	Unsafe 1	1.5	67
Sophy programmable pressure valve Model SP3 Sophysa Orsay, France	Unsafe 1	1.5	67
Sophy programmable pressure valve Model SU8 Sophysa Orsay, France	Unsafe 1	1.5	67
Spectris MR Injection System Medrad, Inc. Indianola, PA	Safe	1.5	
Spiegelberg System Bolt stainless steel Aesculap, Inc. Center Valley, PA	Safe	1.5, 3	83
Sponge Forcep Ti6Al-4V Johnson & Johnson Professional, Inc. Raynham, MA	Safe	1.5	62
Stereotactic headframe with removable mouthpiece aluminum, 8-18 SS Delrin, Titanium Compass International, Inc. Rochester, MN	Safe	1.5	
Swedish Adjustable Gastric Band (SAGB) Titanium, silicone Ethicon Endo-Surgery, Inc. Cincinnati, OH	Safe	1.5, 3	
Suction/Irrigation Handle for Sinuscope Greatbatch Scientific Clarence, NY	Safe	1.5	

Object	Status	Field Strength (T)	Reference

Miscellaneous (continued)

Object	Status	Field Strength (T)	Reference
Super ArrowFlex PSI 10 Fr. x 65 cm 304V SS Arrow International Inc. Reading, PA	Unsafe 2	1.5	
Super ArrowFlex PSI 9 Fr. x 11 cm 304 V SS Arrow International Inc. Reading, PA	Unsafe 2	1.5	
SynchroMed, implantable drug infusion device Medtronic Inc. Minneapolis, MN	Conditional 5	1.5	
Tantalum powder	Safe	1.39	1
Tears Naturale Port Punctal Occluder Alcon Research, Ltd. Fort Worth, Texas	Safe	1.5	
TheraCath 304 V SS Arrow International Inc. Reading, PA	Unsafe 2	1.5	
TheraSeed Radioactive Seed Implant Titanium, graphite, lead Theragenics Corporation Buford, GA	Safe	1.5, 3	
Tweezers, Ceramic prototype, ceramic MicroSurgical Techniques, Inc. Fort Collins, CO	Safe	1.5	48
Urolume Endoprosthesis AMS American Medical Systems, Minnetonka, MN	Safe	1.5, 3	
Vascular marker, O-ring washer 302 SS PIC Design Middlebury, CT	Conditional 1	1.5	

Object	Status	Field Strength (T)	Reference
Vitallium implant	Safe	1.5	
VOCARE Bladder System	Conditional 5	1.5	
Implantable Functional			
Neuromuscular Stimulator (FNS)			
NeuroControl Corporation			
Valley View, OH			
Winged infusion set	Safe	1.5	58
MRI compatible			
E-Z-EM, Inc.			
Westbury, NY			
Woodson Elevator	Safe	1.5	62
304 SS			
Johnson & Johnson			
Professional, Inc.			
Raynham, MA			

Ocular Implants and Devices

Object	Status	Field Strength (T)	Reference
Clip 250, double tantalum clip	Safe	1.5	29
tantalum			
ocular			
Mira Inc.			
Clip 50, double tantalum clip	Safe	1.5	29
tantalum			
ocular			
Mira Inc.			
Clip 51, single tantalum clip	Safe	1.5	29
tantalum			
ocular			
Mira Inc.			
Clip 52, single tantalum clip	Safe	1.5	29
tantalum			
ocular			
Mira Inc.			
Double tantalum clip	Safe	1.5	29
tantalum			
ocular			
Storz Instrument Co.			
Double tantalum clip style 250	Safe	1.5	29
tantalum			
ocular			
Storz Instrument Co.			

Object	Status	Field Strength (T)	Reference

Ocular Implants and Devices (continued)

Object	Status	Field Strength (T)	Reference
Fatio eyelid spring/wire ocular	Unsafe 1	1.5	30
Gold eyelid spring ocular	Safe	1.5	
Intraocular lens implant Binkhorst, iridocapsular lense platinum-iridium loop platinum, iridium ocular	Safe	1.0	31
Intraocular lens implant Binkhorst, iridocapsular lense platinum-iridium loop ocular	Safe	1.5	31
Intraocular lens implant Binkhorst, iridocapsular lense Titanium loop Titanium ocular	Safe	1.0	31
Intraocular lens implant Worst, platinum clip lense ocular	Safe	1.0	31
Retinal tack 303 SS ocular Bascom Palmer Eye Institute	Safe	1.5	32
Retinal tack 303 SS ocular Duke	Safe	1.5	32
Retinal tack aluminum textraoxide ocular Ruby	Safe	1.5	32
Retinal tack cobalt, nickel ocular Greishaber Fallsington, PA	Safe	1.5	32

Object	Status	Field Strength (T)	Reference

Retinal tack
    martensitic SS
    ocular
    Western European
	Unsafe 1	1.5	32

Retinal tack
    Titanium alloy
    ocular
    Coopervision
    Irvine, CA
	Safe	1.5	32

Retinal tack, Norton staple
    platinum, rhodium
    ocular
    Norton
	Safe	1.5	32

Single tantalum clip
    tantalum
    ocular
	Safe	1.5	29

Troutman magnetic ocular implant ocular	Unsafe 1	1.5	
Unitech round wire eye spring ocular	Unsafe 1	1.5	

Wide Angle IMT
    lens implant
    VisionCare Ophthalmic
    Technologies, Inc.
    Saratoga, CA
	Unsafe 1	1.5	

Orthopedic Implants, Materials, and Devices

Object	Status	Field Strength (T)	Reference

AML femoral component bipolar
    hip prothesis
    orthopedic implant
    Zimmer
    Warsaw, IN
	Safe	1.5	3

BioPro Great Toe M-P Joint
    Cobalt Chrome
    BioPro, Inc.
    Port Huron, MI
	Safe	1.5	

Bryan Cervical Disc Prosthesis Spinal
    C. P. Titanium, Titanium alloy
    orthopedic implant
    Dynamics Corporation
    Mercer Island, WA
	Safe	1.5, 3	

Object	Status	Field Strength (T)	Reference

Orthopedic Implants, Materials, and Devices (continued)

Object	Status	Field Strength (T)	Reference
Cannulated cancellous screw 6.5 x 50 mm Titanium alloy orthopedic implant DePuy ACE Medical Co. El Segundo, CA	Safe	1.5	
Captured screw assembly, 100 mm Titanium alloy orthopedic implant DePuy ACE Medical Co. El Segundo, CA	Safe	1.5	
Cervical wire, 18 gauge 316L SS orthopedic implant	Safe	0.3	33
Charnley-Muller hip prosthesis Protasyl-10 alloy orthopedic implant	Safe	0.3	
Cobalt Chrome Staple Cobalt Chrome, ASTM F75 Smith & Nephew, Inc. Orthopedic Division Memphis, TN	Safe	1.5, 3	83
Compression Hip Screw Plate and Lag Screw (tested as assembly) 316L SS Smith & Nephew, Inc. Orthopedic Division Memphis, TN	Safe	1.5, 3	83
Cortical bone screw 4.5 x 36 mm Titanium alloy orthopedic implant DePuy ACE Medical Co. El Segundo, CA	Safe	1.5	
Cortical bone screw large, Titanium alloy orthopedic implant Zimmer Warsaw, IN	Safe	1.5	34

Object	Status	Field Strength (T)	Reference
Cortical bone screw small, Titanium alloy orthopedic implant Zimmer Warsaw, IN	Safe	1.5	34
Cotrel rod SS-ASTM, grade 2 orthopedic implant	Safe	1.5	
Cotrel rods with hooks 316L SS orthopedic implant	Safe	0.3	33
Drummond wire 316L SS orthopedic implant	Safe	0.3	33
DTT, device for transverse traction 316L SS orthopedic implant	Safe	0.3	33
Endoscopic noncannulated interference screw Titanium orthopedic implant Acufex Microsurgical Norwood, MA	Safe	1.5	34
Fixation staple cobalt- chromium alloy orthopedic implant Richards Medical Co. Memphis, TN	Safe	1.5	34
GII Titanium Anchor Titanium Mitek Products Norwood, MA	Safe	1.5	
Halifax clamps orthopedic implant American Medical Electronics Richardson, TX	Safe	1.5	
Harrington compression rod with hooks and nuts 316L SS orthopedic implant	Safe	0.3	33

Object	Status	Field Strength (T)	Reference

Orthopedic Implants, Materials, and Devices (continued)

Object	Status	Field Strength (T)	Reference
Harrington distraction rod with hooks 316L SS orthopedic implant	Safe	0.3	33
Harris hip prosthesis orthopedic implant Zimmer Warsaw, IN	Safe	1.5	3
Hip implant austenitic SS orthopedic implant DePuy Inc. Warsaw, IN	Safe	1.5, 3	83
Jewett nail orthopedic implant Zimmer Warsaw, IN	Safe	1.5	3
Kirschner intermedullary rod orthopedic implant Kirschner Medical Timonium, MD	Safe	1.5	3
L plate, 6-hole Titanium alloy orthopedic implant DePuy ACE Medical Co. El Segundo, CA	Safe	1.5	
L Rod cobalt-nickel alloy orthopedic implant Richards Medical Co. Memphis, TN	Safe	1.5	
Luque Wire orthopedic implant	Safe	0.3	33
Panalok Anchor clear PLA polymer anchor Mitek Products Norwood, MA	Safe	1.5	

Object	Status	Field Strength (T)	Reference
Moe spinal instrumentation orthopedic implant Zimmer Warsaw, IN	Safe	1.5	
Oxidized Zirconium Knee Femoral Component Smith & Nephew, Inc. Orthopedic Division Memphis, TN	Safe	1.5, 3	83
Perfix interence screw 17-4 SS orthopedic implant Instrument Makar Okemos, MI	Conditional 1	1.5	34
Rusch Rod orthopedic implant	Safe	1.5	
Side plate, 6-hole Titanium alloy orthopedic implant DePuy ACE Co. El Segundo, CA	Safe	1.5	
Spinal L-Rod orthopedic implant DePuy Warsaw, IN	Safe	1.5	
Stainless steel mesh orthopedic implant Zimmer Warsaw, IN	Safe	1.5	3
Stainless steel plate orthopedic implant Zimmer Warsaw, IN	Safe	1.5	3
Stainless steel screw orthopedic implant Zimmer Warsaw, IN	Safe	1.5	3
Stainless steel wire orthopedic implant Zimmer Warsaw, IN	Safe	1.5	3

Object	Status	Field Strength (T)	Reference

Orthopedic Implants, Materials, and Devices (continued)

Object	Status	Field Strength (T)	Reference

Staple plate, large
    Zimaloy
    orthopedic implant
    Zimmer
    Warsaw, IN
	Safe	1.5	3

Synthes AO DCP 2, 3, 4, 5 hole plate orthopedic implant	Safe	1.5	

Teno Fix Tendon Repair System
    Ortheon Medical, LLC
    Winter Park, FL
	Safe	1.5	

Tibial nail, 9 mm
    Titanium alloy
    orthopedic implant
    DePuy ACE Medical Co.
    El Segundo, CA
	Safe	1.5	

Titanium Intramedullary Nail
    Titanium alloy
    Smith & Nephew, Inc.
    Orthopedic Division
    Memphis, TN
	Safe	1.5, 3	83

UltraFix RC Suture Anchor
    316L SS
    Linvatec Corporation
	Safe	1.5	

Universal Reconstruction Ribbon
    Titanium
    orthopedic implant
    DePuy ACE Medical Co.
    El Segundo, CA
	Safe	1.5	

Zielke rod with screw
    washer and nut
    316L SS
    orthopedic implant
	Safe	0.3	33

Object	Status	Field Strength (T)	Reference

Otologic Implants

Object	Status	Field Strength (T)	Reference

Austin tytan piston
  Titanium
  otologic implant
  Treace Medical
  Nashville, TN
	Safe	1.5	35

Berger V bobbin ventilation tube
  Titanium
  otologic implant
  Richards Medical Co.
  Memphis, TN
	Safe	1.5	35

Causse Flex H/A
  partial ossicular prosthesis
  Titanium
  otologic implant
  Microtek Medical, Inc.
  Memphis, TN
	Safe	1.5	36

Causse Flex H/A
  total ossicular prosthesis
  Titanium
  otologic implant
  Microtek Medical Inc.
  Memphis, TN
	Safe	1.5	36

Cochlear implant
  Combi 40/40+
  Multichannel system
  otologic implant
  MedEl
  Innsbrook, Austria
	Unsafe 2	0.2, 1.5	65

Cochlear implant
  Nucleus 24 Cochlear
  Implant System
  otologic implant
  Cochlear Corporation
  Engelwood, CO
	Conditional 5	1.5	

Cochlear implant
  Nucleus Mini 20-channel
  otologic implant
  Cochlear Corporation
  Engelwood, CO
	Unsafe 1	1.5	38

Object	Status	Field Strength (T)	Reference

Otologic Implants (continued)

Object	Status	Field Strength (T)	Reference
Cochlear implant otologic implant 3M/House	Unsafe 1	0.6	37
Cochlear implant otologic implant 3M/Vienna	Unsafe 1	0.6	37
Cody tack otologic implant	Safe	0.6	37
Ehmke hook stapes prosthesis platinum otologic implant Richards Medical Co. Memphis, TN	Safe	1.5	35
Fisch piston Teflon, SS otologic implant Richards Medical Co. Memphis, TN	Safe	1.5	38
Flex H/A notched offset total ossicular prosthesis 316L SS otologic implant Microtek Medical, Inc. Memphis, TN	Safe	1.5	36
Flex H/A offset partial ossicular prosthesis 316L SS Microtek Medical, Inc. Memphis, TN	Safe	1.5	36
House double loop ASTM-318- 76 Grade 2 SS otologic implant Storz St. Louis, MO	Safe	1.5	35
House double loop tantalum otologic implant Storz St. Louis, MO	Safe	1.5	35

Object	Status	Field Strength (T)	Reference

House single loop
    ASTM-318- 76, Grade 2 SS
    otologic implant
    Storz
    St. Louis, MO
	Safe	1.5	31

House single loop
    tantalum
    otologic implant
    Storz
    St. Louis, MO
	Safe	1.5	35

House wire
    SS
    otologic implant
    Otomed
	Safe	0.5	39

House wire
    tantalum
    otologic implant
    Otomed
	Safe	0.5	39

House-type incus prosthesis otologic implant	Safe	0.6	

House-type stainless steel
    piston and wire
    ASTM-318-76 Grade 2 SS
    otologic implant
    Xomed-Treace Inc.
    A Bristol-Myers Squibb Co.
	Safe	1.5	35

House-type wire loop stapes prosthesis
    316L SS
    otologic implant
    Richards Medical Co.
    Memphis, TN
	Safe	1.5	35

McGee piston stapes prosthesis
    316L, SS
    otologic implant
    Richards Medical Co.
    Memphis, TN
	Safe	1.5	35

McGee piston stapes prosthesis
    platinum, 316L SS
    otologic implant
    Richards Medical Co.
    Memphis, TN
	Safe	1.5	35

Object	Status	Field Strength (T)	Reference

Otologic Implants (continued)

Object	Status	Field Strength (T)	Reference
McGee piston stapes prosthesis platinum, chromium-nickel alloy, SS otologic implant Richards Medical Co. Memphis, TN	Unsafe 1	1.5	38
McGee Sheperd's Cook stapes prosthesis 316L SS otologic implant Richards Medical Co. Memphis, TN	Safe	1.5	35
Plasti-pore piston 316L SS, Plasti-pore material otologic implant Richards Medical Co. Memphis, TN	Safe	1.5	35
Platinum ribbon loop stapes prosthesis platinum otologic implant Richards Medical Co. Memphis, TN	Safe	1.5	35
Reuter bobbin ventilation tube 316L SS otologic implant Richards Medical Co. Memphis, TN	Safe	1.5	35
Reuter drain tube otologic implant	Safe	1.5	35
Richards bucket handle stapes prosthesis 316L SS otologic implant Richards Medical Co. Memphis, TN	Safe	1.5	35
Richards piston stapes prosthesis platinum, fluoroplastic otologic implant Richards Medical Co. Memphis, TN	Safe	1.5	35

Object	Status	Field Strength (T)	Reference
Richards Plasti-pore with Armstrong-style platinum ribbon platinum otologic implant Richards Medical Co. Memphis, TN	Safe	1.5	35
Richards platinum Teflon piston 0.6 mm Teflon, platinum otologic implant Richards Medical Co. Memphis, TN	Safe	1.5	38
Richards platinum Teflon piston 0.8 mm Teflon, platinum otologic implant Richards Medical Co. Memphis, TN	Safe	1.5	38
Richards Shepherd's crook platinum otologic implant Richards Medical Co. Memphis, TN	Safe	0.5	39
Richards Teflon piston Teflon otologic implant Richards Medical Co. Memphis, TN	Safe	1.5	38
Robinson incus replacement prosthesis ASTM-318-76 Grade 2 SS otologic implant Storz St. Louis, MO	Safe	1.5	35
Robinson stapes prosthesis ASTM-318-76 Grade 2 SS otologic implant Storz St. Louis, MO	Safe	1.5	35

Object	Status	Field Strength (T)	Reference

Otologic Implants (continued)

Object	Status	Field Strength (T)	Reference
Robinson-Moon offset stapes prosthesis ASTM-318-76 Grade 2 SS otologic implant Storz St. Louis, MO	Safe	1.5	35
Robinson-Moon-Lippy offset stapes prosthesis ASTM-318-76 Grade 2 SS otologic implant Storz St. Louis, MO	Safe	1.5	35
Ronis piston stapes prosthesis 316L SS, fluoroplastic otologic implant Richards Medical Co. Memphis, TN	Safe	1.5	35
Schea cup piston stapes prosthesis platinum, fluoroplastic otologic implant Richards Medical Co. Memphis, TN	Safe	1.5	35
Schea malleus attachment piston Teflon otologic implant Richards Medical Co. Memphis, TN	Safe	1.5	38
Schea stainless steel and Teflon wire prosthesis Teflon, 316 L SS otologic implant Richards Medical Co. Memphis, TN	Safe	1.5	38
Scheer piston Teflon, 316L SS otologic implant Richards Medical Co. Memphis, TN	Safe	1.5	33

Object	Status	Field Strength (T)	Reference

Scheer piston stapes prosthesis
  316L SS, fluoroplastic
  otologic implant
  Richards Medical Co.
  Memphis, TN
	Safe	1.5	35

Schuknecht gelfoam and wire
  prosthesis, Armstrong style
  316L SS
  otologic implant
  Richards Medical Co.
  Memphis, TN
	Safe	1.5	40

Schuknecht piston stapes prosthesis
  316L SS, fluoroplastic
  otologic implant
  Richards Medical Co.
  Memphis, TN
	Safe	1.5	35

Schuknecht Tef-wire incus attachment
  ASTM-318-76 Grade 2 SS
  otologic implant
  Storz
  St. Louis, MO
	Safe	1.5	35

Schuknecht Tef-wire malleus attachment
  ASTM-318-76 Grade 2 SS
  otologic implant
  Storz
  St. Louis, MO
	Safe	1.5	35

Schuknecht Teflon wire piston
  0.6 mm
  Teflon, 316L SS
  otologic implant
  Richards Medical Co.
  Memphis, TN
	Safe	1.5	38

Schuknecht Teflon wire piston
  0.8 mm
  Teflon, 316L SS
  otologic implant
  Richards Medical Co.
  Memphis, TN
	Safe	1.5	38

Object	Status	Field Strength (T)	Reference

Otologic Implants (continued)

Object	Status	Field Strength (T)	Reference
Sheehy incus replacement ASTM-318-76 Grade 2 SS otologic implant Storz St. Louis, MO	Safe	1.5	35
Sheehy incus strut 316L SS otologic implant Richards Medical Co. Memphis, TN	Safe	1.5	38
Sheehy-type incus replacement strut Teflon, 316L SS otologic implant Richards Medical Co. Memphis, TN	Safe	1.5	35
Silverstein malleus clip, ventilation tube Teflon, 316L SS otologic implant Richards Medical Co. Memphis, TN	Safe	1.5	38
SOUNDTEC Direct Drive Hearing System SOUNDTEC, Inc. Oklahoma City, OK	Unsafe 1	1.5	
Spoon bobbin ventilation tube 316L SS otologic implant Richards Medical Co. Memphis, TN	Safe	1.5	35
Stapes, fluoroplastic/platinum, piston otologic implant Microtek Medical, Inc. Memphis, TN	Safe	1.5	36
Stapes, fluoroplastic/stainless steel piston 316L SS otologic implant Microtek Medical, Inc. Memphis, TN	Safe	1.5	36

Object	Status	Field Strength (T)	Reference
Tantalum wire loop stages prosthesis tantalum otologic implant Richards Medical Co. Memphis, TN	Safe	1.5	35
Tef-platinum piston platinum otologic implant Xomed-Treace Inc. A Bristol-Myers Squibb Co.	Safe	1.5	35
Total ossibular replacement prosthesis (TORP) 316L SS otologic implant Richards Medical Co. Memphis, TN	Safe	1.5	38
Trapeze ribbon loop stapes prosthesis platinum otologic implant Richards Medical Co. Memphis, TN	Safe	1.5	35
Vibrant Soundbridge Symphonix Devices, Inc. San Jose, CA	Unsafe 1	1.5	
Williams microclip 316L SS otologic implant Richards Medical Co. Memphis, TN	Safe	1.5	35
Xomed Baily stapes implant otologic implant	Safe	1.5	35
Xomed ceravital partial ossicular prosthesis otologic implant	Safe	1.5	
Xomed stapes prosthesis Robinson-style otologic implant Richard's Co. Nashville, TN	Safe	1.5	35

Object	Status	Field Strength (T)	Reference

Otologic Implants (continued)

Object	Status	Field Strength (T)	Reference
Xomed stapes ASTM-318-76 Grade 2 SS otologic implant Xomed-Treace Inc. A Bristol-Myers Squibb Co.	Safe	1.5	35

Patent Ductus Arteriosus (PDA), Atrial Septal Defect (ASD), and Ventricular Septal Defect (VSD) Occluders

Object	Status	Field Strength (T)	Reference
Atrial Septal Defect Occluder Guardian Angel 12 mm occluder Microvena Corporation White Bear Lake, MN	Safe	1.5	
Atrial Septal Defect Occluder Guardian Angel 40 mm occluder Microvena Corporation White Bear Lake, MN	Safe	1.5	
Bard Clamshell Septal Umbrella 17 mm, occluder MP35N C.R. Bard, Inc. Billerica, MA	Safe	1.5	41
Bard Clamshell Septal Umbrella 23 mm, occluder MP35N C.R. Bard, Inc. Billerica, MA	Safe	1.5	41
Bard Clamshell Septal Umbrella 28 mm, occluder MP35N C.R. Bard, Inc. Billerica, MA	Safe	1.5	41

Object	Status	Field Strength (T)	Reference
Bard Clamshell Septal Umbrella 33 mm, occluder MP35N C.R. Bard, Inc. Billerica, MA	Safe	1.5	41
Bard Clamshell Septal Umbrella 40 mm, occluder MP35N C.R. Bard, Inc. Bellerica, MA	Safe	1.5	41
CardioSEAL Septal Occluder 17 mm MP35N NMT Medical, Inc. Boston, MA	Safe	1.5	
CardioSEAL Septal Occluder 23 mm MP35N NMT Medical, Inc. Boston, MA	Safe	1.5	
CardioSEAL Septal Occluder 28 mm MP35N NMT Medical, Inc. Boston, MA	Safe	1.5	
CardioSEAL Septal Occluder 33 mm MP35N NMT Medical, Inc. Boston, MA	Safe	1.5	
CardioSEAL Septal Occluder 40 mm MP35N NMT Medical, Inc. Boston, MA	Safe	1.5	
HELEX ASD closure device occluder size 15 mm Nitinol W. L. Gore and Associates, Inc. Flagstaff, AZ	Safe	1.5	

Object	Status	Field Strength (T)	Reference

Patent Ductus Arteriosus (PDA), Atrial Septal Defect (ASD), and Ventricular Septal Defect (VSD) Occluders (continued)

Object	Status	Field Strength (T)	Reference

HELEX
    ASD closure device
    occluder
    size 35 mm
    Nitinol
    W. L. Gore and Associates, Inc.
    Flagstaff, AZ
	Safe	1.5	

Lock Clamshell Septal Occlusion
    Implant
    17 mm, occluder
    304 V SS
    C.R. Bard, Inc.
    Billerica, MA
	Conditional 2	1.5	41

Lock Clamshell Septal Occlusion
    Implant
    23 mm, occluder
    304 V SS
    C.R. Bard, Inc.
    Billerica, MA
	Conditional 2	1.5	41

Lock Clamshell Septal Occlusion
    Implant
    28 mm, occluder
    304 V SS
    C.R. Bard, Inc.
    Billerica, MA
	Conditional 2	1.5	41

Lock Clamshell Septal Occlusion
    Implant
    33 mm, occluder
    304 V SS
    C.R. Bard, Inc.
    Billerica, MA
	Conditional 2	1.5	41

Lock Clamshell Septal Occlusion
    Implant
    40 mm, occluder
    304 V SS
    C.R. Bard, Inc.
    Billerica, MA
	Conditional 2	1.5	41

Object	Status	Field Strength (T)	Reference

Rashkind PDA Occlusion Implant
    12 mm, occluder
    304V SS
    C.R. Bard, Inc.
    Billerica, MA
	Conditional 2	1.5	41

Rashkind PDA Occlusion Implant
    17 mm, occluder
    304 V SS
    C.R. Bard, Inc.
    Billerica, MA
	Conditional 2	1.5	41

Pellets and Bullets

Object	Status	Field Strength (T)	Reference
BBs (Crosman)	Unsafe 1	1.5	
BBs (Daisy)	Unsafe 1	1.5	

Bullet, .357 inch
    aluminum, lead
    pellets and bullets
    Winchester
	Safe	1.5	42

Bullet, .357 inch
    bronze, plastic
    pellets and bullets
    Patton-Morgan
	Safe	1.5	42

Bullet, .357 inch
    copper, lead
    pellets and bullets
    Cascade
	Safe	1.5	42

Bullet, .357 inch
    copper, lead
    pellets and bullets
    Hornady
	Safe	1.5	42

Bullet, .357 inch
    copper, lead
    pellets and bullets
    Patton-Morgan
	Safe	1.5	42

Bullet, .357 inch
    lead
    pellets and bullets
    Remington
	Safe	1.5	42

Object	Status	Field Strength (T)	Reference

Pellets and Bullets (continued)

Object	Status	Field Strength (T)	Reference
Bullet, .357 inch nickel, copper, lead pellets and bullets Winchester	Safe	1.5	42
Bullet, .357 inch nylon, lead pellets and bullets Smith & Wesson	Safe	1.5	42
Bullet, .357 inch steel, lead pellets and bullets Fiocchi	Safe	1.5	42
Bullet, .380 inch copper, nickel, lead pellets and bullets Winchester	Unsafe 1	1.5	42
Bullet, .380 inch copper, plastic, lead pellets and bullets Glaser	Safe	1.5	42
Bullet, .44 inch Teflon, bronze pellets and bullets North American Ordinance	Safe	1.5	42
Bullet, .45 inch copper, lead pellets and bullets Samson	Safe	1.5	42
Bullet, .45 inch steel, lead pellets and bullets Evansville Ordinance	Unsafe 1	1.5	42
Bullet, 7.62 x 39 mm copper, steel pellets and bullets Norinco	Unsafe 1	1.5	42
Bullet, 9 mm copper, lead pellets and bullets Norma	Unsafe 1	1.5	42

Object	Status	Field Strength (T)	Reference
Bullet, 9 mm	Safe	1.5	42
copper, lead			
pellets and bullets			
Remington			
Shot, 00 buckshot	Safe	1.5	42
lead			
pellets and bullets			
Shot, 12 gauge, size: 00	Safe	1.5	42
copper, lead			
pellets and bullets			
Federal			
Shot, 4	Safe	1.5	42
lead			
pellets and bullets			
Shot, 7 1/2	Safe	1.5	42
lead			
pellets and bullets			

Penile Implants

Object	Status	Field Strength (T)	Reference
Penile implant, 700 Ultrex Plus	Safe	1.5, 3	
AMS			
American Medical Systems			
Minnetonka, MN			
Penile implant, AMS 700 Ultrex	Safe	1.5, 3	
Preconnected			
AMS			
American Medical Systems,			
Minnetonka, MN			
Penile implant, AMS 700 CX Inflatable	Safe	1.5	43
AMS			
American Medical Systems			
Minnetonka, MN			
Penile implant, AMS 700 CX/CXM	Safe	1.5, 3	
AMS			
American Medical Systems			
Minnetonka, MN			
Penile implant, AMS 700 Ultrex	Safe	1.5, 3	
AMS			
American Medical Systems			
Minnetonka, MN			

Object	Status	Field Strength (T)	Reference

Penile Implants (continued)

Object	Status	Field Strength (T)	Reference
Penile implant, AMS Ambicor AMS American Medical Systems Minnetonka, MN	Safe	1.5, 3	
Penile implant, AMS Dynaflex AMS American Medical Systems Minnetonka, MN	Safe	1.5, 3	
Penile implant, AMS Hydroflex AMS American Medical Systems Minnetonka, MN	Safe	1.5, 3	
Penile implant, AMS Malleable 600 AMS American Medical Systems Minnetonka, MN	Safe	1.5, 3	43
Penile implant, AMS Malleable 600M AMS American Medical Systems Minnetonka, MN	Safe	1.5, 3	
Penile implant, AMS Malleable 650 AMS American Medical Systems Minnetonka, MN	Safe	1.5, 3	
Penile Implant, DURA II Penile Prosthesis AMS American Medical Systems Minnetonka, MN	Safe	1.5, 3	
Penile implant, Duraphase	Unsafe 1	1.5	
Penile implant, Flex-Rod II (Firm) Surgitek, Medical Engineering Corp. Racine, WI	Safe	1.5	43
Penile implant, Flexi-Flate Surgitek, Medical Engineering Corp. Racine, WI	Safe	1.5	43

Object	Status	Field Strength (T)	Reference
Penile implant, Flexi-Rod (Standard) Surgitek, Medical Engineering Corp. Racine, WI	Safe	1.5	43
Penile implant, Jonas Dacomed Corp. Minneapolis, MN	Safe	1.5	43
Penile implant, Mentor Flexible Mentor Corp Minneapolis, MN	Safe	1.5	43
Penile implant, Mentor Inflatable Mentor Corp. Minneapolis, MN	Safe	1.5	43
Penile implant, OmniPhase Dacomed Corp. Minneapolis, MN	Unsafe 1	1.5	43
Penile implant, Osmond, external	Safe	1.5	
Penile implant, Uniflex 1000	Safe	1.5	

Sutures

Object	Status	Field Strength (T)	Reference
Suture Biosyn Needle removed glycomer 631 United States Surgical North Haven, CT	Safe	1.5, 3	83
Suture Chromic gut Needle removed gut United States Surgical North Haven, CT	Safe	1.5, 3	83
Suture Flexon Needle removed stainless steel coated with FEP United States Surgical North Haven, CT	Safe	1.5, 3	83

Object	Status	Field Strength (T)	Reference

Sutures (continued)

Object	Status	Field Strength (T)	Reference
Suture 　Maxon 　Needle removed 　polyglyconate 　United States Surgical 　North Haven, CT	Safe	1.5, 3	83
Suture 　Monosof 　Needle removed 　nylon, lead weight 　with latex bolster 　United States Surgical 　North Haven, CT	Safe	1.5, 3	83
Suture 　Novafil 　Needle removed 　polybutester 　United States Surgical 　North Haven, CT	Safe	1.5, 3	83
Suture 　Plain gut 　Needle removed 　gut 　United States Surgical 　North Haven, CT	Safe	1.5, 3	83
Suture 　Polysorb 　Needle removed 　lactomer 9-1 　United States Surgical 　North Haven, CT	Safe	1.5, 3	83
Suture 　SecureStrand 　Needle removed 　UHMW polyethylene 　United States Surgical 　North Haven, CT	Safe	1.5, 3	83

Object	Status	Field Strength (T)	Reference
Suture	Safe	1.5, 3	83
Sofsilk			
Needle removed			
silk			
United States Surgical			
North Haven, CT			
Suture	Safe	1.5, 3	83
Steel			
Needle removed			
316L SS			
United States Surgical			
North Haven, CT			
Suture	Safe	1.5, 3	83
Surgilon			
Needle removed			
braided nylon			
United States Surgical			
North Haven, CT			
Suture	Safe	1.5, 3	83
Surgipro			
Needle removed			
polypropylene			
United States Surgical			
North Haven, CT			

Vascular Access Ports, Infusion Pumps, and Catheters

Object	Status	Field Strength (T)	Reference
A Port Implantable Access System	Safe	1.5	44
Titanium			
Therex Corporation			
Walpole, MA			
Access Implantable	Safe	1.5	44
Titanium, plastic			
Celsa			
Cedex, France			
AccuRx Constant Flow	Safe	1.5	
Implantable Pump			
Titanium, silicone			
Advanced Neuromodulation Systems			
Plano, TX			

Object	Status	Field Strength (T)	Reference

Vascular Access Ports, Infusion Pumps, and Catheters (continued)

Object	Status	Field Strength (T)	Reference
Broviac catheter single lumen silicone, barium sulfate Bard Access Systems Salt Lake City, UT	Safe	1.5	45
Button Vascular Access Port polysulfone polymer, silicone Infusaid Inc. Norwood, MA	Safe	1.5	44
CathLink LP Titanium Bard Access Systems Salt Lake City, UT	Safe	1.5	45
CathLink SP Titanium Bard Access Systems Salt Lake City, UT	Safe	1.5	45
Celsite Port and Catheter Titanium B. Braun Medical Bethlehem, PA	Safe	1.5	44
Cozmo Pump Insulin Pump Deltec Diabetes Division St. Paul, MN	Unsafe 1	1.5	
Dome Port Vascular Access Port Titanium Davol Inc., Subsidiary of C.R. Bard, Inc. Salt Lake City, UT	Safe	1.5	44
Dual MacroPort polysulfone polymer, silicone Infusaid Inc. Norwood, MA	Safe	1.5	44

Object	Status	Field Strength (T)	Reference
Dual MicroPort	Safe	1.5	44
Vascular Access Port			
polysulfone polymer, silicone			
Infusaid Inc.			
Norwood, MA			
DuraCath Intraspinal Catheter	Safe	1.5	
316L SS, silicone			
Advanced Neuromodulation Systems			
Plano, TX			
GRIPPER PLUS Safety Needle	Safe	1.5, 3	
Deltec, Inc.			
Minneapolis, MN			
Groshong Catheter, dual lumen, 9.5 Fr.	Safe	1.5	45
silicone, barium sulfate, tungsten			
Bard Access Systems			
Salt Lake City, UT			
Groshong Catheter, single lumen, 8 Fr.	Safe	1.5	45
silicone, barium sulfate, tungsten			
Bard Access Systems			
Salt Lake City, UT			
Groshong Catheter	Conditional 1	1.5	
Hickman Catheter, dual lumen, 10.0 Fr.	Safe	1.5	45
silicone, barium sulfate			
Bard Access Systems			
Salt Lake City, UT			
Hickman Catheter, single lumen, 3.0 Fr.	Safe	1.5	45
Bard Access Systems			
Salt Lake City, UT			
Hickman Port	Conditional 1	1.5	44
Vascular Access Port			
316L SS			
Davol Inc., Subsidiary of			
C.R. Bard, Inc.			
Salt Lake City, UT			
Hickman Port, Pediatric	Safe	1.5	44
Vascular Access Port			
Titanium			
Davol, Inc., Subsidiary of			
C.R. Bard, Inc.			
Salt Lake City, UT			

Object	Status	Field Strength (T)	Reference

Vascular Access Ports, Infusion Pumps, and Catheters (continued)

Object	Status	Field Strength (T)	Reference
Hickman subcutaneous port Vascular Access Port SS, Titanium, plastic Davol, Inc., Subsidiary of C.R. Bard, Inc. Salt Lake City, UT	Safe	1.5	45
Hickman subcutaneous port attachable catheter Titanium Davol, Inc., Subsidiary of C.R. Bard, Inc. Salt Lake City, UT	Safe	1.5	45
Hickman subcutaneous port Vascular Access Port venous catheter Titanium Davol Inc., Subsidiary of C.R. Bard, Inc. Salt Lake City, UT	Safe	1.5	44
HMP-Port Vascular Access Port plastic Horizon Medical Products Atlanta, GA	Safe	1.5	44
Implantable Infusion Pump Model 3000-16 Titanium Codman Raynham, MA	Safe	1.5, 3	83
Implantable Infusion Pump Model 3000-30 Titanium Codman Raynham, MA	Safe	1.5, 3	83
Implantable Infusion Pump Model 3000-50 Titanium Codman Raynham, MA	Safe	1.5, 3	83

Object	Status	Field Strength (T)	Reference
Implantofix II	Safe	1.5	44
Vascular Access Port			
polysulfone			
Burron Medical Inc.			
Bethlehem, PA			
Infusaid, Model 400	Safe	1.5	44
Vascular Access Port			
Titanium			
Infusaid Inc.			
Norwood, MA			
Infusaid, Model 600	Safe	1.5	44
Vascular Access Port			
Titanium			
Infusaid Inc.			
Norwood, MA			
Infuse-A-Kit	Safe	1.5	44
plastic			
Infusaid			
Norwood, MA			
LifePort Vascular Access System	Safe	1.5	44
attachable catheter			
plastic			
Strato Medical Group			
Beverly, MA			
LifePort Vascular Access System	Safe	1.5	44
attachable catheter and bayonet			
lock ring			
plastic			
Strato Medical Group			
Beverly, MA			
Lifeport, Model 1013	Safe	1.5	44
Vascular Access Port			
Titanium			
Strato Medical Corp.			
Beverly, MA			
LifePort, Model 6013	Safe	1.5	44
Vascular Access Port			
Delrin			
Strato Medical Corporation			
Beverly, MA			

Object	Status	Field Strength (T)	Reference

Vascular Access Ports, Infusion Pumps, and Catheters (continued)

Object	Status	Field Strength (T)	Reference
Low Profile MRI Port Vascular Access Port Delrin Davol, Inc., Subsidiary of C.R. Bard, Inc. Salt Lake City, UT	Safe	1.5	45
Low Profile MRI Port Vascular Access Port Titanium Davol, Inc. Subsidiary of C.R. Bard, Inc. Salt Lake City, UT	Safe	1.5	45
Macroport Vascular Access Port polysulfone, Titanium Infusaid Inc. Norwood, MA	Safe	1.5	
Mediport Vascular Access Port Cormed	Safe	1.5	
Medtronic MiniMed 2007 Implantable Insulin Pump System Medtronic MiniMed Northridge, CA	Unsafe 1	1.5	
MicroPort Vascular Access Port polysulfone, polymersilicone Infusaid Inc. Norwood, MA	Safe	1.5	44
MRI Hard Base Implanted Port Vascular Access Port plastic Davol, Inc., Salt Lake City, UT	Safe	1.5	44

Object	Status	Field Strength (T)	Reference
MRI Port	Safe	1.5	44
Vascular Access Port			
Delrin, silicone			
Davol, Inc., Subsidiary of			
C.R. Bard, Inc.			
Salt Lake City, UT			
Non-Coring (Huber) Needle	Safe	1.5, 3	83
Medi-tech			
Boston Scientific			
Watertown, MA			
Norport-AC	Safe	1.5	44
Vascular Access Port			
Titanium			
Norfolk Medical			
Skokie, IL			
Norport-DL	Safe	1.5	44
Vascular Access Port			
316L SS			
Norfolk Medical			
Skokie, IL			
Norport-LS	Safe	1.5	44
Vascular Access Port			
316L SS			
Norfolk Medical			
Skokie, IL			
Norport-LS	Safe	1.5	44
Vascular Access Port			
polysulfone			
Norfolk Medical			
Skokie, IL			
Norport-LS	Safe	1.5	44
Vascular Access Port			
Titanium			
Norfolk Medical			
Skokie, IL			
Norport-PT	Safe	1.5	44
Vascular Access Port			
Titanium			
Norfolk Medical			
Skokie, IL			

Object	Status	Field Strength (T)	Reference

Vascular Access Ports, Infusion Pumps, and Catheters (continued)

Object	Status	Field Strength (T)	Reference
Norport-SP Vascular Access Port polysulfone, silicone rubber, Dacron Norfolk Medical Skokie, IL	Safe	1.5	44
OmegaPort Access System Vascular Access Port Titanium, 316L SS Norfolk Medical Skokie, IL	Safe	1.5	44
OmegaPort-SR Access System Vascular Access Port Titanium, 316L SS Norfolk Medical Skokie, IL	Safe	1.5	44
Open-ended Catheter, single lumen, 6 Fr. (ChronoFlex) Davol, Inc., Subsidiary of C.R. Bard, Inc. Salt Lake City, UT	Safe	1.5	45
Open-ended Catheter, single lumen, 8 Fr. (ChronoFlex) Davol, Inc., Subsidiary of C.R. Bard, Inc. Salt Lake City, UT	Safe	1.5	45
OptiPort Catheter, single lumen silicone Simms-Deltec St. Paul, MN	Safe	1.5	45
P.A.S. PORT Elite with PolyFlow Polyurethane Catheter Vascular Access Systems Division Deltec, Inc. St. Paul, MN	Safe	1.5, 3	83
PeriPort Vascular Access Port polysulfone, Titanium Infusaid, Inc. Norwood, MA	Safe	1.5	44

Object	Status	Field Strength (T)	Reference
Phantom 　Vascular Access Port 　Norfolk Medical 　Skokie, IL	Safe	1.5	44
Plastic Port 　Vascular Access Port 　polysulfone, Titanium 　Cardial 　Saint-Etienne, France	Safe	1.5	45
PORT-A-CATH II Dual-Lumen 　Low Profile with PolyFlow 　Polyurethane Catheter 　Deltec, Inc. 　St. Paul, MN	Safe	1.5, 3	83
PORT-A-CATH II Dual-Lumen 　with Silicone Catheter 　Vascular Access Systems Division 　Deltec, Inc. 　St. Paul, MN	Safe	1.5, 3	83
PORT-A-CATH II Single-Lumen 　Low Profile with PolyFlow 　Polyurethane Catheter 　Deltec, Inc. 　St. Paul, MN	Safe	1.5, 3	83
PORT-A-CATH II Single-Lumen 　with PolyFlow 　Polyurethane Catheter 　Deltec, Inc. 　St. Paul, MN	Safe	1.5, 3	83
PORT-A-CATH GRIPPER Needle 　Vascular Access Systems Division 　Deltec, Inc. 　St. Paul, MN	Safe	1.5, 3	83
PORT-A-CATH Needle 　Vascular Access 　Systems Division 　Deltec, Inc. 　St. Paul, MN	Conditional 5	1.5, 3	83

Object	Status	Field Strength (T)	Reference

Vascular Access Ports, Infusion Pumps, and Catheters (continued)

Object	Status	Field Strength (T)	Reference
PORT-A-CATH Titanium Dual Lumen Portal Titanium Pharmacia Deltec St. Paul, MN	Safe	1.5	44
PORT-A-CATH Titanium Peritoneal Portal Titanium Pharmacia Deltec St. Paul, MN	Safe	1.5	44
PORT-A-CATH Titanium Venous Low Profile Portal Titanium Pharmacia Deltec St. Paul, MN	Safe	1.5	44
PORT-A-CATH Titanium Venous Portal Titanium Pharmacia Deltec St. Paul, MN	Safe	1.5	44
PORT-A-CATH, P.S.A. Port Portal Titanium Pharmacia Deltec St. Paul, MN	Safe	1.5	44
Porto-cath Pharmacin, NUTECH Pharmacia Deltec St. Paul, MN	Safe	1.5	44
Q-Port Vascular Access Port 316L SS Quinton Instrument Co. Seattle, WA	Conditional 1	1.5	44
R-Port Premier Vascular Access Port silicone, plastic, SS Medi-tech Boston Scientific Watertown, MA	Safe	1.5, 3	83

Object	Status	Field Strength (T)	Reference

S.E.A.
  Vascular Access Port
  Titanium
  Harbor Medical Devices, Inc.
  Boston, MA
	Safe	1.5	44

Snap-Lock
  Vascular Access Port
  Titanium, polysulfone polymer,
  silicone
  Infusaid Inc.
  Norwood, MA
	Safe	1.5	44

Synchromed
  Model 8500-1
  Titanium, thermoplastic, silicone
  Medtronic, Inc.
  Minneapolis, MN
	Conditional 5	1.5	44

TitanPort
  Titanium
  Vascular Access Port
  Norfolk Medical
  Skokie, IL
	Safe	1.5	

Triple Lumen
  Arrow International, Inc.
  Reading, PA
	Safe	1.5	

Vascular Access Catheter With Repair Kit	Safe	1.5	

Vasport
  Vascular Access Port
  Titanium, fluoropolymer
  Gish Biomedical, Inc.
  Santa Ana, CA
	Safe	1.5	44

Vaxess
  Vascular Access Port
  19 gauge x 1/2, 90° hub
  plastic, polyurethane
  Medi-tech
  Watertown, MA
	Safe	1.5, 3	83

Object	Status	Field Strength (T)	Reference

Vascular Access Ports, Infusion Pumps, and Catheters (continued)

Object	Status	Field Strength (T)	Reference

Vaxess
    Vascular Access Port
    plastic, polyurethane
    Medi-tech
    Boston Scientific
    Watertown, MA
	Safe	1.5, 3	83

Vaxess, Titanium
    Vascular Access Port
    Titanium, polyurethane
    Medi-tech
    Boston Scientific
    Watertown, MA
	Safe	1.5, 3	83

Vaxess, Titanium Mini-Port
    with silicone catheter
    Vascular Access Port
    Titanium, silicone
    Medi-tech
    Boston Scientific
    Watertown, MA
	Safe	1.5, 3	83

Vital-Port
    Vascular Access Port
    polysulfone, Titanium
    Cook Corp.
    Leechburg, PA
	Safe	1.5	45

Vital-Port, Dual
    Vascular Access Port
    polysulfone, Titanium
    Cook Corp.
    Leechburg, PA
	Safe	1.5	45

THE LIST

REFERENCES

1. New PFJ, Rosen BR, Brady TJ, et al. Potential hazards and artifacts of ferromagnetic and nonferromagnetic surgical and dental materials and devices in nuclear magnetic resonance imaging. Radiology 1983;147:139-148.

2. Becker RL, Norfray JF, Teitelbaum GP, et al. MR imaging in patients with intracranial aneurysm clips. Am J Roentgenol 1988;9:885-889.

3. Shellock FG, Crues JV. High-field strength MR imaging and metallic biomedical implants: an ex vivo evaluation of deflection forces. Am J Roentgenol 1988;151:389-392.

4. Brown MA, Carden JA, Coleman RE, et al. Magnetic field effects on surgical ligation clips. Magn Reson Imaging 1987;5:443-453.

5. Dujovny M, Kossovsky N, Kossowsky R, et al. Aneurysm clip motion during magnetic resonance imaging: in vivo experimental study with metallurgical factor analysis. Neurosurgery 1985;17:543-548.

6. Barrafato D, Henkelman RM. Magnetic resonance imaging and surgical clips. Can J Surg 1984;27:509-512.

7. Moscatel M, Shellock FG, Morisoli S. Biopsy needles and devices: assessment of ferromagnetism and artifacts during exposure to a 1.5 Tesla MR system. J Magn Reson Imaging 1995;5:369-372.

8. Shellock FG, Shellock VJ. Additional information pertaining to the MR-compatibility of biopsy needles and devices. J Magn Reson Imaging 1996;6:441.

9. Hathout G, Lufkin RB, Jabour B, et al. MR-guided aspiration cytology in the head and neck at high field strength. J Magn Reson Imaging 1992;2:93-94.

10. Fagan LL, Shellock FG, Brenner RJ, Rothman B. Ex vivo evaluation of ferromagnetism, heating, and artifacts of breast tissue expanders exposed to a 1.5 T MR system. J Magn Reson Imaging 1995;5:614-616.

423

11. Teitelbaum GP, Lin MCW, Watanabe AT, et al. Ferromagnetism and MR imaging: safety of carotid vascular clamps. Am J Neuroradiol 1990;11:267-272.

12. Gegauff A, Laurell KA, Thavendrarajah A, et al. A potential MRI hazard: forces on dental magnet keepers. J Oral Rehabil 1990;17:403-410.

13. Shellock FG. Ex vivo assessment of deflection forces and artifacts associated with high-field strength MRI of "mini-magnet" dental prostheses. Magn Reson Imaging 1989;7 (Suppl 1):38.

14. Shellock FG, Slimp G. Halo vest for cervical spine fixation during MR imaging. Am J Roentgenol 1990;154:631-632.

15. Clayman DA, Murakami ME, Vines FS. Compatibility of cervical spine braces with MR imaging. A study of nine nonferrous devices. Am J Neuroradiol 1990;11:385-390.

16. Shellock FG. MR imaging and cervical fixation devices: assessment of ferromagnetism, heating, and artifacts. Magn Reson Imaging 1996;14:1093-1098.

17. Soulen RL, Budinger TF, Higgins CB. Magnetic resonance imaging of prosthetic heart valves. Radiology 1985;154:705-707.

18. Shellock FG, Morisoli SM. Ex vivo evaluation of ferromagnetism, heating, and artifacts for heart valve prostheses exposed to a 1.5 Tesla MR system. J Magn Reson Imaging 1994;4:756-758.

19. Hassler M, Le Bas JF, Wolf JE, et al. Effects of magnetic fields used in MRI on 15 prosthetic heart valves. J Radiol 1986;67:661-666.

20. Frank H, Buxbaum P, Huber L, et al. In vitro behavior of mechanical heart valves in 1.5 T superconducting magnet. Eur J Radiol 1992;2:555-558.

21. Teitelbaum GP, Bradley WG, Klein BD. MR imaging artifacts, ferromagnetism, and magnetic torque of intravascular filters, stents, and coils. Radiology 1988;166:657-664.

22. Marshall MW, Teitelbaum GP, Kim HS, et al. Ferromagnetism and magnetic resonance artifacts of platinum embolization microcoils. Cardiovasc Intervent Radiol 1991;14:163-166.

23. Watanabe AT, Teitelbaum GP, Gomes AS, et al. MR imaging of the bird's nest filter. Radiology 1990;177:578-579.

24. Leibman CE, Messersmith RN, Levin DN, et al. MR imaging of inferior vena caval filter: safety and artifacts. Am J Roentgenol 1988;150:1174-1176.

25. Shellock FG, Detrick MS, Brant-Zawadski M. MR-compatibility of the Guglielmi detachable coils. Radiology 1997;203:568-570.

26. Kiproff PM, Deeb DL, Contractor FM, Khoury MB. Magnetic resonance characteristics of the LGM vena cava filter: technical note. Cardiovasc Intervent Radiol 1991;14:254-255.

27. Teitelbaum GP, Raney M, Carvlin MJ, et al. Evaluation of ferromagnetism and magnetic resonance imaging artifacts of the Strecker tantalum vascular stent. Cardiovasc Intervent Radiol 1989;12:125-127.

28. Girard MJ, Hahn P, Saini S, Dawson SL, Goldberg MA, Mueller PR. Wallstent metallic biliary endoprosthesis: MR imaging characteristics. Radiology 1992;184:874-876.

29. Shellock FG, Myers SM, Schatz CJ. Ex vivo evaluation of ferromagnetism determined for metallic scleral "buckles" exposed to a 1.5 T MR scanner. Radiology 1992;185:288-289.

30. de Keizer RJ, Te Strake L. Intraocular lens implants (pseudophakoi) and steel-wire sutures: a contraindication for MRI? Doc Ophthalmol 1984;61:281-284.

31. Albert DW, Olson KR, Parel JM, et al. Magnetic resonance imaging and retinal tacks. Arch Ophthalmol 1990;108:320-321.

32. Joondeph BC, Peyman GA, Mafee MF, et al. Magnetic resonance imaging and retinal tacks [Letter]. Arch Ophthalmol 1987;105:1479-1480.

33. Lyons CJ, Betz RR, Mesgarzadeh M, et al. The effect of magnetic resonance imaging on metal spine implants. Spine 1989;14:670-672.

34. Shellock FG, Mink JH, Curtin S, et al. MRI and orthopedic implants used for anterior cruciate ligament reconstruction: assessment of ferromagnetism and artifacts. J Magn Reson Imaging 1992;2:225-228.

35. Shellock FG, Schatz CJ. High-field strength MR imaging and metallic otologic implants. Am J Neuroradiol 1991;12:279-281.

36. Nogueira M, Shellock FG. Otologic bioimplants: ex vivo assessment of ferromagnetism and artifacts at 1.5 Tesla. Am J Roentgenol 1995;163:1472-1473.

37. Mattucci KF, Setzen M, Hyman R, et al. The effect of nuclear magnetic resonance imaging on metallic middle ear prostheses. Otolaryngol Head Neck Surg 1986;94:441-443.

38. Applebaum EL, Valvassori GE. Further studies on the effects of magnetic resonance fields on middle ear implants. Ann Otol Rhinol Laryngol 1990;99:801-804.

39. White DW. Interaction between magnetic fields and metallic ossicular prostheses. Am J Otol 1987;8:290-292.

40. Leon JA, Gabriele OF. Middle ear prosthesis: significance in magnetic resonance imaging. Magn Reson Imaging 1987;5:405-406.

41. Shellock FG, Morisoli SM. Ex vivo evaluation of ferromagnetism and artifacts for cardiac occluders exposed to a 1.5 Tesla MR system. J Magn Reson Imaging 1994;4:213-215.

42. Teitelbaum GP, Yee CA, Van Horn DD, et al. Metallic ballistic fragments: MR imaging safety and artifacts. Radiology 1990;175:855-859.

43. Shellock FG, Crues JV, Sacks SA. High-field magnetic resonance imaging of penile prostheses: in vitro evaluation of deflection forces and imaging artifacts [Abstract]. In: Book of Abstracts, Society of Magnetic Resonance in Medicine. Berkeley, CA, Society of Magnetic Resonance in Medicine 1987;3:915.

44. Shellock FG, Nogueira M, Morisoli S. MR imaging and vascular access ports: ex vivo evaluation of ferromagnetism, heating, and artifacts at 1.5 T. J Magn Reson Imaging 1995;4:481-484.

45. Shellock FG, Shellock VJ. Vascular access ports and catheters tested for ferromagnetism, heating, and artifacts associated with MR imaging. Magn Reson Imaging 1996;14:443-447.

46. Kanal E, Shaibani A. Firearm safety in the MR imaging environment. Radiology 1994;193:875-876.

47. Hess T, Stepanow B, Knopp MV. Safety of intrauterine contraceptive devices during MR imaging. Eur Radiol 1996;6:66-68.

48. Shellock FG, Shellock VJ. Evaluation of MR compatibility of 38 bioimplants and devices. Radiology 1995;197:174.

49. Shellock FG, Shellock VJ. Ceramic surgical instruments: ex vivo evaluation of compatibility with MR imaging. J Magn Reson Imaging 1996;6:954-956.

50. Shellock FG, Shellock VJ. Evaluation of cranial flap fixation clamps for compatibility with MR imaging. Radiology 1998;822-825.

51. Lufkin R, Jordan S, Lylcyk M. MR imaging with topographic EEG electrodes in place. Am J Neuroradiol 1988;9:953-954.

52. Marra S, Leonetti JP, Konior RJ, Raslan W. Effect of magnetic resonance imaging on implantable eyelid weights. Ann Otol Rhinol Laryngol 1995;104:448-452.

53. Zhang J, Wilson CL, Levesque MF, Behnke EJ, Lufkin RB. Temperature changes in nickel-chromium intracranial depth electrodes during MR scanning. Am J Neuroradiol 1993;14:497-500.

54. Mark AS, Hricak H. Intrauterine contraceptive devices: MR imaging. Radiology 1987;162:311-314.

55. Go KG, Kamman RL, Mooyaart EL. Interaction of metallic neurosurgical implants with magnetic resonance imaging at 1.5 Tesla as a cause of image distortion and of hazardous movement of the implant. Clin Neurosurg 1989;91:109-115.

56. Fransen P, Dooms G, Thauvoy. Safety of the adjustable pressure ventricular valve in magnetic resonance imaging: problems and solutions. Neuroradiology 1992;34:508-509.

57. ECRI, Health devices alert. A new MRI complication? May 27, 1988.

58. To SYC, Lufkin RB, Chiu L. MR-compatible winged infusion set. Comput Med Imaging Graph 1989;13:469-472.

59. Shellock FG. MR-compatibility of an endoscope designed for use in interventional MRI procedures. Am J Roentgenol 1998;71:1297-1300.

60. Shellock FG, Shellock VJ. Cardiovascular catheters and accessories: Ex vivo testing of ferromagnetism, heating, and artifacts associated with MRI. J Magn Reson Imaging 1998;8:1338-1342.

61. Shellock FG, Shellock VJ. Metallic marking clips used after stereotactic breast biopsy: ex vivo testing of ferromagnetism, heating, and artifacts associated with MRI. Am J Roentgenol 1999;172:1417-1419.

62. Shellock FG. MRI safety of instruments designed for interventional MRI: assessment of ferromagnetism, heating, and artifacts. Workshop on New Insights into Safety and Compatibility Issues Affecting In Vivo MR, Syllabus 1998; pp. 39.

63. Shellock FG, Hatfield M, Simon BJ, Block S, Wamboldt J, Starewicz PM, Punchard WFB. Implantable spinal fusion stimulator: assessment of MRI safety. J Magn Reson Imaging 2000;12:214-223.

64. Shellock FG, Shellock VJ. Stents: Evaluation of MRI safety. Am J Roentgenol 1999;173:543-546.

65. Teissl C, Kremser C, Hochmair ES, Hochmair-Desoyer IJ. Cochlear implants: in vitro investigation of electromagnetic interference at MR imaging-compatibility and safety aspects. Radiology 1998;208:700-708.

66. Teissl C, Kremser C, Hochmair ES, Hochmair-Desoyer IJ. Magnetic resonance imaging and cochlear implants: compatibility and safety aspects. J Magn Reson Imaging 1999;9:26-38.

67. Ortler M, Kostron H, Felber S. Transcutaneous pressure-adjustable valves and magnetic resonance imaging: an ex vivo examination of the Codman-Medos programmable valve and the Sophy adjustable pressure valve. Neurosurgery 1997;40:1050-1057.

68. Shellock FG, Shellock VJ. MR-compatibility evaluation of the Spetzler titanium aneurysm clip. Radiology 1998;206:838-841.

69. Jost C, Kuman V. Are current cardiovascular stents MRI safe? J Invasive Cardiol 1998;10:477-479.

70. Shellock FG, Shellock VJ. MRI Safety of cardiovascular implants: evaluation of ferromagnetism, heating, and artifacts. Radiology 2000;214:P19H.

71. Weishaupt D, Quick HH, Nanz D, Schmidt M, Cassina PC, Debatin JF. Ligating clips for three-dimensional MR angiography at 1.5 T: In vitro evaluation. Radiology 2000;214:902-907.

72. Hug J, Nagel E, Bornstedt A, Schackenburg B, Oswald H, Fleck E. Coronary arterial stents: safety and artifacts during MR imaging. Radiology 2000;216:781-787.

73. Taal BG, Muller SH, Boot H, Koop W. Potential risks and artifacts of magnetic resonance imaging of self-expandable esophageal stents. Gastrointestinal Endoscopy 1997;46:424-429.

74. Shellock FG. Metallic surgical instruments for interventional MRI procedures: evaluation of MR safety. J Magn Reson Imaging 2001;13:152-157.

75. Edwards, M-B, Taylor KM, Shellock FG. Prosthetic heart valves: evaluation of magnetic field interactions, heating, and artifacts at 1.5 Tesla. J Magn Reson Imaging 2000;12:363-369.

76. Shellock FG. New metallic implant used for permanent female contraception: evaluation of MR safety. Am J Roentgenol 2002;178:1513-1516.

77. Shellock FG. Prosthetic heart valves and annuloplasty rings: assessment of magnetic field interactions, heating, and artifacts at 1.5-Tesla. Journal of Cardiovascular Magnetic Resonance. 2001;3:159-169.

78. Shellock FG. Metallic neurosurgical implants: evaluation of magnetic field interactions, heating, and artifacts at 1.5 Tesla. J Magn Reson Imaging 2001;14:295-299.

79. Rezai AR, Finelli D, Ruggieri P, Tkach J, Nyenhuis JA, Shellock FG. Neurostimulators: Potential for excessive heating of deep brain stimulation electrodes during MR imaging. J Magn Reson Imaging 2001;14:488-489.

80. Finelli DA, Rezai AR, Ruggieri P, Tkach J, Nyenhuis J, Hridlicka G, Sharan A, Stypulkowski PH, Shellock FG. MR-related heating of deep brain stimulation electrodes: an in vitro study of clinical imaging sequences. Am J Neuroradiol 2002;23:1795-1802.

81. Greatbatch W, Miller V, Shellock FG. Magnetic resonance safety testing of a newly-developed, fiber-optic cardiac pacing lead. J Magn Reson 2002;16:97-103.

82. Shellock FG, Tkach JA, Ruggieri PM, Masaryk T, Rasmussen P. Aneurysm clips: evaluation of magnetic field interactions and translational attraction using "long-bore" and "short-bore" 3.0-Tesla MR systems. Am J Neuroradiol 2003;24:463-471.

83. Shellock FG. Biomedical implants and devices: assessment of magnetic field interactions with a 3.0-Tesla MR system. J Magn Reson Imaging 2002;16:721-732.

84. Medtronic Heart Valves, Medtronic, Inc., Minneapolis, MN, Permission to publish 3-Tesla MR testing information for Medtronic Heart Valves provided by Kathryn M. Bayer, Senior Technical Consultant, Medtronic Heart Valves, Technical Service, 1-800-328-2518, extension -42861 or 763-514-2861.

85. Shellock FG, Tkach JA, Ruggieri PM, Masaryk TJ. Cardiac pacemakers, ICDs, and loop recorder: Evaluation of translational attraction using conventional ("long-bore") and "short-bore" 1.5- and 3.0-Tesla MR systems. Journal of Cardiovascular Magnetic Resonance 2003;5:387-397.

86. Product Information, Style 133 Family of Breast Tissue Expanders with Magna-Site Injection Sites, McGhan Medical/INAMED Aesthetics, Santa Barbara, CA.

APPENDIX I

MEDICAL DEVICES AND ACCESSORIES DEVELOPED FOR USE IN THE MR ENVIRONMENT AND FOR INTERVENTIONAL MRI PROCEDURES*

Various vendors and manufacturers, prompted by recommendations and requests from MR healthcare professionals, have recognized the need to develop specialized medical devices, accessories, and instruments necessary for use in the MR environment and for interventional MR procedures. Similar to other equipment used in the MR environment, *ex vivo* testing and evaluation is required to demonstrate the safe use and operation of these devices and accessories before utilization. In general, the test procedures include an assessment of magnetic field interactions, heating, induced electrical current (for certain devices), and artifacts. In addition, it may be necessary to assess the functional or operational aspects of the device to ensure proper operation in the MR environment.

Medical devices and accessories are typically characterized with respect to being "MR-safe" or "MR-compatible" in association with the specific electromagnetic fields of the MR environment used for the evaluation. These terms are defined, as follows:

MR-safe: A device is considered "MR-safe" if, when placed in the MR environment, it presents no additional risk to a patient, individual, or MR system operator from magnetic field interactions, heating, or induced voltages, but may effect the quality of the diagnostic information on MR scans.

MR-compatible: A device is considered "MR-compatible" if it is MR safe, its use in the MR environment does not adversely impact the image quality, and it performs its intended function when used in the MR environment according to its specifications in a safe and effective manner.

Importantly, the above terminology is currently undergoing review and, as such, may change in the near future.

This appendix lists various vendors and manufacturers that provide devices, accessories, and instruments that designed or modified for the MR environment or for use in interventional MR procedures.

Note: Some of these medical devices, accessories, and instruments may be pre-product prototypes that are not approved for use by the U.S. Food and Drug Administration (FDA) or other similar agencies. Some products may not be available in all countries. No claims are made regarding patient or staff safety, MR-safety, MR-compatibility, or clinical capability of any device or product included in this list.

Before introduction of any medical product into the MR environment, the product should be inspected and tested by qualified personnel. The product's "magnetic properties" and its clinical operation in the MR environment should be verified before it is used in the MR environment, during an MR procedure or during an interventional MR procedure. The use of these medical products for animal or human procedures must comply with applicable government or local hospital safety and animal/human studies committee requirements. MR healthcare professionals should contact the vendors directly for technical specifications, commercial availability, and pricing of the medical products.

AESCULAP, INC.

3773 Executive Center Pkwy
Center Valley, PA 18034

Products

General surgical instruments

Contact

Phone: 800-282-9000

DAUM CORPORATION

200 S. Wacker Drive
Suite 3100
Chicago, IL 60606

Products

MRI Body Punctures
MRI Neurology
MRI Orthopedics
MRI Device Technology
Coaxial Needles
NeuroGate Trocar

Contact

Phone: 312-674-4529
Fax: 312-674-4528
Email: daum.usa@daum.de

DRAEGER MEDICAL, INC.

3135 Quarry Road
Telford, PA 18969

Products

MRI anesthesia equipment

Contact

Phone: 800-437-2437

E-Z-EM, INC

717 Main Street
Westbury, NY 11590

Products

Lufkin 22 gauge cytology biopsy needles
MRI histology biopsy needles
MRI core biopsy guns, 14 and 18 gauge
MRI lesion marking systems (20 gauge needle with Kopans
style localization wire)

Contact

Phone: 800-544-4624
Website: www.ezem.com

GALIL MEDICAL USA

400 West Cummings Park, Suite 4600
Woburn, MA 01801

Products

MR-compatible cryosurgery Systems
CryoHit precision cryo-ablation system
Cryotherapy Equipment

Contact

Phone: 781-933-2828
Fax: 781-933-9093

GE MEDICAL DIAGNOSTIC IMAGING ACCESSORIES

P.O. Box 414, W-520
Milwaukee, WI 53201

Products

General MR imaging accessories, magnet & coil drapes

Contact

Phone: 800-472-3666

IMAGE-GUIDED NEUROLOGICS, INC.

2290 W. Eau Gallie Blvd.
Melbourne, FL 32935

Products

Develops proprietary devices that allow real-time, magnetic resonance
(MR) image-guided surgery and drug delivery for treating neurological diseases.

Contact

Phone: 321-757-8900 ext. 270
Fax: 321-757-8616

INTEGRA NEUROSCIENCES

(formerly Elekta, Ruggles NMT)
105 Morgan Lane
Plainsboro, NJ 08536

Products

Neuro Biopsy Needle 651-601
Fukishima Retractor Arms
Genesis Retractor Arms
Suction
Malcolm Rand Retractor Bars

Contact

Phone: 609-275-0500
Fax: 609-799-3297

INVIVO CORPORATION

12601 Research Pkwy.
Orlando, FL 32826

Products

Vital sign monitoring equipment capable of monitoring ECG, respiratory, invasive and noninvasive blood pressure

Contact

Phone: 800-331-3220
Website: www.invivocorp.com

J&J ETHICON, INC.

Route 22
Somerville, NJ 08876

Products

Needles with sutures

Contact

Phone: 908-218-2297
Fax: 908-218-2531

JOHNSON & JOHNSON PROFESSIONAL, INC. (CODMAN DIVISION)

325 Paramount Dr.
Raynham, MA 02767

Products

CMC3 Irrigation Bipolar, Bookwalter Arm and Table Post, Rhoton Microsurgical, Hudson Twist Drill

Contact

Phone: 508-880-8100

LIFE INSTRUMENTS

14 Wood Road, Suite 002
Braintree, MA 02184

Products

Penfields, Curettes, Mini Cobbs, Hudson Twist Drill, Drill Bits, Periosteal Elevators

Contact

Phone: 781-849-0109 or 800-925-2995
Fax: 781-849-0128

MAGMEDIX

158 R Main Street
Gardner, MA 01440

Products

Large selection of nonmagnetic products
Respiratory equipment
MRI facility start-up kits
Physiologic monitoring equipment
Patient comfort and positioning devices
MRI tools
Patient transport equipment
Cryogen accessories
MRI accessories
MRI carts and maintenance devices
MRI danger signs

Contact

Bill Husted or Paula Burgess
Phone: 866-646-3349; 978-630-5580
Fax: 978-630-5583
Website: www.Magmedix.com

MALLINCKRODT, INC.

675 McDonnell Blvd.
St. Louis, MO 63134

Product

OptiStar MR Contrast Delivery System

Contact

Phone: 314-654-3981 or 314-654-2000

MEDRAD

One Medrad Dr.
Indianola, PA 15051-0780

Products

Model 6500 Vital Signs Monitor
Includes: Pulse Oximetry, noninvasive blood pressure and Fiber
Optic ECG
Model 9500 Vital Signs/Multigas Monitor
Includes: Pulse Oximetry, noninvasive blood pressure and Fiber
Optic ECG, Capnography
Model F Music System
Spectris MR Injection System

Contact

Phone: 800-633-7231 or 412-767-2400
Fax: 412-963-1964
Website: www.Medrad.com

MINRAD, INC.

847 Main St.
Buffalo, New York 14203

Products

Monitor Stand
Fiber-optic light sources
Laryngoscope blades and accessories
Endoscopic equipment
Headlamp

Contact

Phone: Customer Service 800-832-3303 or
716-855-1068
Fax: 716-855-1078
Website: www.minrad.com

MRI DEVICES CORPORATION

1515 Paramount Drive
Waukesha, WI 53186

Products

Biopsy needles
Biopsy positioning devices
Biopsy localization systems
Specialize equipment and accessories

Contact

Phone: 800-524-1476
Website: www.MRIDevices.com

OMNIVENT

1720 Sublette Avenue
St. Louis, MO 63110

Products

Patient ventilation and ventilator monitors

Contact

Phone: 314-771-2400
Fax: 800-477-7701

RESONANCE TECHNOLOGY, INC.

18121 Parthenia St.
Northridge, CA 91325

Products

MRI audio/video systems
fMRI products

Contact

Phone: 818-882-1997
Websites: www.fmri.com, www.mrivideo.com

SYNERGETICS, INC
17466 Chesterfield Airport Road
Chesterfield, MO 63005

Products
Deep Neuro Dissection Set
Synerturn N000-160

Contact
Phone: 636-939-5100
Fax: 636-939-6885

[*Special thanks to Ms. Karen Streit of General Electric Medical Systems, Milwaukee, WI for providing the majority of the information used for this compilation.]

APPENDIX II

GUIDANCE FOR INDUSTRY AND FDA STAFF

CRITERIA FOR SIGNIFICANT RISK INVESTIGATIONS OF MAGNETIC RESONANCE DIAGNOSTIC DEVICES

Document issued on: July 14, 2003

This document supersedes "Guidance for Magnetic Resonance Diagnostic Devices - Criteria for Significant Risk Investigations" issued on September 29, 1997

For questions regarding this document, contact Loren A. Zaremba, Ph.D., at 301-594-1212, ext. 137 or by e-mail at lzz@cdrh.fda.gov

U.S. Department of Health and Human Services
Food and Drug Administration
Center for Devices and Radiological Health

Radiological Devices Branch
Division of Reproductive, Abdominal, and Radiological Devices
Office of Device Evaluation

PREFACE

Public Comment

Written comments and suggestions may be submitted at any time for Agency consideration to Division of Dockets Management, Food and Drug Administration, 5630 Fishers Lane, Room 1061, (HFA-305), Rockville, MD, 20852. When submitting comments, please refer to the exact title of this guidance document. Comments may not be acted upon by the Agency until the document is next revised or updated.

Additional Copies

Additional copies are available from the Internet at: http://www.fda.gov/cdrh/ode/guidance/793.pdf or to receive this document by fax, call the CDRH Facts-On-Demand system at 800-899-0381 or 301-827-0111 from a touch-tone telephone. Press 1 to enter the system. At the second voice prompt, press 1 to order a document. Enter the document number (793) followed by the pound sign (#). Follow the remaining voice prompts to complete your request.

This guidance represents the Food and Drug Administration's (FDA's) current thinking on this topic. It does not create or confer any rights for or on any person and does not operate to bind FDA or the public. You can use an alternative approach if the approach satisfies the requirements of the applicable statutes and regulations. If you want to discuss an alternative approach, contact the FDA staff responsible for implementing this guidance. If you cannot identify the appropriate FDA staff, call the appropriate number listed on the title page of this guidance.

INTRODUCTION

This guidance describes the device operation conditions for magnetic resonance diagnostic devices that FDA considers significant risk for the purposes of determining whether a clinical study requires Agency approval of an Investigation Device Exemption (IDE). Magnetic resonance diagnostic devices are class II devices described under 21 CFR 892.1000. The product codes for these devices are:

LNH Magnetic Resonance Imaging System

LNI Magnetic Resonance Spectroscopic System

This guidance supersedes **Guidance for Magnetic Resonance Diagnostic Devices - Criteria for Significant Risk Investigations**, issued September 29, 1997. We have revised our recommendation for the main static magnetic field strength, increasing it to 8 Tesla for most populations. This is based on ongoing experience in the field and numerous literature reviews (1, 2).

FDA's guidance documents, including this guidance, do not establish legally enforceable responsibilities. Instead, guidances describe the Agency's current thinking on a topic and should be viewed only as recommendations, unless specific regulatory or statutory requirements are cited. The use of the word *should* in Agency guidances means that something is suggested or recommended, but not required.

THE LEAST BURDENSOME APPROACH

We believe we should consider the least burdensome approach in all areas of medical device regulation. This guidance reflects our careful review of the relevant scientific and legal requirements and what we believe is the least burdensome way for you to comply with those requirements. However, if you believe that an alternative approach would be less burdensome, please contact us so we can consider your point of view. You may send your written comments to the contact person listed in the preface to this guidance or to the CDRH Ombudsman. Comprehensive information on CDRH's Ombudsman, including ways to contact him, can be found on the Internet at http://www.fda.gov/cdrh/ombudsman/

STUDIES OF MAGNETIC RESONANCE DIAGNOSTIC DEVICES

If a clinical study is needed to demonstrate substantial equivalence, i.e., conducted prior to obtaining 510(k) clearance of the device, the study must be conducted under the IDE regulation (21 CFR Part 812). FDA believes that a magnetic resonance diagnostic device used under any one of the operating conditions listed below is a significant risk device as defined in 21 CFR 812.3(m)(4) and, therefore, that studies involving such a device do not qualify for the abbreviated IDE requirements of 21 CFR 812.2(b). In addition to the requirement of having an FDA-approved IDE, sponsors of significant risk studies must comply with the regulations governing institutional review boards (21 CFR Part 56) and informed consent (21 CFR Part 50).

SIGNIFICANT RISK MAGNETIC RESONANCE DIAGNOSTIC DEVICES

You should consider the following operating conditions when assessing whether a study may be considered significant risk:

- main static magnetic field
- specific absorption rate (SAR)
- gradient fields rate of change
- sound pressure level

Generally, FDA deems magnetic resonance diagnostic devices significant risk when used under any of the operating conditions described below.

Main Static Magnetic Field

Population	Main static magnetic field greater than (Tesla)
adults, children, and infants aged > 1 month	8
neonates i.e., infants aged 1 month or less	4

Specific Absorption Rate (SAR)

Site	Dose	Time (min) equal to or greater than:	SAR (W/kg)
whole body	averaged over	15	4
head	averaged over	10	3
head or torso	per gram of tissue	5	8
extremities	per gram of tissue	5	12

Gradient Fields Rate of Change

Any time rate of change of gradient fields (dB/dt) sufficient to produce severe discomfort or painful nerve stimulation

Sound Pressure Level

Peak unweighted sound pressure level greater than 140 dB.

A-weighted root mean square (rms) sound pressure level greater than 99 dBA with hearing protection in place.

These criteria apply only to device operating conditions. Other aspects of the study may involve significant risks and the study, therefore, may require IDE approval regardless of operating conditions. See Blue Book Memorandum entitled **Significant Risk and Non-significant Risk Medical Device Studies**, http://www.fda.gov/cdrh/d861.html for further discussion.

After FDA determines that the device is substantially equivalent, clinical studies conducted in accordance with the indications reviewed in the 510(k), including clinical design validation studies conducted in accordance with the quality systems regulation, are exempt from the investigational device exemptions (IDE) requirements. However, such studies must be performed in conformance with 21 CFR 56 and 21 CFR 50.

(1) Kangarlu A, Burgess RE, Zu H, et al. Cognitive, cardiac and physiological studies in ultra high field magnetic resonance imaging. *Magnetic Resonance Imaging*, 1999;17:1407-1416.

(2) Schenck John F, Safety of strong, static magnetic fields. *Journal of Magnetic Resonance Imaging*, 2000;12:2-19.

APPENDIX III

WEB SITES FOR MRI SAFETY, BIOEFFECTS, AND PATIENT MANAGEMENT

There are two web sites devoted to MRI safety, bioeffects, and patient management: www.MRIsafety.com and www.IMRSER.org

www.MRIsafety.com

The international information resource for MRI safety, bioeffects, and patient management.®

This web site provides up-to-date, concise information for healthcare providers and patients seeking answers to questions on MRI safety-related topics. The latest information is also provided for screening patients with implants, materials, and medical devices.

Key Features

✎ *The List*: a searchable database that contains over 1,200 implants and other objects tested for MRI safety.

✎ *Safety Information*: concise information that pertains to the latest recommendations and guidelines for patient care and management in the MR environment.

442

✎ *Research Summary*: a presentation and summary of over 100 peer-reviewed articles on MRI bioeffects and safety.

✎ *Screening Forms*: information and forms for screening patients and other individuals available to imaging facilities in a "downloadable" format.

www.IMRSER.org

The web site of the Institute for Magnetic Resonance Safety, Education, and Research (IMRSER)

The Institute for Magnetic Resonance Safety, Education, and Research (IMRSER) was formed in response to the growing need for information on matters pertaining to magnetic resonance (MR) safety. The IMRSER is the first independent, multidisciplinary, professional organization devoted to promoting awareness, understanding, and communication of MR safety issues through education and research.

The functions and activities of the IMRSER involve development of MR safety guidelines and dissemination of this information to the MR community. This is accomplished through the efforts of two Advisory Boards: the Medical, Scientific, and Technology Advisory Board and the Corporate Advisory Board.

The Medical, Scientific, and Technology Advisory Board consists of recognized leaders in the field of MR including diagnostic radiologists, clinicians, research scientists, physicists, MRI technologists, MR facility managers, and other allied healthcare professionals involved in MR technology and safety. The members of the Medical, Scientific, and Technology Advisory Board represent academic, private, research, and institutional MR facilities utilizing scanners operating at static magnetic field strengths ranging from 0.2-Tesla (including dedicated-extremity and interventional MR systems) to 8.0-Tesla. In addition, the Food and Drug Administration has assigned a Federal Liaison to the IMRSER.

The Corporate Advisory Board is comprised of representatives from the medical industry including MR system manufacturers, contrast agent pharmaceutical companies, RF coil manufacturers, MR accessory vendors, medical product manufacturers, and other related corporate organizations.

The Institute for Magnetic Resonance Safety, Education, and Research develops MR safety guidelines utilizing the pertinent peer-reviewed literature and by relying on each board member's extensive clinical, research, or other appropriate experience. Notably, documents developed by the IMRSER consider and incorporate MR safety guidelines and recommendations created by the International Society for Magnetic Resonance in Medicine (ISMRM), the American College of Radiology (ACR), the Food and Drug Administration (FDA), the National Electrical Manufacturers Association (NEMA), the Medical Devices Agency (MDA), the International Electrotechnical Commission (IEC), and other organizations.

Frank G. Shellock, Ph.D. is a physiologist with more than 17 years of experience conducting laboratory and clinical investigations in the field of magnetic resonance imaging. He is an Adjunct Clinical Professor of Radiology at the Keck School of Medicine, University of Southern California and the Founder of the Institute for Magnetic Resonance Safety, Education, and Research (www.IMRS-ER.org). As a commitment to the field of MRI safety, bioeffects, and patient management, he created and maintains the internationally popular web site, www.MRIsafety.com. This web site currently has over 35,000 registered users.

Dr. Shellock is a Member of the Guidelines and Standards Committee (Body MRI), Commission on Neuroradiology and Magnetic Resonance for the American College of Radiology, the Safety Committee for the International Society for Magnetic Resonance Imaging, and a member of the Sub-Committee on MR Safety and Compatibility for the American Society for Testing and Materials. Additionally, he serves in advisory roles to government, industry, and other policy-making organizations. Dr. Shellock is an Associate Editor for the Journal of Magnetic Resonance Imaging and a Reviewing Editor for several medical and scientific journals.

Dr. Shellock's memberships in professional societies include the American College of Radiology, the International Society for Magnetic Resonance in Medicine, the Radiological Society of North America, the California Radiological Society, the Hawaii Radiological Society, and the Society for Cardiovascular Magnetic Resonance. He is a member and Fellow of the American College of Sportsmedicine.

Dr. Shellock is a recipient of a National Research Service Award from the National Institutes of Health, National Heart, Lung, and Blood Institute and has received numerous grants from governmental agencies and private organizations. He is currently participating in MR safety research at the Cleveland Clinic Foundation that is funded by a grant awarded from the National Institutes of Health and conducting MRI research on Ironman Triathletes at North Hawaii Community Hospital.

Dr. Shellock is frequently invited to lecture, both nationally and internationally. He has provided plenary lectures on MR safety to numerous organizations including the Radiological Society of North America, the International Society for Magnetic Resonance in Medicine, the American College of Radiology, American Roentgen Ray Society, the American Society of Neuroradiology, the Environmental Protection Agency, the Oklahoma Heart Institute, the Head and Neck Radiology Society, the Center for Devices and Radiological Health (FDA), the Magnetic Resonance Managers Society, the American Heart Association, the American Society of Neuroimaging, and the Society for Cardiovascular Magnetic Resonance.

Dr. Shellock's company, Magnetic Resonance Safety Testing Services, specializes in the assessment of MR safety for implants and devices as well as the evaluation of electromagnetic field-related bioeffects and the development of new clinical MR imaging applications for low-field (0.2-Tesla), high-field, and very-high-field strength (8.0-Tesla) MR systems (www.MagneticResonanceSafetyTesting.com).

NOTES

NOTES

NOTES

NOTES